Topology of Foliations: An Introduction

Recent Titles in This Series

(*Continued in the back of this publication*)

Translations of
MATHEMATICAL
MONOGRAPHS

Volume 97

Topology of Foliations: An Introduction

Itiro Tamura

Translated by
Kiki Hudson

American Mathematical Society
Providence, Rhode Island

葉層のトポロジー

YOUSOU NO TOPOROJII (Topology of Foliations)

by Itiro Tamura

Originally published in Japanese by Iwanami Shoten, Publishers, Tokyo in 1976

Translated from the Japanese by Kiki Hudson

2000 *Mathematics Subject Classification.* Primary 57R30;
Secondary 58–XX.

ABSTRACT. This book presents an introduction to the theory of foliations, especially those with compact leaves, and it covers all of the important basic results in this field. The book includes many graphical representations of examples in low dimensions. Specific topics presented include dynamical systems on the torus and on the three-sphere, local and global stability theorems for foliations, the existence of compact leaves on three-spheres, and foliated cobordisms of three-spheres. Chapter II and §§26 and 27 of Chapter VII form a mini-introduction to the theory of differentiable manifolds.

Library of Congress Cataloging-in-Publication Data

Tamura, Ichirō, 1926–
[Yōsō no toporojī. English]
Topology of foliations: an introduction/Itiro Tamura; translated by Kiki Hudson.
p. cm. — (Translations of mathematical monographs; v. 97)
Translation of: Yōsō no toporojī.
Includes bibliographical references and index.
ISBN 0-8218-4543-8 (hardcover; alk. paper); ISBN 0-8218-4200-5 (softcover; alk. paper)
1. Foliations (Mathematics) I. Title. II. Series.

QA613.62.T3513 1992 91-37135
514.72—dc20 CIP

Printed in the United States of America.

Information on copying and reprinting can be found in the back of this volume.
∞ The paper used in this book is acid-free and falls within the guidelines
established to ensure permanence and durability.
Visit the AMS home page at `http://www.ams.org/`

10 9 8 7 6 5 4 3 2   11 10 09 08 07 06 05 06

Contents

Foreword

Topology of Foliations was first published in Japanese in 1976, with the principal aim of introducing the theory of foliations to young students. The author, Professor Itiro Tamura, passed away on February 21, 1991, during the preparation of the English translation of this successful book.

Professor Tamura worked on differential topology in the late 1950's. His results included establishing the non-homotopy-invariance of the Pontrjagin classes, demonstrating the existence of a topological eight-manifold with no differential structure, etc. He began working on the theory of foliations in the late 1960's and made an important contribution to the problem of the existence of foliations on closed manifolds in 1971: he constructed codimension-one foliations on every odd-dimensional homotopy sphere. This was done by showing the existence of spinnable structures on such manifolds. This work stimulated the study of foliations, which grew very rapidly in the 1970's.

At almost the same time as the publication of the original version of this book, the problem of the existence of foliations on closed manifolds was solved by W. Thurston. Much work in foliation theory has been done since 1976. An important part of this work was done by Tamura's students, who had read the original version of *Topology of Foliations*, as the author had hoped.

The author himself was active in the theory of foliations. He initiated the theory of transverse foliations. In the last several years of his life, he was studying the Seifert conjecture and the minimal flow conjecture. I regret very much that he could not continue his research.

The *Topology of Foliations* was translated into Russian in 1979. Now an English translation has been completed and many more students of mathematics can touch the spirit of Itiro Tamura.

The author begins this book with very elementary facts. This part, Chapters I and II, can be used as an introduction to manifold theory and dynamical-system theory. A typical course on manifold theory seldom extends to the Frobenius Theorem, but the reader will find a nice exposition of this theorem here in Chapter VII.

In the body of this book, the author explains important results which motivate further research. Chapter III leads the reader to the Seifert Conjecture or to the minimal flow conjecture. There are now works by Harrison and Fathi-Herman on this subject. Chapters IV and V lead to the architecture of foliations or the theory of depth or height, which was developed by Nishimori, Hector, Cantwell, and Conlon. Chapter VI, which contains the theorem of Novikov, leads to the theory of embedded surfaces in foliations due to Rosenberg, Roussarie, Thurston, and Gabai. Chapter VIII leads to the theory of characteristic classes of foliations. This theory was first developed by Bott-Haefliger and Gelfand-Fuks; since then there have been many results in this direction, with too many names to be cited here. The Godbillon-Vey invariant is still one of the most interesting characteristic classes, commanding the attention of many researchers. The reader can easily find appropriate references for this further research by using the names of authors mentioned above.

Even now, fifteen years after the publication of the original version of this book, I strongly recommend it as a first step into research on the theory of foliations. I am sure that the theory of foliations will be continue to be developed by the readers of this translation.

Takashi Tsuboi

Tokyo, 1991

Preface

The orbits of a nonsingular vector field form a pattern of stripes on the torus. This pattern is a decomposition of the torus into a union of the orbits (Figure 1.10). Such a pattern is an example of the foliation of a manifold which is decomposed thereby into arcwise-connected subsets called the leaves of the foliation. The leaves locally form layers of Euclidean spaces (Figures 4.2, 4.4), and each orbit in the above example is a leaf. The leaves of a foliation of a manifold are submanifolds of a common dimension; the difference of the dimensions of the manifold and a leaf is called the codimension of the foliation.

Thus all foliations with a given codimension of a manifold have a common local structure. Yet their patterns vary globally from one foliation to another. The topology of foliations examines the global nature of foliations from a topological viewpoint, asking questions such as 'Do compact leaves exist?' , 'Do locally dense leaves exist?' , 'Are all the leaves homeomorphic?' , etc.

Aside from foliations there are other structures defined on a manifold by its local characterization—differential structures, combinatorial structures, and complex structures, to name a few—and global investigations of these structures form the core of current research in geometry. The purpose of this book is to introduce present-day geometrical thinking to a broader mathematical community. We focus our attention on the foliations of manifolds. Starting with the origin of this branch of geometry, we have picked out several major works during its forty years of development in a more-or-less chronological order, and tried to give as instructive an analysis as possible at each stage. Any reader with a background in mathematics at the lower-division level in college should be able to follow the arguments. Each concept is explained in the text and supplemented by notes given in the appendices when necessary.

The initial usage of foliations was local in nature, arising from their approach to solutions of systems of differential equations. We see the first global consideration of foliations in various studies concerning trajectories on the torus, although the theory of foliations did not establish itself as an independent field of research until after Reeb's monograph was published. More on Reeb's work later.

During the 1880's Poincaré published a series of papers each entitled *Mémoire sur Les Courbes Définies par une Équation Différentielle,* for the first time airing his topological ideas. These ideas were extended by Denjoy in his paper on the topological study of the orbits on the torus, published in 1932; the topic expanded further with Siegel's work published in 1945. Despite the fact that the environment was limited to the torus, these papers presented strikingly fresh results and made effective use of the topological methods epitomizing research in foliations. They can be regarded as the starting point of the topology of foliations. In Chapter I we develop the Poincaré-Denjoy-Siegel theorem in detail with many illustrations. The study of tangent vectors, tangent planes, and so forth on the torus should prepare the reader psychologically to meet these concepts in the more general environment of manifolds in Chapter II and beyond.

Chapter II deals with basic concepts of C^r manifolds. We limited our discussion to what is needed in the later chapters and hence barely touched the surface of the field; however combined with the sections on differential forms in Chapter VII, this chapter might serve as a basic abridged introduction to the theory of manifolds. In §28 we present the theorem of Frobenius to highlight its fundamental relationship to the theory of foliations.

When a foliation has a compact leaf, under certain conditions one can remove the compact leaf through a local change of the foliation (§23); this technique is discussed in conjunction with the dynamical systems of Wilson (1966) and Schweitzer (1972) in Chapter III.

In the monograph *Sur certains propriétés topologiques des variétés feuilletées* (1952), Reeb gave a precise definition of foliations as geometric objects; they had previously been treated in analytic environments. In Chapter V we present the main results of his book: the local and global stability theorems inspired by the comparison between C^r bundles, which had been developed by Reeb's teacher Ehresmann and others, and foliations.

In Chapter VI we discuss the theorem of Novikov (1964) which asserts the existence of a compact leaf in a codimension one foliation of the three-sphere. This result added to his portfolio of achievement, including contributions in differential topology and algebraic topology and Novikov won a Field medal in 1970.

In 1971 Godbillon and Vey discovered a singular characteristic class, the first of its kind. This discovery is having a major impact on the field of geometry in the large. In Chapter VIII we give the definition of Godbillon-Vey classes and present the calculation of Thurston which shows that the cardinality of the three-dimensional foliated cobordism group is the continuum (1972). The theorems of Chapter VII are repeatedly applied in Chapter VIII.

One outstanding trait of foliations is the accessibility to their examples in low-dimensional manifolds: the torus in Chapter I, the Seifert conjecture on codimension two foliations of S^3 in §14, Novikov's theorem in §25,

Thurston's calculation in Chapter VIII, and so on. We were particular in choosing results in low-dimensional manifolds so that the figures enhance the explanations—the figures should almost be self-standing.

Each chapter of this book deals with an independent topic, although all are related to one another. We believe that a reader accustomed to reading mathematics will have little trouble in starting at any chapter, with of course the use of the numbered theorems cited as references.

We have added a bibliography of papers and monographs on foliations including many topics which are not covered in the text for those readers who wish to further their knowledge in this subject. The bibliography, though not complete, includes research results up to present (see the translator's afterword).

The author thanks Tadayoshi Mizutani, Shigeo Ichiraku, Shigeyuki Morita, Toshiyuki Nishimori, Nobuo Tsuchiya, and Takashi Inaba for their valuable comments and their contribution in compiling the bibliography. Finally special thanks go to Mr. Hideo Arai of the Iwanami Shoten Publishers without whose assistance the book would not have materialized at all.

Itiro Tamura
University of Tokyo
February, 1976

CHAPTER I

Nonsingular Dynamical Systems on the Torus

§1. 'On the curves defined by a differential equation'

During the years 1881 through 1886 Poincaré pursued a novel direction in his study of differential equations: geometrical investigations of the solution curves. This was presented in four papers, each entitled *Mémoire sur Les Courbes Définies par une Équation Différentielle* (see Appendix I). At the beginning of the first paper he stated the following:

Une théorie complète des fonctions définies par les équations différentielles serait d'une grande utilité dans un grand nombre de questions de Mathématiques pures ou de Mécanique. Malheureusement, il est évident que, dans la grande, généralité des cas qui se présentent, on ne peut pas intégrer ces équations à l'aide des fonctions déjà connues, par exemple à l'aide des fonctions définies par les quadratures. Si l'on voulait donc se restreindre aux cas que l'on peut étudier avec des intégrales définies ou indéfinies, le champ de nos recherches serait singulièrement diminué, et l'immense majorité des questions qui se présentent dans les applications demeureraient insolubles.

Il est donc nécessaire d'étudier les fonctions définies par des équations différentielles en elles-mêmes et sans chercher à les ramener à des fonctions plus simples, ...

Here Poincaré proposes a qualitative investigation of functions defined as the solutions of differential equations.

L'étude complète d'une fonction comprend deux parties:

1^0 *Partie qualitative (pour ainsi dire), ou étude géométrique de la courbe définie par la fonction;*

2^0 *Partie quantitative, ou calcul numérique des valeures de la fonction.*

Ainsi, par exemple, pour étudier une équation algébrique, on commence par rechercher, à l'aide du théorème de Sturm, quel est le nombre des racines réelles, c'est la partie qualitative, puis on calcule la valeur numérique de ces racines, ce qui constitue l'étude quantitative de l'équation. De même, pour étudier une courbe algébrique, on commence par construire *cette courbe ..., c'est-à-dire qu'on cherche quelles sont les branches de courbes fermées, branches infinies, etc. Après cette étude qualitative de la courbe, on peut déterminer exactement un certain nombre de points.*

C'est naturellement par la partie qualitative qu'on doit aborder la théorie de toute fonction et c'est pourquoi le problème qui se présent en premier lieu est le suivant:

Construire les courbes définies par des équations différentielles.

Poincaré then goes on to point out the importance of qualitative studies for the numerical calculations of functions and states his plan in the following:

D'ailleures, cette étude qualitative aura par elle-même un intérêt du premier ordre. Diverses questions fort importantes d'Analyse et de Mécanique peuvent en effet s'y ramener. Prenons pour exemple le problème des trois corps: ne peut-on pas se demander si l'un des corps restera toujours dans une certaine région du ciel ou bien s'il pourra s'éloigner indéfiniment; si la distance de deux des corps augmentera, ou diminuera à l'infini, ou bien si elle restera comprise entre certaines limites? Ne peut-on pas se poser mille questions de ce genre qui seront toutes résolues quand on saura construire qualitativement les trajectoires des trois corps? .... Tel est le vaste champ de découvertes qui s'ouvre devant les géomètres.

The qualitative study of Poincaré belongs to the realm of *topology*—in today's language—in which one tries to understand the characteristics of curves defined by differential equations not through numerical, but through topological investigations. He developed his plan further, culminating in the theory of 'dynamical systems' (see Appendix I). In the last chapter of the third paper of the four mentioned above, Poincaré discusses the curves defined by differential equations on the torus. We shall show that there exist systems of nonsingular differential equations—*dynamical systems*—on the torus, and that these solution curves define *foliations* of the torus. In this context Poincaré's four papers serve as the starting point of 'Topology of Foliations'. In this chapter we discuss the Poincaré's work and the major contributions of Denjoy and Siegel that followed it.

§2. Vector fields and their trajectories

By specifying a coordinate—that is the origin 0, a positive direction, and a unit segment—we can associate to each point on a line a unique real number. Similarly, a Cartesian coordinate system consisting of two fixed perpendicular lines in the plane, the x-axis and y-axis, enables us to represent any point P in the plane by the pair (x, y) of real numbers x and y, and to assign to each pair (x, y) of real numbers the point $P(x, y)$ in the plane. We shall denote by $\mathbf{R}$ the set of all real numbers. In set notation, $\{P(x, y); x, y \in \mathbf{R}\}$ represents the plane in a given Cartesian coordinate system. More generally, if $C(a)$ is the statement that a has a certain property C, then $\{a; C(a)\}$ is the set of all a with the property C.

Let $P(x, y)$ and $P'(x', y')$ be two points in the plane. The usual formula for the *distance* between the points P and P',

$$\rho(P, P') = \sqrt{(x - x')^2 + (y - y')^2},$$

satisfies the following '*metric axioms*', and ρ is called the standard *metric* in the plane R^2:

For points P, P', and P'' in the plane:

(i) $\rho(P, P') \geq 0$, and $\rho(P, P') = 0$ if and only if $P = P'$,

(ii) $\rho(P, P') = \rho(P', P)$,

(iii) $\rho(P, P') \leq \rho(P, P'') + \rho(P'', P')$, the *triangle inequality*.

Let P be a point in the plane. Given a number $\epsilon > 0$ denote by $U_\epsilon(P)$ the set of all points in the plane whose respective distances from P are less than ϵ:

$$U_\epsilon(P) = \{P'; \rho(P, P') < \epsilon\}.$$

The set $U_\epsilon(P)$ is called the *ϵ-neighborhood* of P. A point P is an *interior point* of A if there exists $\epsilon > 0$ such that $U_\epsilon(P) \subset A$. A point P' is a *cluster point* of A if for every $\epsilon > 0$ $A \cap U_\epsilon(P') \neq \varnothing$, where $\varnothing$ denotes the empty set. We say that P' is an *accumulation point* of A if for every $\epsilon > 0$, $A \cap U_\epsilon(P')$ contains infinitely many points. For instance, let $A = \{P(x, 0); x \text{ a rational number}\}$: the point $P(\sqrt{2}, 0)$ is both a cluster point and an accumulation point of A, but no $P(x, 0)$ is an interior point of A.

The *interior* of A, denoted by $\operatorname{Int} A$, is defined to be the set of all interior points of A. If $A = \operatorname{Int} A$ we say that A is an *open* set. The *closure* of the set A, which we shall denote by $\bar{A}$, is defined to be the set of all cluster points of A. We say that A is a *closed* set if $A = \bar{A}$.

Let $S^1 = \{(x, y); x^2 + y^2 = 1\}$ be the unit circle in the plane. Define a subset A of S^1 by $A = \{(\cos\theta, \sin\theta); \theta \text{ rational}\}$; then $\bar{A} = S^1$.

The set $U_\epsilon(P)$ is open. To see this, for $P' \in U_\epsilon$ let $\epsilon' = \epsilon - \rho(P, P')$. We then have

$$U_{\epsilon'}(P') \subset U_\epsilon(P).$$

For a sequence of points $P_1, P_2, \ldots, P_n, \ldots$, if there exists P_0 such that

$$\lim_{n\to\infty} \rho(P_n, P_0) = 0,$$

we write $\lim_{n\to\infty} P_n = P_0$, and say that P_0 is the *limit point* of this sequence.

An ordered pair of points P and Q in the plane determines the directed segment $\overrightarrow{PQ}$, called a *vector based at* P (Figure 1.1). We define a vector in the usual sense to be the whole of the directed segments which are translates of each other; so if a directed segment $\overrightarrow{PQ}$ represents a vector v, then by the pair (P, v) we mean the vector $\overrightarrow{PQ}$ based at P. The pair (P, v) will be denoted by v_P (Figure 1.1). Sometimes we say that v_P is a *fixed vector* in order to distinguish it from the usual vector v. Identifying $(0, v)$ with the vector v, where 0 is the origin, is another common practice.

A fixed vector v_P is determined by a point P together with the x-coordinate ξ and the y-coordinate η of a vector v. The numbers ξ and η will be called the *x-component* and *y-component* respectively of v_P. In particular, v_P is the *zero vector* at P, if $\xi = \eta = 0$.

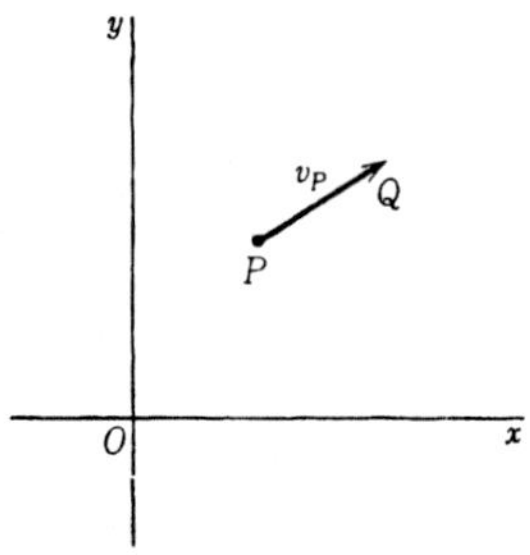

FIGURE 1.1

Let A be a set in the plane. A *vector field*, denoted $X(A)$ or X, assigns to each point $P \in A$ a vector based at P, that is, $X(A)$ is the set $\{v_P; P \in A\}$. Let v_P be a vector at $P(x, y)$ and let $\xi(x, y)$ be the x-component and $\eta(x, y)$ the y-component respectively of v_P; then $\xi(x, y)$ and $\eta(x, y)$ are functions defined for all (x, y) such that $P(x, y) \in A$. Such functions $\xi(x, y)$ and $\eta(x, y)$ determine a vector field X on A, which we shall sometimes refer to as the component functions of X (Figure 1.2).

A vector field $X = \{v_P; P \in A\}$ is *nonsingular* if every $v_P \in X$ is a nonzero vector.

Let X be a vector field on A. Let $\xi(x, y)$ and $\eta(x, y)$ be the component functions of X. We say that X is *differentiable of class* C^r or simply C^r at P if $\xi(x, y)$ and $\eta(x, y)$ are differentiable of class C^r. This means that all partial derivatives of $\xi(x, y)$ and $\eta(x, y)$ of orders up to r exist and are continuous at P. If A is open and X is C^r at each point of A, we say that X is a C^r *vector field*, $r = 0, 1, 2, \ldots, \infty$. A C^0 vector field has continuous component functions $\xi(x, y)$ and $\eta(x, y)$; for a C^∞ vector field the rth partial derivatives of $\xi(x, y)$ and $\eta(x, y)$ exist for all nonnegative integers r.

The characterization C^∞ is often replaced by the adjective 'smooth'.

Let α, β be real numbers, $\alpha < \beta$, and let φ be a map from an open

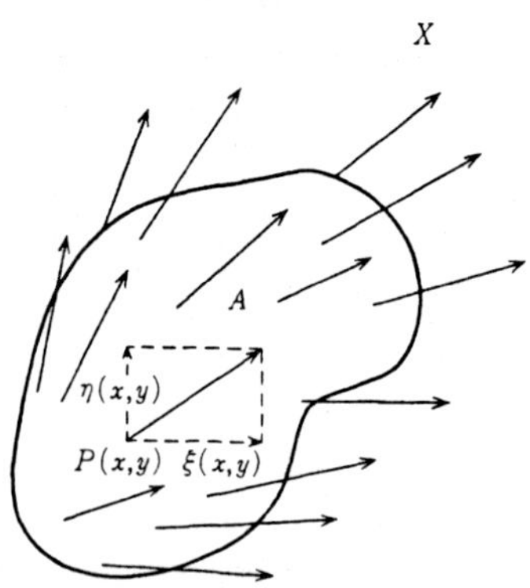

FIGURE 1.2

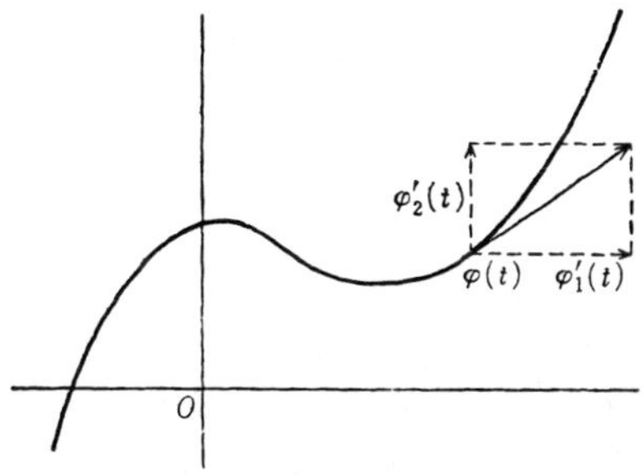

FIGURE 1.3

interval $(\alpha, \beta) = \{t \in \mathbf{R}; \alpha < t < \beta\}$ to the plane. The coordinates $(\varphi_1(t), \varphi_2(t))$ of $\varphi(t)$ in the plane define real-valued functions $\varphi_1(t)$ and $\varphi_2(t)$ on (α, β). When $\varphi_1(t)$ and $\varphi_2(t)$ are of class C^r, $r = 0, 1, 2, \ldots, \infty$, we say that φ is a C^r *curve* defined on the open interval (α, β). Similarly there are C^r curves defined on intervals $[\alpha, \beta] = \{t \in \mathbf{R}; \alpha \le t \le \beta\}$, $[\alpha, \beta) = \{t \in \mathbf{R}; \alpha \le t < \beta\}$, $(\alpha, \beta]$, (α, ∞), $(-\infty, \beta)$, $(-\infty, \beta]$.

Let φ be a C^r curve, $r \ge 1$, defined on an open interval (α, β). The *tangent vector* to φ at $\varphi(t)$ is defined to be the vector at $\varphi(t)$ whose x-component and y-component are $\varphi_1'(t)$ and $\varphi_2'(t)$ respectively, where $\varphi_i'(t)$ is the derivative of φ with respect to t, $i = 1, 2$ (Figure 1.3).

Let A be an open set in the plane and $X = \{v_P; P \in A\}$ a vector field on A. We say that a C^{r+1} curve φ defined on an open interval (α, β) is an *integral curve* of X if it satisfies for all t, $\alpha < t < \beta$, the following:

(i) $\varphi(t) \in A$, and

(ii) the tangent vector of φ at $\varphi(t)$ is equal to $v_{\varphi(t)}$ (see Figure 1.4).

We may also consider integral curves defined on intervals $[\alpha, \beta]$, $[\alpha, \infty)$, *etc.* Terms such as *solution curves* or *trajectories* are often used in place of integral curves.

Let $\xi(x, y)$ and $\eta(x, y)$ be the x and y-components of a vector $v_P \in X$ at $P(x, y)$. Then φ is an integral curve of X if and only if the coordinate functions $\varphi_1(t)$ and $\varphi_2(t)$ satisfy the system of ordinary differential equations:

$$\frac{d\varphi_1}{dt}(t) = \xi(\varphi_1(t), \varphi_2(t)), \qquad \frac{d\varphi_2}{dt}(t) = \eta(\varphi_1(t), \varphi_2(t)). \tag{1}$$

We now assume that X is a C^r vector field in the entire plane $r \ge 1$. Then given a point $P_0(x_0, y_0)$ in the plane and a real number t_0 there exist, by the existence and uniqueness theorem for ordinary differential equations, unique C^{r+1} functions φ_1 and φ_2 defined on some interval $(t_0 - \tau, t_0 + \tau)$ which satisfy the following:

(i) The system of differential equations (1) is valid for $t_0 - \tau < t < t_0 + \tau$, and

(ii) $\varphi_1(t_0) = x_0$, $\varphi_2(t_0) = y_0$

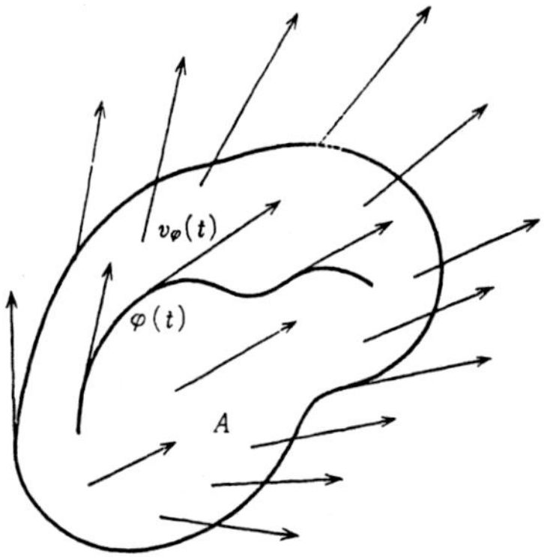

FIGURE 1.4

(see Appendix I, Note 1). Hence there exists a unique integral curve for X defined on $(t_0 - \tau, t_0 + \tau)$ with $\varphi(t_0) = P_0$. We note in particular that if

$$|\xi(x, y)| < M_0, \qquad |\eta(x, y)| < M_0, \qquad M_0 \text{ a positive number,}$$

in the neighborhood $U_\epsilon(P_0)$ of P_0, we may choose $\epsilon/2M_0$ for τ.

Next we let $t_1 \in (t_0 - \tau, t_0 + \tau)$ be a number arbitrarily close to $t_0 + \tau$. Suppose that there is a positive number M_1 such that

$$|\xi(x, y)| < M_1, \quad |\eta(x, y)| < M_1,$$

in the ϵ'-neighborhood $U_{\epsilon'}(\varphi(t_1))$ of $\varphi(t_1)$. Then taking $\tau' = \epsilon'/2M_1$ as discussed above, we obtain a unique integral curve $\overline{\varphi}$ of X defined on the open interval $(t_1 - \tau', t_1 + \tau')$ with $\overline{\varphi}(t_1) = \varphi(t_1)$. If $t_0 + \tau < t_1 + \tau'$, by the uniqueness of integral curves, φ agrees with $\overline{\varphi}$ on the interval $(t_1, t_0 + \tau)$; hence we may define φ by $\varphi(t) = \overline{\varphi}(t)$ for $t \in (t_1, t_1 + \tau')$ and we have an integral curve φ defined on $(t_0 - \tau, t_1 + \tau)$ such that $\varphi(t_0) = P_0$.

Under suitable conditions, for example if there exists a positive number M such that $|\xi(x, y)| < M$, $|\eta(x, y)| < M$ for every (x, y) in the plane, we can extend φ in both the positive and negative directions of t_0 to obtain a unique integral curve φ defined on $(-\infty, \infty)$ such that $\varphi(t_0) = P_0$.

We often take $t_0 = 0$. Then $\varphi(0) = P_0$ and we say that φ is the *integral curve* of X *starting at* P_0 or *through* P_0, and we denote it by $\varphi(t, P_0)$, with parameter t.

An integral curve $\varphi(t, P(x, y))$ of X starting at $P(x, y)$—that is through $P(x, y)$—is a function of t as well as x and y. We observed earlier that the curve $\varphi(t, P(x, y))$ is of class C^{r+1} with respect to t; however, it is of class C^r with respect to x and y according to the theorem for the differentiability of solutions as functions of the initial conditions.

§3. Vector fields and trajectories on the torus

For points $P(x, y)$ and $P'(x', y')$ in the plane, define a relation $P \sim P'$ if $x - x'$ and $y - y'$ are integers. The relation $\sim$ is an equivalence relation, i.e.,

(i) $P \sim P$,
(ii) if $P \sim P'$ then $P' \sim P$,
(iii) if $P \sim P'$ and $P' \sim P''$, then $P \sim P''$.

The set $\{Q; Q \sim P\}$ of the points in the plane which are related to P by $\sim$ is called the *equivalence class* determined by P and is denoted by $[P]$. We say that P is a *representative* of the equivalence class. A class $[P(x, y)]$ is the set of all points $(x+m, y+n)$, $m, n = 0, \pm 1, \pm 2, \dots$, in the plane.

The points in the plane decompose into equivalence classes; the quotient set of the plane by the equivalence relation $\sim$ is the *torus*, here and below frequently denoted by T. We shall denote by p the equivalence class $[P]$, by p' the equivalence class $[P']$, and so forth; p, p', *etc.* are points of T.

Consider the square $ABCD$ with vertices at $A(0, 1)$, $B(0, 0)$, $C(1, 0)$, and $D(1, 1)$ in the plane. For an arbitrary point $P(x, y)$ in the plane choose integers m, n such that $0 \le x - m \le 1$, $0 \le y - n \le 1$; $Q(x - m, y - n)$ is in square $ABCD$ and satisfies

$$P(x, y) \sim Q(x - m, y - n).$$

We also notice that if P', P'' are distinct points of the square $ABCD$, then $P' \sim P''$ only if both P' and P'' are on the sides of the square and then they must be of the form $P'(x, 0) \sim P''(x, 1)$, $P'(0, y) \sim P''(1, y)$, *etc.* Therefore if we paste side AB onto side DC, and side AD onto side BC according to the relation $\sim$, the result will be the torus (Figure 1.5). Or again one may think of the torus as the result of identifying the end circles of the cylinder (Figure 1.5 (ii)).

Let $p = [P]$, $p' = [P']$ be points on T. Let $\rho(p, p')$ be the minimum of $\rho(Q, Q')$ with $Q \in [P]$ and $Q' \in [P']$:

$$\rho(p, p') = \min_{\substack{Q \in [P] = p, \\ Q' \in [P'] = p'}} \rho(Q, Q').$$

We adopt $\rho(p, p')$ as the distance between p and p' on T. The following lemma shows that T with the distance ρ is a metric space.

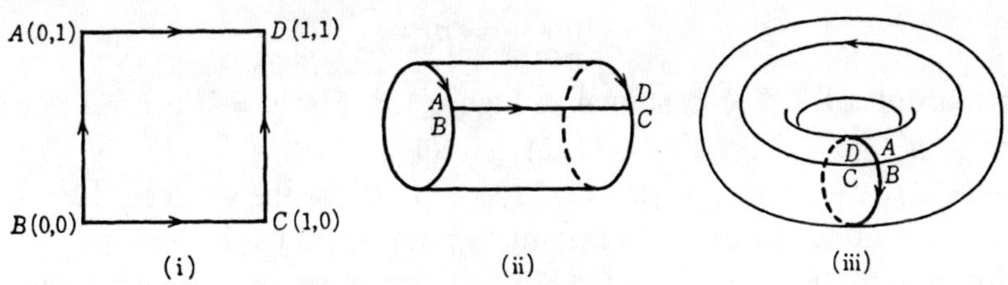

FIGURE 1.5

LEMMA 1.1. *The distance* ρ *satisfies the following*:

(i) $\rho(p, p') \geq 0$, *and* $p = p'$ *if and only if* $\rho(p, p') = 0$.

(ii) $\rho(p, p') = \rho(p', p)$.

(iii) $\rho(p, p') \leq \rho(p, p'') + \rho(p'', p')$.

Here $p, p', p'' \in T$.

PROOF. (i) and (ii) are obvious. By definition there exist $Q \in [P] = p$, $Q'' \in [P''] = p''$, $Q''' \in [P''] = p''$, and $Q' \in [P'] = p'$ satisfying

$$\rho(p, p'') = \rho(Q, Q''), \qquad \rho(p'', p') = \rho(Q''', Q').$$

If Q', Q'', Q''' are $Q'(x', y')$, $Q''(x'', y'')$, $Q'''(x''', y''')$ respectively, then

$$Q'(x', y') \sim \bar{Q}'(x' + x'' - x''', y' + y'' - y'''),$$

as $x'' - x'''$ and $y'' - y'''$ are integers. Hence

$$\begin{aligned}\rho(p, p') &\leq \rho(Q, \bar{Q}') \leq \rho(Q, Q'') + \rho(Q'', \bar{Q}') \\ &= \rho(Q, Q'') + \rho(Q''', Q') = \rho(p, p'') + \rho(p'', p');\end{aligned}$$

hence we have proven (iii).

Let p be a point on T. The *ϵ-neighborhood* $U_\epsilon(p)$ of p is

$$U_\epsilon(p) = \{p' \in T;\ \rho(p', p) < \epsilon\},$$

where $\epsilon > 0$. The following definitions are exactly analogous with those defined for the plane. Let A be a subset of T. We say that p is an *interior point* of A if $U_\epsilon(p) \subset A$, for some $\epsilon > 0$. A point $p' \in T$ is a *cluster point* of A if $U_\epsilon(p') \cap A \neq \varnothing$, for every $\epsilon > 0$. If $U_\epsilon(p') \cap A$ is an infinite set for every $\epsilon > 0$, p' is an *accumulation point* of A. The set of all interior points of A, denoted by $\operatorname{Int} A$, is the *interior* of A. The set A is *open* if $A = \operatorname{Int} A$. The *closure* of the set A, denoted by $\bar{A}$, is the set of all cluster points of A. We say that A is *closed* if $A = \bar{A}$. The set A is *dense* in T if $\bar{A} = T$. For example, the set $A = \{[P(x, y)];\ x, y \text{ rational}\}$ is dense in T.

A point p_0 is a *limit point* of a sequence $p_1, p_2, \dots$ on T if

$$\lim_{n\to\infty} \rho(p_n, p_0) = 0;$$

in this case we simply write

$$\lim_{n\to\infty} p_n = p_0.$$

An example of $U_\epsilon(p)$ is shown in Figure 1.6. For $p = [P(x, y)]$ consider square $A'B'C'D'$, with $A'(x - (1/2), y + (1/2))$, $B'(x - (1/2), y - (1/2))$, $C'(x+(1/2), y-(1/2))$, $D'(x+(1/2), y+(1/2))$ as the vertices. Identifying the opposite sides of this square point to point gives us the torus T. Therefore for any ϵ, $0 < \epsilon < 1/2$, $U_\epsilon(P(x, y))$ becomes the ϵ-neighborhood of p on the torus T.

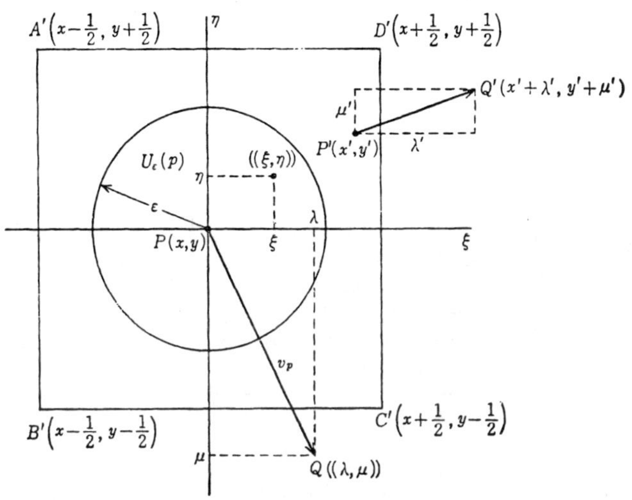

FIGURE 1.6

Motivated by the above example, we now introduce a local coordinate system on $U_\epsilon(p)$. First consider a new coordinate system in the plane: $P(x, y)$ is the origin, the ξ-coordinate is the line through $P(x, y)$ parallel to the x-axis, and the η-coordinate is the line through $P(x, y)$ parallel to the y-axis. We shall use $((\,,\,))$ to represent the points in this new coordinate system, which will be denoted by $(P; \xi, \eta)$. Now passing to T, we can represent points in $U_\epsilon(p)$, $\epsilon < 1/2$, by $((\xi, \eta))$. Here $\sqrt{\xi^2 + \eta^2} < \epsilon$ (Figure 1.6). We shall call $(P; \xi, \eta)$ a *local coordinate system* on $U_\epsilon(p)$, $\epsilon < 1/2$. The plane with coordinate system $(P; \xi, \eta)$ is called the *tangent plane* of T at p. We may think of $U_\epsilon(p)$ as a subset of the tangent plane; the coordinates $((\xi, \eta))$ with $\sqrt{\xi^2 + \eta^2} < \epsilon$ are the local coordinates on $U_\epsilon(p)$.

Throughout the rest of this chapter, 'local coordinate system' on the ϵ-neighborhood $U_\epsilon(p)$ of a point p in the torus T refers to the one we have just developed.

A point $((\lambda, \mu))$ in the tangent plane of T at p is called a tangent vector to T at p; a generic notation for this vector is v_p. Observe that v_p can be regarded as a directed segment $\overrightarrow{P((0, 0))Q((\lambda, \mu))}$ or $\overrightarrow{P(x, y)Q(x+\lambda, y+\mu)}$; as a vector $(P(x, y), (\lambda, \mu))$ based at $P(x, y)$ or $(((0, 0)), ((\lambda, \mu)))$ based at $((0, 0))$; or as a vector $((\lambda, \mu))$. We can even mix the two coordinate systems: $v_p = \overrightarrow{P(x, y)Q((\lambda, \mu))}$ (Figure 1.6). Hence v_p is determined by p and a pair of numbers $((\lambda, \mu))$. We shall call λ and μ the ξ- and η-components respectively of v_p. When $\lambda = 0, \eta = 0$, we say that v_p is a *zero vector*.

The vector $v_{p'}$ at p' whose ξ, η-components are λ', η' can be represented by the directed segment $\overrightarrow{P'(x', y')Q'(\xi' + \lambda', y' + \mu')}$ (Figure 1.6).

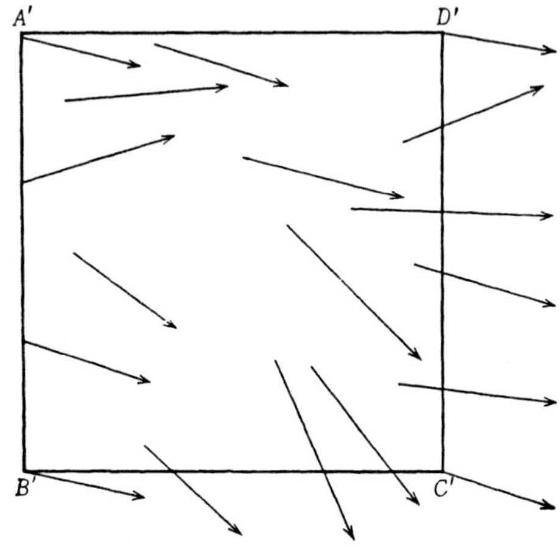

FIGURE 1.7

Let X assign to each point p on T the vector v_p; then as a set, X can be written: $X = \{ v_p ; p \in \mathrm{T} \}$. We call such a set a *vector field* on T. When none of the v_p in X is the zero vector, X is called a *nonsingular vector field*. Transfer each vector of X to a directed segment in square $A'B'C'D'$ as indicated above. We then get the vector field $\hat{X}$ in square $A'B'C'D'$ as defined in §2—notice that the vectors of $\hat{X}$ at the points A', B', C' and D' are parallel translates of each other (Figure 1.7).

A vector field X on T is *of class* C^r at p if $\hat{X}$ is of class C^r at $P(x, y)$ in the sense defined in §2, and X is *of class* C^r if it is of class C^r at every point on T. If X is a C^r vector field on T, then it is clear from definition that $\hat{X}$ is of class C^r in $U_\epsilon(p)$.

Consider a map $\varphi\colon (\alpha, \beta) \to T$, from the open interval (α, β) to the torus T. We say that φ is *continuous* at $\bar{t}$, $\alpha < \bar{t} < \beta$, if for any ϵ-neighborhood $U_\epsilon(\varphi(\bar{t}))$ of $\varphi(\bar{t})$ there exists a number $\delta > 0$ such that

$$\varphi\big((\bar{t} - \delta, \bar{t} + \delta)\big) \subset U_\epsilon(\varphi(\bar{t})).$$

The map φ is a *continuous curve* on T defined on (α, β) if it is continuous at every point of (α, β).

Let $\varphi\colon (\alpha, \beta) \to T$ be a continuous curve; in particular, suppose that for some real number $\delta > 0$ and for each t, $\bar{t} - \delta < t < \bar{t} + \delta$, $\varphi(t) \in U_\epsilon(\varphi(\bar{t}))$. Assume $\epsilon < 1/2$, and consider the local coordinates $((\varphi_1(t), \varphi_2(t)))$ of $\varphi(t)$ in $U_\epsilon(\varphi(\bar{t}))$. Then φ_1, φ_2 are continuous functions defined on the open interval $(\bar{t} - \delta, \bar{t} + \delta)$ since φ is a continuous curve. Here we say that the continuous curve φ is *differentiable of class* C^r at $\bar{t}$ if both φ_1 and φ_2 are differentiable of class C^r at $\bar{t}$. A continuous curve on the torus defined on an open interval (α, β) is called differentiable of class C^r when it is differentiable of class C^r at every point of (α, β).

If φ is a C^r curve, it follows immediately from the definition of local coordinate systems that coordinate functions φ_1 and φ_2 are both C^r on the interval $(\bar{t}-\delta, \bar{t}+\delta)$.

Similarly we can define continuous or C^r curves on T for intervals $[\alpha, \beta]$, (α, ∞), $(-\infty, \infty)$, etc.

A C^r curve on T, $\varphi\colon(-\infty, \infty) \to T$, is called *closed* if there exists a positive number u such that

$$\varphi(t+u) = \varphi(t), \qquad -\infty < t < \infty. \tag{1}$$

If $\bar{u}$ denotes the smallest among the u satisfying (1), the image $\varphi((-\infty, \infty))$ becomes

$$\varphi((-\infty, \infty)) = \varphi([0, \bar{u}]),$$

and φ is determined simply by the values it takes on $[0, \bar{u}]$. In such a situation, a C^r closed curve may be redefined as a C^r map $\varphi\colon [0, \bar{u}] \to T$ with $\varphi(0) = \varphi(\bar{u})$. This means that the right derivatives of the coordinate functions φ_1, φ_2 at $t = 0$ of orders up to r agree with their respective left derivatives at $t = \bar{u}$ of orders up to r, where $((\varphi_1(t), \varphi_2(t)))$ are the local coordinates of $\varphi(t)$, $0 \le t < \delta$ or $\bar{u} - \delta < t \le \bar{u}$, in the ϵ-neighborhood of $\varphi(0) = \varphi(\bar{u})$, $U_\epsilon(\varphi_1(0)) = U_\epsilon(\varphi(\bar{u}))$. The image $\varphi([0, \bar{u}])$ under a closed curve φ is a closed subset of T.

A closed C^r curve φ on T defined on $[0, \bar{u}]$ as above is *simple* if it satisfies $\varphi(t_1) \ne \varphi(t_2)$, for every pair t_1, t_2 such that $0 \le t_1 < t_2 < \bar{u}$. The image $L = \varphi([0, \bar{u}])$ under φ is also called a *simple closed C^r curve.*

Let φ be a C^r curve on T defined over the interval (α, β), $r \ge 1$. We use the local coordinates $((\varphi_1(t), \varphi_2(t)))$ for $\varphi(t)$, $\bar{t}-\delta < t < \bar{t}+\delta$, in $U_\epsilon(\varphi(\bar{t}))$ and define the *tangent vector* to φ at $\varphi(\bar{t})$, denoted $(d\varphi/dt)(\bar{t})$, to be the tangent vector $((\varphi_1'(\bar{t}), \varphi_2'(\bar{t})))$ of T at the point $\varphi(\bar{t})$ (Figure 1.8).

Let X be a C^r vector field on T. A C^{r+1} curve φ defined on an open interval (α, β) is called an *integral curve* of X if the tangent vector $(d\varphi/dt)(t)$

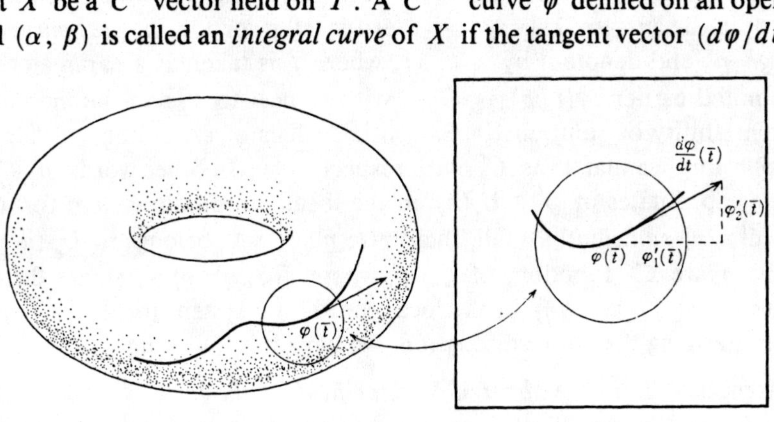

FIGURE 1.8

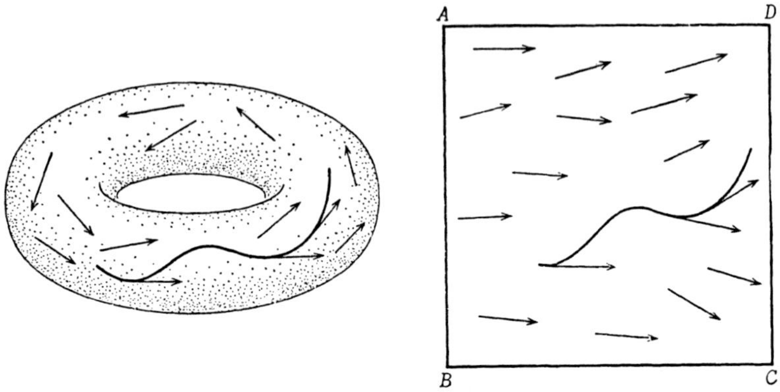

FIGURE 1.9

of φ at $\varphi(t)$ is equal to $v_{\varphi(t)} \in X$ for every t with $\alpha < t < \beta$ (Figure 1.9).

Suppose now that $((\xi, \eta))$ are the local coordinates of $p' \in U_\epsilon(p)$ as before, and denote by $\lambda(\xi, \eta)$ and $\mu(\xi, \eta)$ the ξ and η-components respectively of $v_{p'} \in X$. Then φ is an integral curve of X in $U_\epsilon(p)$ if and only if the following system of ordinary differential equations is satisfied:

$$\frac{d\varphi_1}{dt}(t) = \lambda(\varphi_1(t), \varphi_2(t)), \qquad \frac{d\varphi_2}{dt}(t) = \mu(\varphi_1(t), \varphi_2(t)),$$

where $((\varphi_1(t), \varphi_2(t)))$ are the local coordinates of $\varphi(t)$.

Therefore if $r \geq 1$, our situation here is analogous to that of §2. By the existence and uniqueness theorem for ordinary differential equations there exists a unique integral curve φ of X defined on some open interval $(-\tau, \tau)$ satisfying $\varphi(0) = p$.

Moreover, it is well known that for a C^r vector field X on T, there is a positive number M such that $|\lambda(\xi, \eta)| < M$ and $|\mu(\xi, \eta)| < M$ (see Appendix I, Note 2). Hence adopting the method discussed in §2, we can extend the above φ defined on $(-\tau, \tau)$ to a unique integral curve φ of X defined over the entire real line $(-\infty, \infty)$ satisfying $\varphi(0) = p$; we then say that φ is the *integral curve of* X *starting at* p or the *integral curve of* X *through* p, and denote it by $\varphi(t, p)$, where t is taken as a parameter.

As noted earlier, $\varphi(t, p)$ is C^{r+1} with respect to t, and the theorem on differentiability of solutions in the initial values of an ordinary differential equation implies that φ is C^r with respect to p. In other words, if $((\xi, \eta))$ are the coordinates of $p' \in U_\epsilon(p)$ in the local coordinate system for $U_\epsilon(p)$ with $|\xi|$, $|\eta|$ sufficiently small, then $\varphi(t, p'((\xi, \eta)))$ belongs to $U_{\epsilon'}(\varphi(t, p))$ and $\bar{\xi}$, $\bar{\eta}$ are C^r functions of ξ, η, where $\bar{\xi}(\xi, \eta)$, $\bar{\eta}(\xi, \eta)$ are the coordinates of $\varphi(t, p'((\xi, \eta)))$ in the local coordinate system for $U_{\epsilon'}(\varphi(t, p))$.

Summarizing the above discussion, we have the following

THEOREM 1.2. *Let X be a C^r vector field on the torus T, $r \geq 1$. Then for each point p on T, there exists a unique integral curve $\varphi(t, p)$ of X through p, which is differentiable of class C^{r+1} in t and of class C^r in p.*

If X is a nonsingular vector field, the curve $\varphi(t, p)$ is either closed, or it extends infinitely as t increases or decreases.

Sometimes a vector field is called a *dynamical system*, especially when it is defined as a system of differential equations. In this context the integral curves of the given system of differential equations are called the integral curves of the dynamical system.

Suppose $\bar{p} = \varphi(\bar{t}, p)$ is a point on the integral curve $\varphi(t, p)$ through p of a C^r vector field X on T. Then the curve $\bar{\varphi}$ defined by $\bar{\varphi}(t) = \varphi(t+\bar{t}, p)$ with parameter t is also an integral curve of X, and $\bar{\varphi}(0) = \varphi(\bar{t}, p) = \bar{p}$; so by the uniqueness of integral curves, we have $\bar{\varphi}(t) = \varphi(t, \bar{p})$. Hence

$$\varphi(t, \varphi(\bar{t}, p)) = \varphi(t+\bar{t}, p). \tag{2}$$

We call the subset $\{\varphi(t, p);\ -\infty < t < \infty\}$ of T the *trajectory* or the *orbit* of X through p and denote it by $C(p)$. If $\bar{p} \in C(p)$, then by (2) $C(p) = C(\bar{p})$. From this and the fact that every point of T must belong to a trajectory, we see that T decomposes into the trajectories of X.

We say that the integral curve $\varphi(t, p)$ of X through p is *periodic* if for some t_1 and t_2, $t_1 < t_2$,

$$\varphi(t_1, p) = \varphi(t_2, p) \quad \text{but} \quad \varphi(t, p) \neq \varphi(t_1, p)$$

for some t, $t_1 < t < t_2$. In this case trajectory $C(p)$ is called *periodic* or *closed.*

A periodic integral curve $\varphi(t, p)$ is C^{r+1} and closed, and by the uniqueness of integral curves it must be a simple closed curve.

The trajectory $C(p)$ is *ergodic* on T if it is dense in T, i.e., $\overline{C(p)} = T$. In this case $C(p)$ is obviously not a closed set.

§4. Topological properties of trajectories on T

Let X be a C^r vector field on the torus T, $r \geq 1$, and let $\varphi(t, p)$ be its integral curve through p. The ω-limit set $L^+(p)$ of the trajectory $C(p)$ through p is defined to be

$$L^+(p) = \bigcap_{0 \leq s < \infty} \overline{\{\varphi(t, p);\ s \leq t < \infty\}}.$$

The α-limit set $L^-(p)$ of $C(p)$ is defined by

$$L^-(p) = \bigcap_{-\infty < s \leq 0} \overline{\{\varphi(t, p);\ -\infty < t \leq s\}}.$$

Both of the sets $L^+(p)$ and $L^-(p)$ are nonempty and closed (see Note 3, Appendix I).

Let A be a subset of T. We say that A is an *invariant set* if for each point $p \in A$ the integral curve $\varphi(t, p)$ through p is contained in A:

$$\bigcup_{p \in A} \{\varphi(p, t);\ -\infty < t < \infty\} = A.$$

Theorem 1.3. *Both ω-limit sets and α-limit sets are invariant.*

Proof. Let $x \in L^+(p)$; then by definition there exists a sequence of numbers, $\{t_i\}$, with $\lim_{i\to\infty} t_i = \infty$ such that

$$\lim_{i\to\infty} \varphi(t_i, p) = x.$$

If $\varphi(\hat{t}, x)$ is a point on the trajectory of X through x, then

$$\lim_{i\to\infty} \varphi(t_i + \hat{t}, p) = \lim_{i\to\infty} \varphi(\hat{t}, \varphi(t_i, p)) = \varphi(\hat{t}, x).$$

Hence $\varphi(\hat{t}, x) \in L^+(p)$; so $L^+(p)$ is an invariant set. Similarly we can show that the set $L^-(p)$ is invariant.

We give a few examples of vector fields on T. Let a and b be numbers not both zero, and let $X_{a,b}$ be the vector field which assigns to each point of T the tangent vector whose ξ-component and η-component are a and b respectively. Then $X_{a,b}$ is a smooth nonsingular vector field.

The integral curve $\varphi(t, p)$ of $X_{a,b}$ through p is given as the solution of

$$\frac{d\varphi_1}{dt} = a, \qquad \frac{d\varphi_2}{dt} = b$$

in the local coordinate system of each $U_\epsilon(q)$ ($q \in T$). Setting $p = [P(x, y)]$, we get the integral curve

$$\varphi(t, p) = [Q(x + at, y + bt)]$$

(see Figure 1.10).

The ratio of a and b is either rational or irrational, and we consider each case separately.

(I) *b/a is rational, or $a = 0$.* In this case,

$$a = m\lambda, \qquad b = n\lambda$$

for some integers m, n and a real number λ. Hence

$$\varphi\left(\frac{1}{\lambda}, p\right) = [Q(x + m, y + n)] = p,$$

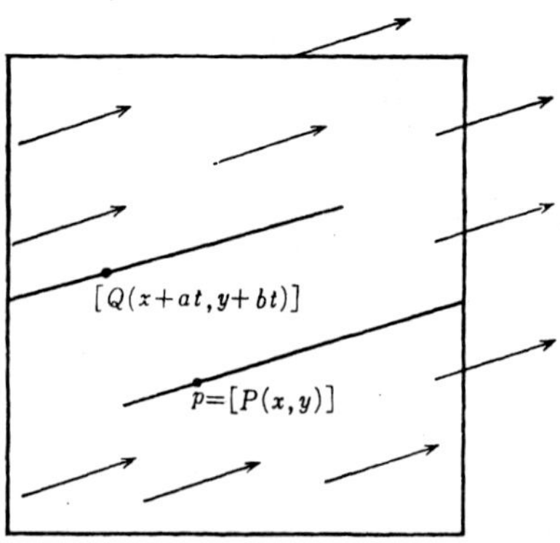

Figure 1.10

and the integral curve starting at p comes back to p at $t = 1/\lambda$. Thus the trajectory $C(p)$ is periodic; $C(p)$ is a simple closed curve on T and is its own α-limit and ω-limit sets.

(II) b/a *is irrational.* Here no trajectory is periodic. For if there were a periodic integral curve $\varphi(t, p)$, there would be numbers t_1, t_2 such that $t_1 \neq t_2$ but $\varphi(t_1, p) = \varphi(t_2, p)$. Letting $p = [P(x, y)]$ we have

$$[P_1(x + at_1, y + bt_1)] = [P_2(x + at_2, y + bt_2)],$$

which would imply that both $a(t_1 - t_2)$ and $b(t_1 - t_2)$ are integers. Hence either $a = 0$, or b/a would have to be rational.

We now let $p' = [P'(x', y')]$ be a point on T. The integral curve of $X_{a,b}$ starting at $p = [P(x, y)]$ passes through

$$p_n = \left[P_n(x', y' + (y - y') + \frac{b}{a}(x' - x) + \frac{b}{a}n)\right],$$

for $n = 0, 1, 2, \ldots$. For a real number λ define $\langle\lambda\rangle$ to be the number in the set $\{\lambda + m; m = 0, \pm 1, \pm 2, \ldots\}$ with the smallest absolute value. Then, as b/a is an irrational number, by the Diophantine approximation the sequence

$$\left\langle y - y' + \frac{b}{a}(x' - x) + \frac{b}{a}n \right\rangle, \qquad n = 0, 1, 2, \ldots$$

has a subsequence which converges to zero (see Note 4, Appendix I). Therefore, given $\epsilon > 0$ there exists a point p_n such that $\rho(p', p_n) < \epsilon$ and hence, $p' \in \overline{C(p)}$. But p' was chosen arbitrarily. Hence

$$\overline{C(p)} = T,$$

that is, C(p) is ergodic. In this case the ω-limit set and the α-limit set of any trajectory $C(p)$ satisfy $L^+(p) = L^-(p) = T$.

Summarizing the above discussion we have the following

THEOREM 1.4. *Every trajectory of $X_{a,b}$ is periodic when $a = 0$, or b/a is a rational number. Every trajectory is ergodic when b/a is an irrational number.*

Thus the trajectories of $X_{a,b}$ must be either all periodic or all nonperiodic. However this is not the case with a nonsingular C^r vector field in general. We shall investigate some vector fields which have both periodic and nonperiodic orbits. But first we prove a lemma (Figure 1.11) which will be frequently used later.

LEMMA 1.5. *There exists a C^∞ function $\Phi(x)$ defined on $(-\infty, \infty)$, satisfying the following*:

(i) $\Phi(x) = \begin{cases} 0, & |x| \geq 2, \\ 1, & |x| \leq 1. \end{cases}$

(ii) $0 \leq \Phi(x) \leq 1$ *for* $1 \leq |x| \leq 2$.

(iii) $\Phi'(x) \geq 0$ *for* $x \leq 0$, $\Phi'(x) \leq 0$ *for* $x \geq 0$.

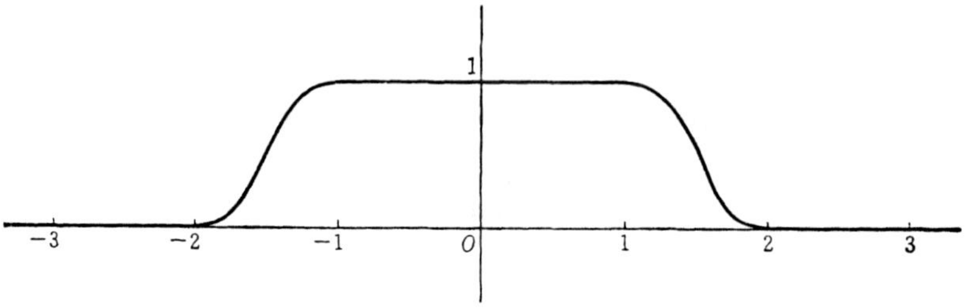

FIGURE 1.11

PROOF. Define $\alpha(x)$ by

$$\alpha(x) = \begin{cases} e^{1/(x(x-1))}, & 0 < x < 1, \\ 0, & x \leq 0 \text{ or } 1 \leq x. \end{cases}$$

Then $\alpha(x)$ is a C^∞ function defined on $(-\infty, \infty)$. Let β be defined by

$$\beta(x) = \int_0^x \alpha(x)\,dx \Big/ \int_0^1 \alpha(x)\,dx.$$

The function $\beta(x)$ is defined on $(-\infty, \infty)$ and is C^∞, such that $\beta(x) = 0$ for $x \leq 0$, $\beta(x) = 1$ for $x \geq 1$, and $0 < \beta(x) < 1$ for $0 < x < 1$. Now define Φ by $\Phi(x) = \beta(x+2)\beta(2-x)$. Then Φ is the desired function.

Let a curve C_s with parameter t, $0 \leq t \leq 1$, be defined on square $ABCD$ with vertices $A(0, 1)$, $B(0, 0)$, $C(1, 0)$, and $D(1, 1)$ by

$$x = t, \qquad y = s + \frac{1}{c}\Phi(2-4s)(\Phi(t+1) - 1),$$

where $0 \leq s \leq 1$, and c is a real number such that

$$c > 4|\Phi'(x)|, \qquad -\infty < x < \infty \tag{3}$$

(see Figure 1.12).

The family of curves $\{C_s;\ 0 \leq s \leq 1\}$ covers square $ABCD$, and as y is monotone increasing in s when t is fixed, $C_s \cap C'_s = \varnothing$ if $s \neq s'$. Furthermore the tangent lines of C_s along the sides of the square $ABCD$ are parallel to the x-axis. Hence the family $\{C_s;\ 0 \leq s \leq 1\}$ gives rise to the C^∞ curves covering the torus T which is obtained from square ABCD via the point-on-point identifications $\overrightarrow{AB}$ with $\overrightarrow{DC}$ and $\overrightarrow{BC}$ with $\overrightarrow{AD}$. Let X be the vector field on T induced by the tangent vectors of $\{C_s; 0 \leq s \leq 1\}$. Then X is a nonsingular C^∞ vector field whose integral curves are $\{C_s;\ 0 \leq s \leq 1\}$ with the above identification. Among the trajectories of X the one corresponding to C_0 or C_1 is closed, and the rest are neither periodic nor ergodic since their closures are different from T.

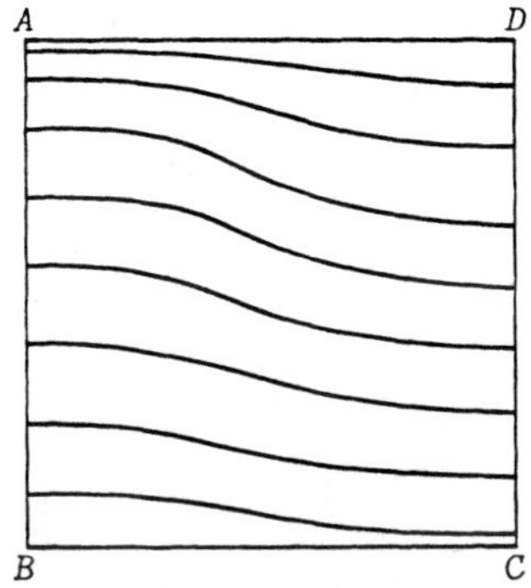

FIGURE 1.12

As we stated in §1, it was Poincaré who proposed investigating the topological properties of trajectories of vector fields on the torus. He posed questions such as 'Are there periodic trajectories?', or 'What are the conditions for every trajectory to be ergodic?'. His approach was radically different from any quantitative method with which onc actually calculated integral curves, etc. What Poincaré investigated was *analytic* vector fields. In 1932 Denjoy, removing the analyticity condition, proved a beautiful theorem concerning the r-dependence of the topological properties of the trajectories of vector fields. We shall state this result in Theorem 1.10 and Theorem 1.12. Later, in 1945, Siegel generalized the result of Denjoy (see Appendix I) and proved the following

THEOREM 1.6 (Denjoy-Siegel). *Let X be a nonsingular C^r vector field on the torus T. If $r \geq 2$, we have one of the following*:

(i) *X has some periodic trajectories.*
(ii) *Every trajectory of X is ergodic.*

In the case $r = 1$, there are examples of vector fields which satisfy neither (i) *nor* (ii).

The proof of the theorem will be given in §§5 through 7.

The vector field $X_{a,b}$ of Theorem 1.4 is an example for case (i) if either $a = 0$, or b/a is rational, and it provides an example for case (ii) if b/a is irrational. The C^∞ vector field defined from $\{C_s; 0 \leq s \leq 1\}$ in (3) is another example of case (i).

§5. Theorem of Denjoy

Denjoy studied a nonsingular C^r vector field $X = \{v_p; p \in T\}$ in which the ξ-component of each v_p is different from zero, $r \geq 2$. Since the ξ-component of v_p is continuous in p, its values are either all positive or all negative. As the proof is identical in each case, we may assume that the ξ-component of each v_p is positive. The torus T is obtained from the cylinder in Figure 1.5 (ii) by identifying the end circles. If we denote by S_0^1

the circle corresponding to side AB and $S_1^{\ 1}$ the circle corresponding to side CD, then $S_0^{\ 1}$ and $S_1^{\ 1}$ represent the same circle on T, but in what follows we think of T as cut along $S_0^{\ 1} = S_1^{\ 1}$ (Figure 1.13).

Let $p \in S_0^{\ 1}$. Then the integral curve $\varphi(t, p)$ of X through p is the solution $((\varphi_1, \varphi_2))$ of

$$\frac{d\varphi_1}{dt}(t) = \lambda(\varphi_1(t), \varphi_2(t)), \qquad \frac{d\varphi_2}{dt} = \mu(\varphi_1(t), \varphi_2(t))$$

in the local coordinate system, so by assumption we have $\lambda(\varphi_1(t), \varphi_2(t)) > 0$. Hence $\varphi(t, p)$ approaches $S_1^{\ 1}$ strictly as t increases from 0. Thus there exists a minimum $\bar{t} > 0$ such that $\varphi(\bar{t}, p) \in S_1^{\ 1}$.

Define the map $f\colon S_0^{\ 1} \to S_1^{\ 1}$ by

$$f(p) = \varphi(\bar{t}, p). \tag{1}$$

More generally, let $g\colon S^1 \to S^1$ be a map of circle S^1 into S^1. If we use complex numbers $e^{2\pi\theta i}$, $\theta \in \mathbf{R}$, to represent the points on S^1 we have $g(e^{2\pi\theta i}) = e^{2\pi\bar{\theta} i}$ for some number $\bar{\theta}$ which is unique up to integers. Set $\bar{g}(\theta) = \bar{\theta}$ so that

$$g(e^{2\pi\theta i}) = e^{2\pi\bar{g}(\theta)i}. \tag{2}$$

Now suppose g is continuous at $e^{2\pi\theta i}$, that is, given $\bar{g}(\theta)$ and $\epsilon > 0$, there exists a $\delta > 0$ such that for $e^{2\pi\theta' i}$ with $|\theta - \theta'| < \delta$ we can select $\bar{g}(\theta')$ which satisfies

$$|\bar{g}(\theta) - \bar{g}(\theta')| < \epsilon.$$

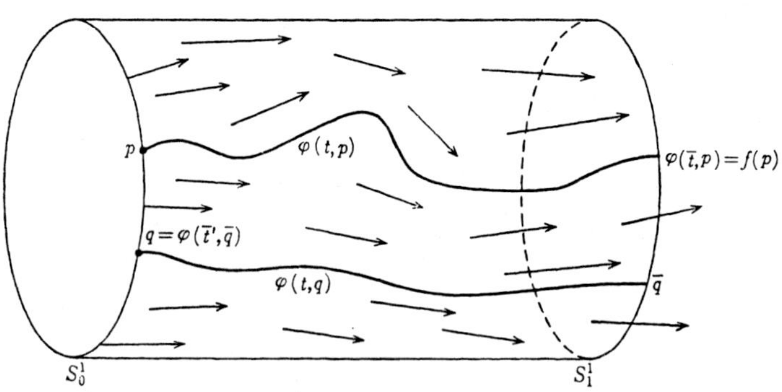

FIGURE 1.13

For θ' with $|\theta - \theta'| < \delta$, a $\bar{g}(\theta')$ satisfying equation (2) is uniquely determined, and so $\bar{g}$ is a continuous function. We say that g is *differentiable of class* C^r at the point $e^{2\pi\theta i}$ if the map $\bar{g}$ is C^r differentiable at θ. The map $g: S^1 \to S^1$ is *differentiable of class* C^r if g is C^r at every point of S^1. The map g is *orientation preserving* if for each point $e^{2\pi\theta i}$ of S^1, we have $\bar{g}(\theta) < \bar{g}(\theta')$ whenever $\theta < \theta' < \theta + \delta$. The map g is a C^r *diffeomorphism* of S^1 if g is one-to-one and onto S^1 such that both g and its inverse $g^{-1}: S^1 \to S^1$ are differentiable of class C^r.

By Theorem 1.2 $\varphi(t, p)$ is C^r in p. The function f in (1) is C^r, because if we set $p = e^{2\pi\theta i}$ and $f(p) = e^{2\pi\bar{f}(\theta)i}$, then $\bar{f}(\theta')$ which is chosen to satisfy $|\bar{f}(\theta) - \bar{f}(\theta')| < \epsilon$, ϵ sufficiently small, whenever $|\theta - \theta'| < \delta$, is a differentiable C^r function of θ'.

The integral curve $\varphi(t, p)$ of X through $\bar{q} \in S_1{}^1$ approaches $S_0{}^1$ strictly as t decreases from zero. Hence there exists a maximum $\bar{t}'$ such that $\bar{t}' < 0$ and $\varphi(\bar{t}', \bar{q}) \in S_0{}^1$. Let $\varphi(\bar{t}') = q$. Then by the uniqueness of the integral curves, we have $f(q) = \bar{q}$ (Figure 1.13), that is, the map which sends $\bar{q}$ to q is the inverse $f^{-1}: S_1{}^1 \to S_0{}^1$ of f. As $\varphi(t, \bar{q})$ is of class C^r with respect to $\bar{q}$, f^{-1} is also of class C^r. Hence $f: S_0{}^1 \to S_1{}^1$ is a C^r diffeomorphism. Further it should be evident from Figure 1.13 that f is orientation preserving.

The integral curve of X through each point of T approaches $S_1{}^1$ strictly as the parameter t increases, and meets $S_1{}^0 = S_1{}^1$. Hence every trajectory of X can be obtained from an integral curve through a point on $S_0{}^1$.

As t increases from zero, the integral curve $\varphi(t, p)$ through $p \in S_0{}^1 = S_1{}^1$ returns to $S_0{}^1 = S_1{}^1$ at $f(p) = \varphi(\bar{t}, p)$. The segment of $\varphi(t, p)$ between p and $f(p)$ is an arc with parameter t, $0 \leq t \leq \bar{t}$ (Figure 1.14). The segment of $\varphi(t, p)$ for $t > \bar{t}$ can be thought of as the integral curve $\varphi(t, f(p))$ defined for $t > 0$; $\varphi(t, p)$ $(t > \bar{t})$ intersects $S_0{}^1 = S_1{}^1$ next at $f(f(p))$ where $\varphi(t, f(p))$ meets $S_0{}^1 = S_1{}^1$ for the first time as t increases from zero. Continuing this iteration we obtain the sequence

$$f(p), f(f(p)), f(f(f(p))), \ldots \tag{3}$$

of points at which the curve $\varphi(t, p)$, $t > 0$, intersects $S_0{}^1 = S_1{}^1$. Similarly we can show that $\varphi(t, p)$, $t < 0$, intersects $S_0{}^1 = S_1{}^1$ at

$$f^{-1}(p), f^{-1}(f^{-1}(p)), f^{-1}(f^{-1}(f^{-1}(p))), \ldots. \tag{4}$$

The topological properties of the trajectory $C(p)$ for $\varphi(t, p)$ are characterized by the intersection points of $C(p)$ and $S_0{}^1 = S_1{}^1$ given in (3) and (4).

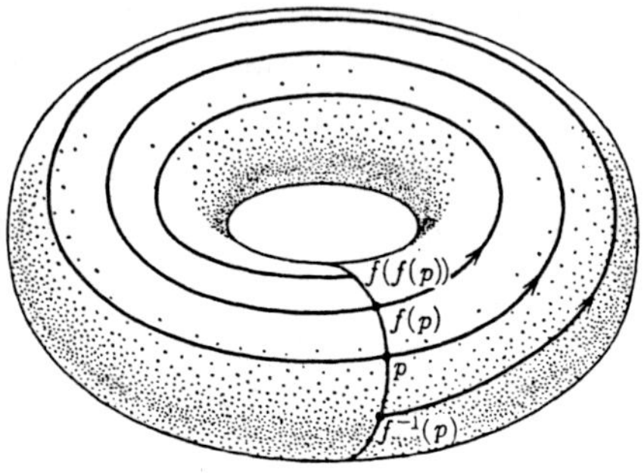

FIGURE 1.14

Given a C^r diffeomorphism $g: S^1 \to S^1$ and $p \in S^1$, we adopt the notation

$$g^m(p) = \overbrace{g(g(g(\cdots(g}^{m}(p))\cdots), \qquad g^{-m}(p) = \overbrace{g^{-1}(g^{-1}(g^{-1}\cdots(g^{-1}}^{m}(p))\cdots),$$

while setting $g^0(p) = p$. A point $p \in S^1$ is *periodic relative to* g if there exists a nonzero integer such that $g^n(p) = p$. If the set $\{g^m(p); m = 0, \pm 1, \pm 2, \dots\}$ is dense in S^1, we say that the point p is *ergodic relative to* g.

If the point $p \in S_0{}^1 = S_1{}^1$ is periodic relative to f, where $f: S_0{}^1 \to S_1{}^1$ was defined in (1), $\varphi(t, p)$ satisfies $\varphi(t', p) = p$ for some $t' \neq 0$; hence $C(p)$ is periodic. On the other hand, if p is ergodic relative to f, then for each point p' on $S_0{}^1 = S_1{}^1$, $\varphi(t, p)$ passes through a point $f^n(p)$ arbitrarily close to p', and so the points $\varphi(t, p')$ and $\varphi(t, f^n(p))$ are arbitrarily close. Hence $\overline{C(p)} = T$. Hence the trajectory $C(p)$ is ergodic.

The function $f: S_0{}^1 \to S_1{}^1$ associated with $X_{a,b}$ in Theorem 1.4, $a \neq 0$, is given by $f(e^{2\pi\theta i}) = e^{2\pi(\theta + b/a)i}$. If b/a is rational then every point is periodic relative to f. If b/a is irrational every point is ergodic.

Denjoy proved the following

THEOREM 1.7 (Denjoy). *Let $g: S^1 \to S^1$ be an orientation-preserving diffeomorphism of class C^r, $r \geq 2$. Then exactly one of the following happens.*

(i) *g has periodic points.*

(ii) *Every point on S^1 is ergodic relative to g.*

The following proof is a streamlined version due to Siegel, which requires two lemmas. Case (i) of Theorem 1.7 does indeed happen, and $X_{a,b}$ with a rational b/a offers an example. So we assume that we are not in case (i) and proceed with our discussion.

For simplicity we write

$$p_n = g^n(p), \qquad n = 0, \pm 1, \pm 2, \dots,$$

where $p \in S^1$.

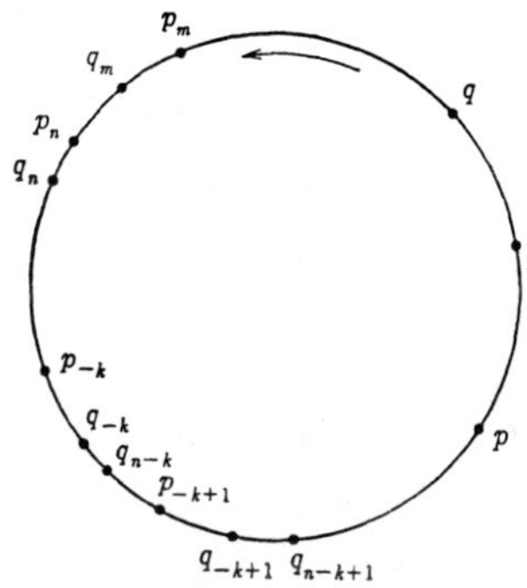

FIGURE 1.15

Choose an orientation for S^1. Given two points $p, q \in S^1$, let pq be the arc from p to q in this orientation. If the point p' is on pq and $p' \neq p, q$ we write

$$p \prec p' \prec q$$

(see Figure 1.15). More generally, if distinct points p', p'', p''', ..., are on the arc pq in this order, we write

$$p \prec p' \prec p'' \prec p''' \prec \cdots \prec q.$$

LEMMA 1.8. *Suppose g has no periodic points. Then given a point p on the unit circle S^1 and a nonzero integer m there exists an integer h such that*

$$p \prec p_h \prec p_m.$$

PROOF. Suppose we have

$$p_1 \prec p_2 \prec \cdots \prec p_l \prec p, \qquad p_1 \prec p_{l+1} \prec p$$

for an integer $l \geq 2$. Then the position of p_{l+1} is either

$$p_l \prec p_{l+1} \prec p \tag{5}$$

or

$$p_k \prec p_{l+1} \prec P_{k+1} \tag{6}$$

for some integer k with $1 \leq k \leq l-1$. If it is (6) then the image of (6) by the map g^{-k} is

$$p \prec p_{l-k+1} \prec p_1 \qquad (2 \leq l-k \leq l),$$

which contradicts the first ordering in our assumption. Hence it must be the position (5), and so we have

$$p_1 \prec p_2 \prec p_3 \prec \cdots \prec p_{l+1} \prec p.$$

Supposing that

$$p_1 \prec p_n \prec p$$

for $n = 2, 3, \ldots$, we repeat the above result to obtain

$$p_1 \prec p_2 \prec p_3 \prec \cdots \prec p.$$

Hence $p_1, p_2, \ldots$ is a bounded monotone sequence, and it converges to a point $\bar{p}$, $\lim_{n\to\infty} p_n = \bar{p}$. But the image of $\bar{p}$ by g is

$$g(\bar{p}) = g(\lim_{n\to\infty} p_n) = \lim_{n\to\infty} g(p_n) = \lim_{n\to\infty} p_{n+1} = \bar{p}.$$

Thus $\bar{p}$ is periodic relative to g, which contradicts the hypothesis that g has no periodic points. Hence we have

$$p \prec p_h \prec p_1$$

for some integer $h > 1$, that is, we have proved Lemma 1.8 for $m = 1$. Replace g by g^m to prove the lemma for a general case.

Let $g: S^1 \to S^1$ be an orientation-preserving C^r diffeomorphism without periodic points. Suppose that a point $u \in S^1$ is not ergodic relative to g. If we denote by K the closure of the set $\{u, g^{\pm 1}(u), g^{\pm 2}(u), \ldots\}$, we have that $K \neq S^1$, and so $S^1 - K$ is a nonempty open set. Hence there exists an arc pq on S^1 with the property that $p, q \in K$ but any other point of pq is outside K. Since u is not periodic relative to g, $p \neq q$. In what follows we will denote by *open* pq the set obtained from the arc pq by removing the points p, q.

From the definition of K, the points

$$p_n = g^n(p), \qquad q_n = g^n(q), \qquad n = 0, \pm 1, \pm 2, \ldots,$$

all belong to K. Furthermore, if a point p' belonging to *open* $p_n q_n$ also belongs to K, then $g^{-n}(p')$ which is in *open* pq must belong to K. Hence the open arc *open* $p_n q_n$ and K have no points in common. Hence for an arbitrary pair of integers m, n ($m \neq n$), *open* $p_n q_n$ and *open* $p_m q_m$ are disjoint (Figure 1.15).

For the above p_n, q_n, $n = 0, \pm 1, \pm 2, \ldots$, we have the following

LEMMA 1.9. *Given a positive integer N, there exists an integer n with $n > N$ which satisfies one of the following*:

(i) *Among the arcs $p_{-k} q_{n-k}$, $k = 1, 2, \ldots, n$, no two have points in common.*

(ii) *Among the arcs $p_{n-k} q_{-k}$, $k = 1, 2, \ldots, n$, no two have points in common.*

PROOF. Consider $2N$ arcs $pp_{j'}$, $j' = \pm 1, \pm 2, \ldots, \pm N$. Suppose arc pp_m is the smallest among them. Then we have

$$p \prec p_m \prec p_{j'}, \qquad j' \neq m.$$

By Lemma 1.8, there is an integer h satisfying

$$p \prec p_h \prec p_m.$$

Evidently $|h| > N$. Among such h's we choose the one with the smallest absolute value and relabel it as h. Then we have

$$p \prec p_h \prec p_j, \qquad 0 < |j| < |h|.$$

Putting $n = |h|$, we get the desired result. For suppose neither (i) nor (ii) holds for $n = |h|$. Then we must have both of the following.

(i)′ Among $p_{-k}q_{|h|-k}$, $k = 1, 2, \dots, n$, there are some which have points in common.

(ii)′ Among $p_{|h|-k}q_{-k}$, $k = 1, 2, \dots, n$, there are some which share common points.

If $h > 0$, (i)′ becomes

(i)″ Among the n arcs $p_{-k}q_{h-k}$, $k = 1, 2, \dots, n$ there are some which share common points.

If $h < 0$, (ii)′ reads:

Among the n arcs $p_{-h-k}q_{-k}$, $k = 1, 2, \dots, n$, there are some which have points in common.

By taking the image $g^h(p_{-h-k}q_{-k}) = p_{-k}q_{h-k}$ of $p_{-h-k}q_{-k}$ by g^h this statement becomes (i)″. Thus regardless of the sign of h, we have (i)″. Therefore for some integers k and l, $k \neq l$, selected from $1, 2, \dots, n$, we must have

$$p_{-k} \prec p_{-l} \prec q_{h-k}. \tag{7}$$

Putting $k - l = j$, we have $0 < |j| < |h|$. The image of (7) by g^k is

$$p \prec p_j \prec q_h.$$

This gives us $p \prec p_j \prec p_h$, since p_j is not on the arc p_hq_h. This contradicts the choice of h. Hence, for $n = |h|$, one of (i) or (ii) must hold.

Proof of Theorem 1.7. Let $g: S^1 \to S^1$ be an orientation-preserving C^r diffeomorphism, $r \geq 2$, which sends a point $e^{2\pi\eta i}$ on S^1 to $g(e^{2\pi\eta i}) = e^{2\pi\bar{g}(\eta)i}$. Then the values of $\bar{g}$ are well defined up to integers. For simplicity, we shall write $e^{2\pi\eta_n i}$ for $g^n(e^{2\pi\eta i})$: that is, $\bar{g}^n(\eta) = \eta_n$. We fix a value of $\bar{g}(\eta)$. If we choose a $\delta > 0$ small enough, for any η' such that $|\eta - \eta'| < \delta$ there is a unique $\bar{g}(\eta')$ satisfying $|\bar{g}(\eta) - \bar{g}(\eta')| < 1/2$. Thus $\bar{g}$ is a C^r function defined on $(\eta - \delta, \eta + \delta)$. Hence

$$\frac{d\bar{g}}{d\eta}(\eta) = \frac{d\eta_1}{d\eta}(\eta)$$

is well defined. Using the chain rule, we get

$$\frac{d\eta_n}{d\eta} = \prod_{k=1}^{n} \bar{g}'(\eta_{n-k}), \qquad \frac{d\eta}{d\eta_{-n}} = \prod_{k=1}^{n} \bar{g}'(\eta_{-k}), \qquad n = 1, 2, \dots. \tag{8}$$

We now assume that neither (i) nor (ii) of Theorem 1.7 holds for g so that there are no periodic points relative to g and that there is a point $u \in S^1$

which is not ergodic relative to g. Using this u we define *open* p_nq_n, $n = 0, \pm 1, \pm 2, \dots$, as before.

Define real numbers α_n, β_n by

$$\textit{open}\ p_nq_n = \{e^{2\pi\eta i} : \alpha_n < \eta < \beta_n\}.$$

The numbers α_n, β_n are unique up to integers. Set

$$\delta_n = \beta_n - \alpha_n.$$

Then $\delta_n > 0$ is well defined, and as no two open arcs among the *open* p_nq_n, $n = 0, \pm 1, \pm 2, \dots$, have points in common, we have

$$\sum_{n=-\infty}^{\infty} \delta_n \leq 1.$$

Hence it follows that

$$\lim_{n\to\infty} \delta_n\delta_{-n} = 0.$$

By the mean value theorem, there exists η', $\alpha_0 < \eta' < \beta_0$, such that

$$\frac{\delta_n}{\delta_0} = \frac{\beta_n - \alpha_n}{\beta_0 - \alpha_0} = \frac{\bar{g}^n(\beta_0) - \bar{g}^n(\alpha_0)}{\beta_0 - \alpha_0} = \frac{d\bar{g}^n}{d\eta}(\eta').$$

Hence by (8) we have

$$\frac{\delta_n}{\delta_0} = \prod_{k=1}^{n} \bar{g}'(\eta'_{n-k}), \tag{9}$$

where $\bar{g}^k(\eta') = \eta'_k$. Similarly from (8) we get

$$\frac{\delta_0}{\delta_{-n}} = \prod_{k=1}^{n} \bar{g}'(\eta''_{-k}), \tag{10}$$

where η'' is some number with $\alpha_0 < \eta'' < \beta_0$, and $\bar{g}^k(\eta'') = \eta''_k$. From (9) and (10) we obtain

$$\log\frac{\delta_0^2}{\delta_n\delta_{-n}} = \log\frac{\delta_0}{\delta_{-n}} - \log\frac{\delta_n}{\delta_0} = \sum_{k=1}^{n}(\log\bar{g}'(\eta''_{-k}) - \log\bar{g}'(\eta'_{n-k}))$$
$$\leq \sum_{k=1}^{n} |\log\bar{g}'(\eta''_{-k}) - \log\bar{g}'(\eta'_{n-k})|.$$

But if $n \to \infty$, then $\delta_n\delta_{-n} \to 0$, and so the left hand side of this expression tends to infinity; hence we must have

$$\lim_{n\to\infty} \sum_{k=0}^{n} |\log\bar{g}'(\eta''_{-k}) - \log\bar{g}'(\eta'_{n-k})| = \infty. \tag{11}$$

Now let $a'_k = e^{2\pi\eta'_k i}$ and $a''_k = e^{2\pi\eta''_k i}$, $k = 0, \pm 1, \pm 2, \dots$. We then have

$$p_{-k} \prec a'_{-k} \prec q_{-k}, \qquad p_{n-k} \prec a''_{n-k} \prec q_{n-k}.$$

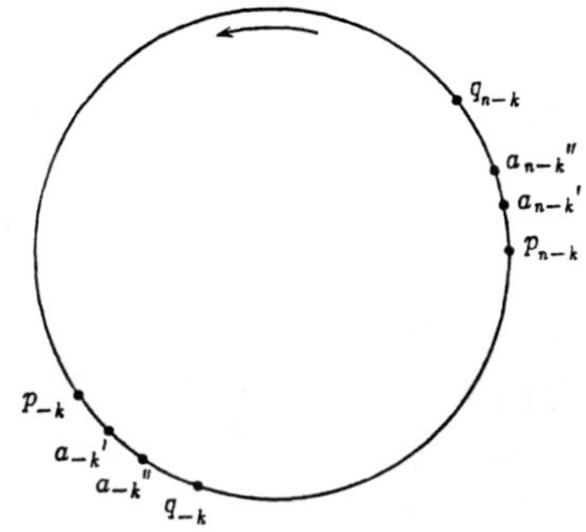

FIGURE 1.16

But we saw earlier that *open* $p_{-k}q_{-k}$ and *open* $p_{n-k}q_{n-k}$ do not intersect. Hence we have

$$\begin{aligned} &p_{-k} \prec a'_{-k} \prec q_{-k} \prec a''_{n-k} \prec q_{n-k} \qquad (k = 1, 2, \ldots, n), \\ &p_{n-k} \prec a''_{n-k} \prec q_{n-k} \prec a'_{-k} \prec q_{-k} \qquad (k = 1, 2, \ldots, n) \end{aligned} \tag{12}$$

(see Figure 1.16).

Since $\bar{g}'(\eta): S^1 \to \mathbf{R}$ is a C^{r-1} function of η and $\bar{g}$ is a C^r diffeomorphism, $r \geq 2$, we have $\bar{g}'(\eta) > 0$. Hence $\log \bar{g}'(\eta)$ is a C^1 function and in particular has bounded variation.

Given an integer N there exists an integer $n > N$ which satisfies one of (i) or (ii) in Lemma 1.9. The relations (12) show that for such an n both sets of arcs $a'_{-k}a''_{n-k}$, $k = 1, 2, \ldots, n$, and $a''_{n-k}a'_{-k}$, $k = 1, 2, \ldots, n$, are pairwise disjoint. Thus $\sum_{k=0}^{n} |\log \bar{g}'(\eta''_{-k}) - \log \bar{g}'(\eta'_{n-k})|$ can be bounded by a constant independent of n, as $\log \bar{g}'(\eta)$ is of bounded variation. This contradicts (11). Hence one of (i) or (ii) of Theorem 1.7 must hold.

It is clear from the proof that Theorem 1.7 still holds if we replace the assumption that g is C^r, $r \geq 2$, by the weaker condition that $\bar{g}'$ has bounded variation.

At the beginning of this section we defined the map $f: S_0^1 \to S_1^1$ by (1) for a nonsingular vector field X on T with the property that the ξ-component of each of its vectors is different from zero, and we observed that if f has a periodic point then X has a periodic trajectory, while if f has an ergodic point X has an ergodic trajectory. Thus the following theorem is an immediate consequence of Theorem 1.7 and a special case of Theorem 1.6; the proof is due to Denjoy.

THEOREM 1.10 (Denjoy). *Let X be a nonsingular vector field on the torus T such that the ξ-component never vanishes. If $r \geq 2$, one of the following holds.*

(i) *X has periodic trajectories.*
(ii) *Every trajectory of X is ergodic.*

§6. Denjoy's C^1 vector field

Denjoy constructed a vector field to show that it is not possible to weaken the differentiability condition $r \geq 2$ to $r \geq 1$ in Theorems 1.7 and 1.10. In this section we discuss this rather fascinating C^1 vector field.

Consider a countable set $\{I_m = [0, l_m];\, m \in \mathbf{Z},\, l_m > 0\}$ of closed intervals which satisfies the following:

(i) $\sum_{m\in\mathbf{Z}} l_m = l$, l finite,
(ii) $\lim_{m\to\pm\infty}(l_m/l_{m+1}) = 1$.

For example we may take $l_m = 1/(1+m^2)$.

Let α be an irrational number and define α_m, $m \in \mathbf{Z}$, by

$$0 \leq \alpha_m < 1, \qquad m\alpha - \alpha_m = \text{an integer}.$$

We construct an interval of length $1+l$ by inserting I_m at α_m, $m \in \mathbf{Z}$, in the interval $[0, 1]$ (Figure 1.17). We obtain a circle of circumference $1+l$ via the identification of the end points of this interval, and denote it by S^1. Think of the following situation:

$$I_m \subset S^1, \qquad m \in \mathbf{Z}, \qquad S^1 = [0, 1] \cup \Big(\bigcup_{m\in\mathbf{Z}} I_m\Big).$$

Construct a C^1 map $f_m\colon I_m \to I_{m+1}$, $m \in \mathbf{Z}$, which satisfies the following:

(i) $df_m/dt > 0$,

(ii) there exists $\delta_m > 0$ such that df_m/dt takes on the value one on $[0, \delta_m)$ and $(l_m - \delta_m, l_m]$, and

(iii)
$$\min\Big\{1, \frac{l_{m+1}}{l_m}\Big\} - \Big(1 - \frac{l_{m+1}}{l_m}\Big)^2 \leq \frac{df_m}{dt} \leq \max\Big\{1, \frac{l_{m+1}}{l_m}\Big\} + \Big(1 - \frac{l_{m+1}}{l_m}\Big)^2.$$

The graph of f_m may be visualized in Figure 1.18, for example, where δ_m is sufficiently small and the broken line $ABCD$ is smoothed out at the corner points B and C by suitable arcs.

Define a map $f_D\colon S^1 \to S^1$ by

$$f_D(x) = \begin{cases} f_m(x) \in I_{m+1} \subset S^1, & x \in I_m, \\ x + \alpha - (\text{integer}) \in [0, 1] \subset S^1, & x \in [0, 1]. \end{cases}$$

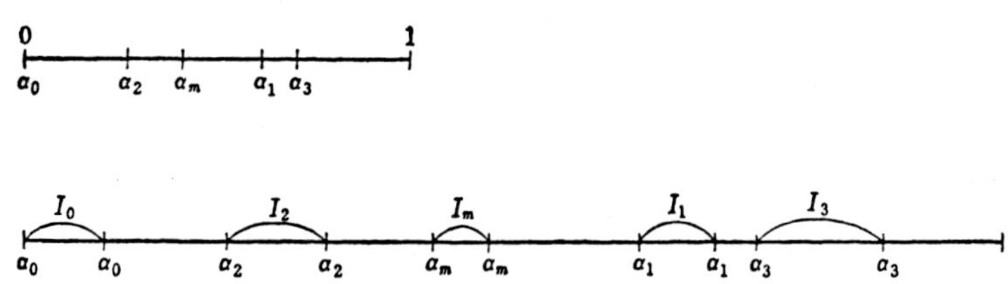

FIGURE 1.17

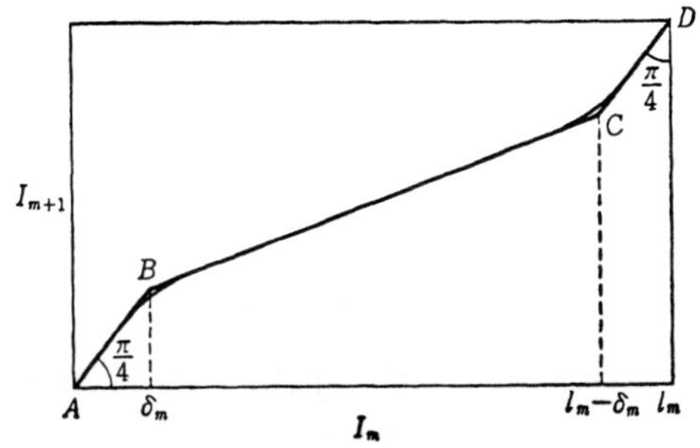

FIGURE 1.18

Evidently f_D is an orientation-preserving C^1 diffeomorphism which has neither periodic nor ergodic points. Thus we have established the following

THEOREM 1.11. *There exists an orientation-preserving C^1 diffeomorphism $f_D: S^1 \to S^1$ which has neither periodic nor ergodic points.*

We use Theorem 1.11 to show that there are C^1 vector fields on T for which Theorem 1.10 does not hold.

For the map f_D in Theorem 1.11, define a C^1 function $\bar{f}_D$ on the closed interval $[0, 1]$ by

$$f_D(e^{2\pi\eta i}) = e^{2\pi\bar{f}_D(\eta)i}, \qquad \eta \in [0, 1],$$

where $e^{2\pi\eta i} \in S^1$. Then $\bar{f}_D$ is monotone increasing.

Define a C^∞ curve $\bar{c}_s: [0, 1] \to S^1 \times I$, $s \in [0, 1)$, by

$$\bar{c}_s(t) = (\exp\{2\pi(s + \Phi(t-2)(\bar{f}_D(s) - s))i\}, t),$$

where Φ is the C^∞ function of Lemma 1.5. From the definition it follows that the family of C^∞ curves on $S^1 \times I$, $\{\bar{c}_s; 0 \le s < 1\}$, satisfies

$$\bar{c}_s(0) = (e^{2\pi si}, 0), \qquad \bar{c}_s(1) = (\exp(2\pi\bar{f}_D(s)i), 1),$$
$$\{\bar{c}_s(t); 0 \le t \le 1\} \cap \{\bar{c}_{s'}(t); 0 \le t \le 1\} = \varnothing, \qquad s \ne s'.$$

We make $S^1 \times I$ into the torus T by identifying $S^1 \times \{0\}$ and $S^1 \times \{1\}$, point by point (Figure 1.5).

Pick a sequence of numbers, s_q, $q = 0, 1, 2, \ldots$, such that

$$0 \le s_q < 1, \qquad (f_D)^q(e^{2\pi si}) = e^{2\pi s_q i}.$$

Then $\bar{c}_s = \bar{c}_{s_0}$ connects to $\bar{c}_{s_1}$, $\bar{c}_{s_1}$ connects to $\bar{c}_{s_2}$, etc., and we can define a curve c_s: $(-\infty, \infty) \to T$.

It follows from the definition of $\bar{c}_s$ that c_s is a C^∞ curve on T.

Denoting by $v_{s,t}$ the tangent vector to the curve c_s at point $c_s(t)$, we see from the definition of c_s that $c_s(t) = c_{s'}(t')$ implies $v_{s,t} = v_{s',t'}$. Hence

$$\{v_{s,t}: 0 \le s < 1, -\infty < t < \infty\} \tag{1}$$

is a vector field on T, which we shall denote by X_D. As f_D is C^1, so is X_D. Since the curve c_s, $0 \le s < 1$, is the integral curve of X_D, it follows from Theorem 1.11 that X_D has neither periodic trajectories nor ergodic trajectories. Thus we have proved

THEOREM 1.12. *There exists a nonsingular C^1 vector field on the torus which has neither periodic nor ergodic trajectories.*

We shall see in §16 that this vector field is an example of a *foliation* in possession of both *proper* and *exceptional leaves.*

§7. Theorem of Siegel

In this section we examine Siegel's proof of Theorem 1.6. For a vector field on T in Theorem 1.10, we were able to select a closed curve, such as $S_0^1 = S_1^1$ of §5, in a natural way so that every integral curve crossed it. We then constructed a C^r diffeomorphism $f\colon S^1 \to S^1$ from this curve and through the behavior of the map f we investigated the topological properties of the trajectories. The existence of such a closed curve for an arbitrary nonsingular vector field is not at all obvious but holds nevertheless. The proof of this is due to Siegel.

Let $X = \{v_p : p \in T\}$ be a nonsingular C^r vector field on the torus T with $r \ge 1$. Let φ be a simple closed $C^{r'}$ curve on T defined over the closed interval $[0, \alpha]$, with $r' \ge 1$. We say that φ or the image of φ is a simple closed curve *transverse* to X if given t, $0 \le t \le \alpha$, there exists no real number λ such that

$$\frac{d\varphi}{dt}(t) = \lambda v_{\varphi(t)}.$$

LEMMA 1.13. *Let X be a nonsingular C^r vector field on T, $r \ge 1$. Then there exists a simple closed C^{r+1} curve $L(X)$ transverse to X.*

PROOF. Let λ_p and μ_p be the ξ-component and the η-component respectively of $v_p \in X$. Denote by $\bar{v}_p$ the tangent vector of T at the point p whose ξ-component is $-\mu_p$ and η-component λ_p. Then $\bar{X} = \{\bar{v}_p : p \in T\}$ is a nonsingular C^r vector field on T (Figure 1.19).

Now fix a point p_0 on T and consider the integral curve $\bar{\varphi}(t, p_0)$ of $\bar{X}$ through p_0. If $\bar{\varphi}(t, p_0)$ is periodic it is a simple closed curve, and clearly

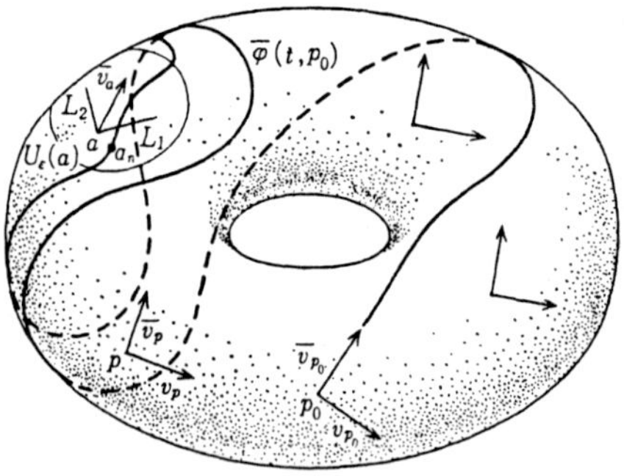

FIGURE 1.19

from the definition of $\bar{X}$ it is transverse to X. Here we have the desired result. On the other hand, if $\overline{\varphi}$ is not periodic we may construct the desired simple closed curve using a segment of $\overline{\varphi}(t, p_0)$ as follows.

Since $\overline{\varphi}(t, p_0)$ is not periodic, the set $\{\overline{\varphi}(n, p_0) : n = 0, 1, 2, \dots\}$ consists of infinitely many points. Thus there exists at least one accumulation point, say a, of this set (see Appendix I, Note 5). We may choose a subsequence $\bar{s}_1, \bar{s}_2, \bar{s}_3, \dots$, of $1, 2, 3, \dots$ so that setting $a_n = \overline{\varphi}(\bar{s}_n, p_0)$, we have $\lim_{n\to\infty} a_n = a$. If we choose a small enough $\epsilon > 0$, the ϵ-neighborhood $U_\epsilon(a)$ of a will satisfy the following.

(i) Denote by $(\bar{\lambda}_p, \bar{\mu}_p)$ the components of vector $\bar{v}_p \in \bar{X}$ at $p \in U_\epsilon(a)$. Then changes in $\bar{\lambda}_p$ and $\bar{\mu}_p$ in U_ϵ (a) are sufficiently small and in particular changes in $\tan^{-1}(\bar{\mu}_p/\bar{\lambda}_p)$ can be less than $\pi/8$.

(ii) The arc $a_n a_{n+1}$ between a_n and a_{n+1} on $\overline{\varphi}(t, p_0)$ is never contained entirely in $U_\epsilon(a)$, $n = 0, 1, 2, \dots$.

Let $(\bar{\lambda}_a, \bar{\mu}_a)$ be the components of the vector $\bar{v}_a \in \bar{X}$ at a. Let L_1, L_2 be subsets of $U_\epsilon(a)$ defined by

$$L_1 = \{(((\bar{\lambda}_a + \bar{\mu}_a)t, (-\bar{\lambda}_a + \bar{\mu}_a)t)); 0 \le t < \epsilon'\}$$
$$L_2 = \{(((\bar{\lambda}_a - \bar{\mu}_a)t, (\bar{\lambda}_a + \bar{\mu}_a)t)); 0 \le t < \epsilon'\},$$

in the local coordinate system on $U_\epsilon(a)$. The sets L_1, L_2 are lines each emanating from a and making angles $\pm\pi/4$ with $\bar{v}_a$ (Figures 1.19 and 1.20).

Since $\lim_{n\to\infty} a_n = a$, $\overline{\varphi}(t, p_0)$ intersects L_1 or L_2 near a_n for some sufficiently large n. We denote the intersection point by $b_n = \overline{\varphi}(s_n, p_0)$. Without loss of generality, we may assume that the arc $a_n b_n$ on $\overline{\varphi}(t, p_0)$ is contained in $U_\epsilon(a)$. Taking a suitable new subsequence, $\bar{s}^1, \bar{s}^2, \bar{s}^3, \dots$, if necessary, we can assume that for each number a_n, $n = 1, 2, \dots$, the b_n as above can be selected. It follows from properties (i), (ii) of $U_\epsilon(a)$ that for each a_n the choice of b_n is unique, the sequence $s_1, s_2, \dots$ is monotone increasing, and the points $b_1, b_2, \dots$ are all distinct. A priori, we have the relation $\lim_{n\to\infty} b_n = a$.

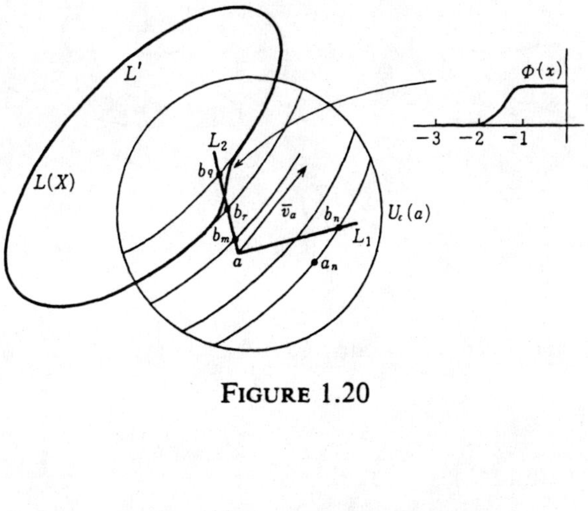

FIGURE 1.20

Since we chose the $b_1, b_2, \ldots, b_n, \ldots$ from $L_1 \cup L_2$, either L_1 or L_2 must contain an infinite number of them. Therefore there exist integers m, q such that $q < m$ and $b_m \in \overline{ab_q}$ (see Figure 1.20). Let b_r be the point at which $\overline{\varphi}(t, p_0)$, $s_q < t \leq s_m$, meets the segment $\overline{b_q b_m}$ for the first time. Then the union of the arc $b_q b_r$ on the trajectory of $\overline{\varphi}(t, p_0)$ and the segment $\overline{b_r b_q}$ on $L_1 \cup L_2$ is a simple closed curve L'. From the construction, L' is obviously transverse to X and is of class C^{r+1} except at b_q, b_r. To smooth out L' at b_q and b_r we use the graph $\{(x, \Phi(x)) : -3 \leq x \leq 0\}$, where $\Phi(x)$ is the C^∞ function constructed in Lemma 1.5. We modify L' by placing the segment

$$\{ (x, \Phi(x)) : -3 \leq x \leq 2, \ -1 \leq x \leq 0 \}$$

on $b_q b_r$ and approximating $\overline{b_q b_r}$ by the segment

$$\{(x, \Phi(x)); -2 \leq x \leq -1\}.$$

The resulting curve, denoted by $L(X)$, is a simple closed C^{r+1} curve transverse to X.

LEMMA 1.14. *The curve* $L(X)$ *of Lemma* 1.13 *does not divide* T *into two domains.*

PROOF. Suppose on the contrary that $L(X)$ divides T into two domains A_1 and A_2 (Figure 1.21 (i)). Noticing that the boundary of A_i is $L(X)$, we construct $M_i = A_i \cup D_i$ by gluing the disk D^2 to the M_i along their boundaries, $i = 1, 2$ (Figure 1.21(ii) & (iii)). Now each M_i is a so-called closed surface (see Appendix I, Note 6).

Since T is a closed orientable surface, the surfaces M_1 and M_2 are also orientable. Moreover, by approximating $L(X)$ by a broken line, we can give each of M_1 and M_2 a simplex cell division. It is well known that a closed orientable surface M can be classified by its Euler number $\chi(M)$; the Euler

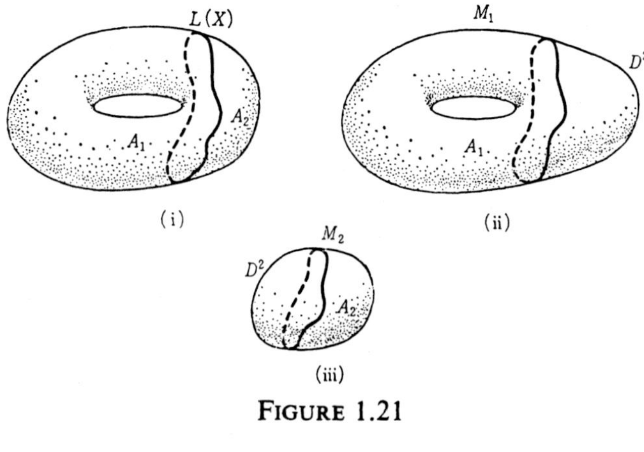

FIGURE 1.21

number $\chi(T)$ is zero, the Euler number $\chi(S^2)$ of the two-sphere S^2 is two, and the Euler number of any other closed orientable surface is negative and even. Suppose neither M_1 nor M_2 is S^2. Then we must have

$$\chi(M_1) \leq 0, \qquad \chi(M_2) \leq 0;$$

hence we have the inequality

$$\chi(A_i) = \chi(M_i) - \chi(D^2) + \chi(S^1) \leq -1, \qquad i = 1, 2,$$

as $\chi(D^2) = 1$ and $\chi(S^1) = 0$. It follows that

$$\chi(T) = \chi(A_1 \cup A_2) = \chi(A_1) + \chi(A_2) - \chi(S^1) \leq -2,$$

but this is a contradiction. So one of M_1 or M_2, say M_1, must be the two-sphere S^2 and A_1 can be obtained by removing the interior of the disk D^2 from the two-sphere S^2. Thus we identify A_1 with D^2; the same argument applies verbatim for the case $M_2 = S^2$.

Denote by X_1 the set $\{v_p ; v_p \in X, p \in A_1\}$; X_1 is a C^r vector field on A_1 and hence on D^2. The vectors of X_1 either all point outward or they all point inward along the boundary S^1 of D^2 since $L(X)$ is transverse to the vector field X (Figure 1.22). Denote by $\hat{v}_p$ the vector of X_1 at p in the disk D^2. Define $f(p)$ to be the point on the boundary S^1 which the line through the origin in the direction of $\hat{v}_p$ passes; f defines a continuous map from D^2 to S^1:

$$f: D^2 \to S^1.$$

Consider the restriction of f to the boundary S^1 of D^2:

$$f|S^1 : S^1 \to S^1.$$

Since either the vectors of X_1 all point outward or they all point inward along the boundary circle S^1, $f|S^1$ is homotopic to the identity map and the mapping degree $\gamma(f|S^1)$ of $f|S^1$ is one (see Appendix I, Note 6). On the other hand, as $f|S^1$ is the restriction of $f: D^2 \to S^1$, we must have that $\gamma(f|S^1) = 0$. But this is a contradiction; hence neither M_1 nor M_2 can be the two-sphere S^2. Hence $L(X)$ does not separate T into two domains.

Lemma 1.15. *Let L be a simple closed C^0 curve on the torus T which does not intersect $L(X)$ of Lemma 1.13. Then L divides $T - L(X)$ into two domains.*

Proof. Cut T along $L(X)$; by Lemma 1.14 we obtain a surface with two boundary components each of which is $L(X)$. We then attach D^2 to this surface along each of the boundary components identifying the boundary of D^2 and $L(X)$ in each case (Figure 1.23). The resulting set M' is a closed

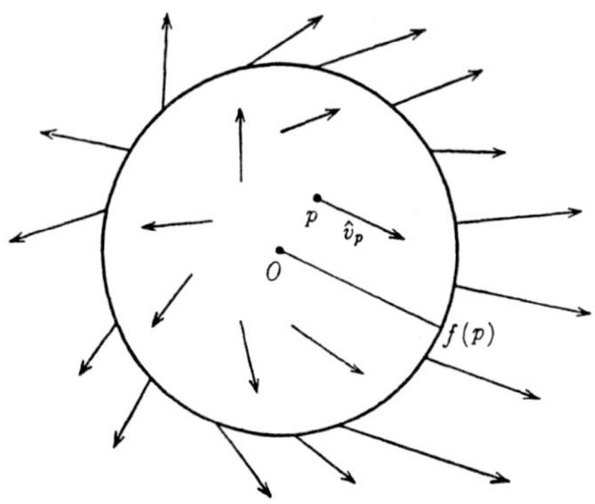

FIGURE 1.22

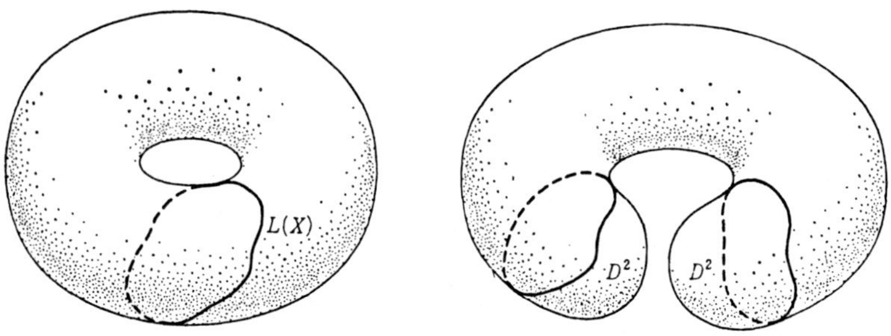

FIGURE 1.23

surface, and as T is orientable so is M'. Since the Euler number $\chi(T)$ is zero we must have

$$\chi(M') = \chi(T) + 2\chi(D^2) - \chi(S^1) = 2,$$

and so according to a classification theorem for closed surfaces, M' must be the two-sphere S^2. But L is a simple closed curve on M'; so L divides M' into two domains by the Jordan curve theorem. Hence L divides the set $T - L(X)$ into two domains.

In the following discussion $X = \{v_p ; p \in T\}$ is a nonsingular C^r vector field with no periodic trajectories.

Let $\varphi(t, q)$ be the integral curve of X through $q \in T$. Assume that the curve $\varphi(t, q)$ does not intersect $L(X)$ for $t > 0$; we want to get a contradiction from this assumption. We saw in §4 that the ω-limit set $L^+(q)$ of the trajectory $C(q)$ passing through q is a nonempty closed set. Since the curve $L(X)$ is transverse to X and $\{\varphi(t, q); 0 \leq t < \infty\} \cap L(X) = \varnothing$, it is obvious that $L^+(q) \cap L(X) = \varnothing$. For $L^+(q)$ we have the following

LEMMA 1.16. *For any point q' in $L^+(q)$, the integral curve $\varphi(t, q')$ of the vector field X through q' is periodic.*

PROOF. Suppose that the integral curve $\varphi(t, q')$ is not periodic. As in the proof of Lemma 1.13, we choose a monotone increasing sequence of positive integers, $t'_1, t'_2, \ldots$, such that the sequence $\varphi(t'_n, q')$, $n = 1, 2, \ldots$, converges to some point c in T. For simplicity we set $a_n = \varphi(t'_n, q')$; then

$$\lim_{n\to\infty} a_n = c.$$

Since L^+ is an invariant set (Theorem 1.3), $a_n \in L^+(q)$, $n = 1, 2, \ldots$; so $c \in L^+(q)$. It follows that $c \notin L(X)$. Thus for a small enough ϵ, the ϵ-neighborhood $U_\epsilon(c)$ of c satisfies the following conditions.

(i) $U_\epsilon(c) \cap L(X) = \varnothing$.

(ii) If $(\lambda_{p'}, \mu_{p'})$ are the coordinates of the vector $v_{p'}$ at $p' \in U_\epsilon(c)$, the change in $(\lambda_{p'}, \mu_{p'})$ is sufficiently small in $U_\epsilon(c)$.

(iii) Compared with $|\lambda_{p'}| + |\mu_{p'}|$, ϵ is sufficiently small.

Now let (λ_c, μ_c) be the coordinates of the vector v_c of X at c, and define the line segment l with respect to the local coordinates of $U_\epsilon(c)$ by

$$l = \{ ((-\mu_c t, \lambda_c t)); -\delta < t < \delta \},$$

where we take δ small enough so that $l \subset U_\epsilon(c)$. The line segment l intersects v_c orthogonally at c (Figure 1.24).

The integral curve $\varphi(t, q')$ intersects l near a_n for a large enough n; denote by $b_n = \varphi(t_n, q')$ this intersection point; here $|t_n - t'_n| < \epsilon'$. Denote by b' the point $\varphi(t', q')$ at which the segment $\varphi(t, q')$, $t > t_n$, intersects l for the first time. Denote by L' the truncated curve $\left\{ \varphi(t, q'); t_n \le t \le t' \right\}$. Then evidently $L' \cap L(X) = \varnothing$. Therefore the union $L = L' \cup \overline{b_n b'}$ of L' and the segment $\overline{b_n b'}$ on l is a simple closed C^0 curve on T, and we have

$$L \cap L(X) = \varnothing$$

(see Figure 1.24). By Lemma 1.15 L divides $T - L(X)$ into two domains: one of them contains $\varphi(t_n - \epsilon', q')$ and will be denoted by A_-, and the other contains $\varphi(t' + \epsilon', q')$ and will be denoted by A_+.

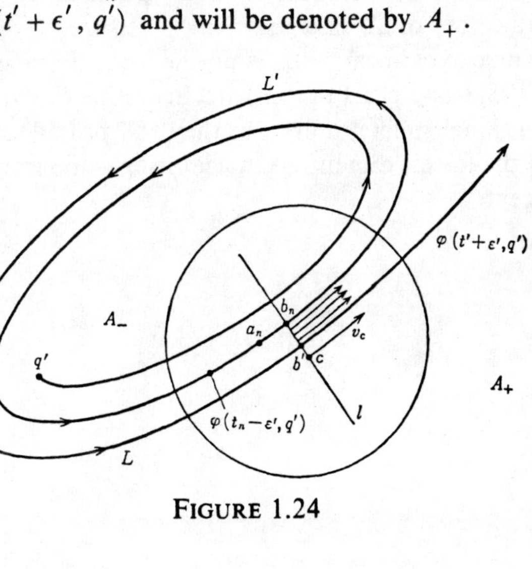

FIGURE 1.24

Since $L^+(q)$ is an invariant set we have that $\varphi(t'+\epsilon', q') \in L^+(q)$; hence for some $\bar{t}$, $\varphi(\bar{t}, q)$ is very close to $\varphi(t'+\epsilon', q')$, and $\varphi(\bar{t}, q) \in A_+$.

We now consider the segment $L_1 = \{\varphi(t, q); \bar{t} \le t < \infty\}$ of $\varphi(t, q)$. By assumption, $L_1 \cap L(X) = \varnothing$. Moreover L_1 does not cross L', and $L_1 \subset A_+$ since the vectors along $\overline{b_n b'}$ are directed from A_- to A_+ (Figure 1.24). Hence we must have that $L^+(q) \subset A_+$, but this contradicts the fact that $\varphi(t_n - \epsilon', q') \in A_-$ is a point of $L^+(q)$. Thus the integral curve $\varphi(t, q')$ is periodic.

We note that Lemma 1.16 remains true if we replace the assumption $t > 0$ by $t < 0$. Now Lemma 1.16 leads to the following

LEMMA 1.17. *Let X be a nonsingular C^r vector field, $r \ge 1$, and assume that none of the integral curves of X is periodic. Then the integral curve $\varphi(t, q)$, starting at an arbitrary point q on T and restricted to $t > 0$, must intersect the simple closed curve $L(X)$ of Lemma 1.13. Here if we replace $t > 0$ by $t < 0$ we will still get the same conclusion.*

In the above lemma the condition that none of the integral curves of X is periodic is necessary; without this condition we may produce a counter example such as the third example in §4.

PROOF OF THEOREM 1.6. We prove Theorem 1.6 using Lemma 1.17 and Theorem 1.7. Suppose that none of the integral curves of X is periodic. The integral curve $\varphi(t, q)$ through q on T intersects $L(X)$ (Lemma 1.17); so it is enough to examine the trajectories through the points on $L(X)$. Denote by $\bar{f}(q)$ the intersection point of $\varphi(t, q)$ with $L(X)$, with the smallest t, $t > 0$: by Lemma 1.17 such $\bar{f}(q)$ always exists. Identify $L(X)$ with S^1; then $\bar{f}(q)$ determines a C^r diffeomorphism $\bar{f}: S^1 \to S^1$, $r \ge 2$. If $\bar{f}$ has a periodic point, we saw in §5 that X has a periodic trajectory; this contradicts our hypothesis. Hence by Theorem 1.7 every point on S^1 must be ergodic. Hence every trajectory of X is ergodic.

The nonsingular dynamical systems on the torus discussed in this chapter will be used in §23, Chapter III to construct Schweitzer's dynamical system on the three-dimensional sphere; moreover they will provide valuable examples of *foliations* of the lowest dimension in acquiring some intuitive understanding of the concept.

CHAPTER II

C^r Manifolds and Tangent Spaces

§8. Topological spaces

In anticipation of defining manifolds later, this section offers a brief review of basic concepts in topology.

Given a set X suppose that a family $\mathcal{O} = \{U_\lambda; U_\lambda \subset X, \lambda \in \Gamma\}$ of subsets of X satisfies the following:

(O_I) $X \in \mathcal{O}$, $\emptyset \in \mathcal{O}$, where $\emptyset$ denotes the empty set.

(O_{II}) If $\{U_\lambda; \lambda \in \Gamma'\}$ is a subset of $\mathcal{O}$, then

$$\bigcup_{\lambda \in \Gamma'} U_\lambda \in \mathcal{O},$$

(O_{III}) If $U_\lambda, U_\mu \in \mathcal{O}$, then $U_\lambda \cap U_\mu \in \mathcal{O}$.

We say that $\mathcal{O}$ is a *topology* or a *system of open sets* on X. Elements of the topology $\mathcal{O}$ are called *open sets* of X, and $(X, \mathcal{O})$ or X a *topological space*.

Elements of X are called *points* of X. For a point x in X, an open set containing x is called a *neighborhood* of x. A subset F of X is called *closed* if $F = X - U$ for some open set U in $\mathcal{O}$. A subset $\mathcal{O}'$ of $\mathcal{O}$ is called a *base* or *basis* for $\mathcal{O}$ if an arbitrary open set $U \in \mathcal{O}$ is a union of some open sets belonging to $\mathcal{O}'$:

$$U = \bigcup_\mu U_\mu, \qquad U_\mu \in \mathcal{O}'.$$

In particular, when $\mathcal{O}'$ is countable we say that $\mathcal{O}'$ is a *countable basis* and that $(X, \mathcal{O})$ is a topological space *with a countable basis.*

Let $(X, \mathcal{O})$ be a topological space and Y a subset of X. Denote by $\mathcal{O}_Y$ the family of open sets

$$\mathcal{O}_Y = \{Y \cap U_\lambda; U_\lambda \in \mathcal{O}\}.$$

Then $\mathcal{O}_Y$ is a topology on Y, and hence $(Y, \mathcal{O}_Y)$ is a topological space. We say that $\mathcal{O}_Y$ is the *relative topology* of Y with respect to $\mathcal{O}$ and that the pair $(Y, \mathcal{O}_Y)$ is a *subspace* of $(X, \mathcal{O})$.

A topological space X is *connected* if X cannot be a union of two disjoint nonempty open sets. A subspace Y of X is *connected* if it is connected in the relative topology of Y. If both subspaces Y_1 and Y_2 of X are connected and $Y_1 \cap Y_2 \neq \emptyset$, it is easy to show that $Y_1 \cup Y_2$ is connected. It follows that for a point p in a topological space X the union C_p of all connected subspaces Y_σ containing p, $C_p = \bigcup_{\sigma \in \Sigma} Y_\sigma$, is connected. We call C_p the *connected component* of X containing p. Now for any pair $p, q \in X$ either $C_p = C_q$ or $C_p \cap C_q = \emptyset$, so X decomposes into its mutually disjoint connected components.

Let $(X, \mathscr{O})$ and $(X', \mathscr{O}')$ be topological spaces. For the set $X \times X'$ there exists a unique family $\tilde{\mathscr{O}}$ whose elements are subsets of $X \times X'$ satisfying the following:

(i) $\tilde{\mathscr{O}}$ is a topology on $X \times X'$.
(ii) If $U_\lambda \in \mathscr{O}$ and $U'_{\lambda'} \in \mathscr{O}'$, then $U_\lambda \times U'_{\lambda'} \in \tilde{\mathscr{O}}$.
(iii) $\tilde{\mathscr{O}}$ is minimal with respect to conditions (i) and (ii).

We call $(X \times X', \tilde{\mathscr{O}})$ the *product space* of $(X, \mathscr{O})$ and $(X', \mathscr{O}')$, and denote it simply by $X \times X'$.

A topological space $(X, \mathscr{O})$ is *discrete* if $\mathscr{O}$ consists of all subsets of X.

A topological space X is called *Hausdorff* if for p, q in X, $p \neq q$, there exist $U, U' \in \mathscr{O}$ such that $p \in U$, $q \in U'$, and $U \cap U' = \emptyset$. A singleton set in a Hausdorff space is closed.

Let A be a set. Suppose that for each pair of elements x, y in A a real number $\rho(x, y)$ is defined and satisfies *the metric axioms*:

(i) $\rho(x, y) \geq 0$, and $x = y$ if and only if $\rho(x, y) = 0$.
(ii) $\rho(x, y) = \rho(y, x)$.
(iii) For $x, y, z \in A$, $\rho(x, y) \leq \rho(x, z) + \rho(z, y)$.

We then say that ρ is a *metric* on A and that A is a *metric space.* A subset B of a metric space A becomes a metric space with the restriction of $\rho(x, y)$ to $x, y \in B$.

Euclidean n-space $\mathbf{R}^n = \{(x_1, x_2, \dots, x_n); x_i \in \mathbf{R}, i = 1, 2, \dots, n\}$ becomes a metric space when we define a metric on $\mathbf{R}^n$ by

$$\rho(x, y) = \sqrt{(x_1 - y_1)^2 + (x_2 - y_2)^2 + \cdots + (x_n - y_n)^2},$$

where $x = (x_1, x_2, \dots, x_n)$ and $y = (y_1, y_2, \dots, y_n)$. It follows that every subset of $\mathbf{R}^n$ is a metric space.

Let A be a metric space. For $x \in A$ and a positive number $\epsilon > 0$, we define the *ϵ-neighborhood* $U_\epsilon(x)$ of x by

$$U_\epsilon(x) = \{y \in A; \rho(x, y) < \epsilon\}.$$

We say that a subset U of A is *open with respect to the metric* ρ if for every x in U, there exists an ϵ such that $U_\epsilon(x) \subset U$. Let $\mathscr{O}_\rho$ be the family of all open sets with respect to ρ; one easily checks that $\mathscr{O}_\rho$ satisfies conditions $(\mathrm{O_I})$, $(\mathrm{O_{II}})$, and $(\mathrm{O_{III}})$; hence $\mathscr{O}_\rho$ is a topology on A. We say that $\mathscr{O}_\rho$ is the *metric topology defined by* ρ, and $(A, \mathscr{O}_\rho)$ is a *topological space determined by metric* ρ. We note that a metric space is Hausdorff.

Let $(X, \mathscr{O})$ be a topological space and A a subset of X. The *closure* of the set A denoted by $\bar{A}$ is the smallest closed set containing A. The largest open set, denoted by $\operatorname{Int} A$, is called the *interior* of A. The *boundary* of A is defined to be $\bar{A} - \operatorname{Int} A$; elements belonging to the boundary are *boundary points* of A. A subset B of X is *dense* in X if $\bar{B} = X$.

Let $p_1, p_2, \dots, p_n, \dots$ be a sequence of points in X. We say that the sequence $p_1, p_2, \dots, p_n, \dots$ *converges* to a point a in X if given a neighborhood U of a there exists a natural number $n(U)$ such that

$$m > n(U) \quad \text{implies} \quad p_m \in U.$$

We write $\lim_{n\to\infty} p_n = a$, and say that a is a *limit point* of this sequence.

In a metric space X, $\lim_{n\to\infty} p_n = a$ if and only if $\lim_{n\to\infty} \rho(p_n, a) = 0$.

A family $\mathfrak{U}$ of open sets of a topological space $(X, \mathscr{O})$

$$\mathfrak{U} = \{ U_\sigma ; U_\sigma \in \mathscr{O}, \sigma \in \Sigma \}$$

is called an *open cover* of X if $\bigcup_\sigma U_\sigma = X$. A subfamily of $\mathfrak{U}$ is a *subcover* if it still covers X. We say that X is a *compact* topological space if every open cover $\mathfrak{U}$ of X has a finite subcover $\{ U_{\sigma_i} ; U_{\sigma_i} \in \mathfrak{U}, i = 1, 2, \dots, s \}$. A bounded closed set in $\mathbf{R}^n$ is compact. Let Y be a closed set in a compact topological space $(X, \mathscr{O})$; then $(Y, \mathscr{O}_Y)$ is compact.

Let $\mathscr{F} = \{ F_\lambda ; \lambda \in \Lambda \}$ be a family of closed sets F_λ in a compact topological space X. Suppose that $\mathscr{F}$ has the *finite intersection property*, that is, $F_{\lambda_1} \cap F_{\lambda_2} \cap \cdots \cap F_{\lambda_n} \neq \varnothing$ for any finite subfamily $\{F_{\lambda_1}, F_{\lambda_2}, \dots, F_{\lambda_n}\}$ of $\mathscr{F}$. Then

$$\bigcap_{\lambda \in \Lambda} F_\lambda \neq \varnothing.$$

For suppose not; then $\{ U_\lambda ; \lambda \in \Lambda \}$, where $U_\lambda = X - F_\lambda$, is an open cover of X, and hence we can choose finitely many U_{λ_i}, $i = 1, 2, \dots, n$, such that $\bigcup_{i=1}^n U_{\lambda_i} = X$. Hence $\bigcap_{i=1}^n F_{\lambda_i} = \varnothing$.

Theorem 2.1. *Let $(A, \mathscr{O}_\rho)$ be a compact metric space and $\{p_i\}_{i=1}^\infty$ a sequence of points in A. Then $\{p_i\}_{i=1}^\infty$ has a subsequence which converges to a point in A.*

Proof. Suppose no convergent subsequence exists. Then for an arbitrary point x in $A - \{p_i\}_{i=1}^\infty$ there is an ϵ-neighborhood $U_\epsilon(x)$ of x such that $U_\epsilon(x) \cap \{p_i\}_{i=1}^\infty = \varnothing$, and so $A - \{p_i\}_{i=1}^\infty$ is an open set in A. Moreover for each p_i there exists $U_{\epsilon_i}(p_i)$ such that $U_{\epsilon_i} \cap \{p_i\}_{i=1}^\infty = \{p_i\}$. Thus although the sets $A - \{p_i\}_{i=1}^\infty$ and the sets $U_{\epsilon_i}(p_i)$, $i = 1, 2, \dots$, form an open cover of A, we cannot select a finite cover of A among them; this contradicts the compactness of A.

Let $(X, \mathscr{O})$ and $(X', \mathscr{O}')$ be topological spaces. A map $f: X \to X'$ is said to be *continuous* if $f^{-1}(U') \in \mathscr{O}$ for every open set $U' \in \mathscr{O}'$. In particular, f is called a *homeomorphism* if f is a one-to-one map of X onto X' and both f and f^{-1} are continuous. We say that X and X' are *homeomorphic* when a homeomorphism f exists between X and X'.

Let X, X', and X'' be topological spaces. If both the map $f: X \to X'$ and the map $f': X' \to X''$ are continuous, so is the map $f' \circ f: X \to X''$.

The image $f([0, 1])$ of a continuous map $f: [0, 1] \to X$ of the closed interval $[0, 1]$ to a topological space X is called a *path* or an *arc* in X. We say that the path $f([0, 1])$ *connects* or *joins* p and q if we have $f(0) = p$ and $f(1) = q$; X is *arcwise connected* or *pathwise connected* if every pair of points x and y in X can be connected by a path in X. It is easy to show that X is connected if it is arcwise connected but the converse is not always true.

For a point p in X, denote by C_p' the union of all arcwise connected subspaces of X containing p. Then C_p' is obviously arcwise connected; C_p' is called the *arcwise connected component* of X containing p. We note that the topological space X decomposes into mutually disjoint arcwise connected components.

§9. C^r manifold

The subspace of $(n+1)$-dimensional Euclidean space $\mathbf{R}^{n+1}$

$$\{(x_1, x_2, \dots, x_{n+1}); x_i \in \mathbf{R}, i = 1, 2, \dots, n+1, x_1^2 + x_2^2 + \cdots + x_{n+1}^2 = 1\}$$

is called the *n-dimensional sphere* or the *n-sphere* and is denoted by S^n. The subspace of $\mathbf{R}^n$

$$\{(x_1, x_2, \dots, x_n); x_i \in \mathbf{R}, i = 1, 2, \dots, n, x_1^2 + x_2^2 + \cdots + x_n^2 \leq 1\}$$

is called the *n-dimensional ball* or *n-ball* and is denoted by D^n. It follows from the definitions that S^{n-1} is a subset of D^n. The *interior* of the n-ball, which we shall denote by $\operatorname{Int} D^n$, is the set $D^n - S^{n-1}$.

A point of S^n is represented by cartesian coordinates $(x_1, x_2, \dots, x_{n+1})$ together with the relation $x_1^2 + x_2^2 + \cdots + x_{n+1}^2 = 1$; these x_i's are not independent. Nevertheless, consider the following subsets of S^n:

$$\begin{aligned} U_i^+ &= \{(x_1, x_2, \dots, x_{n+1}); x_1^2 + x_2^2 + \cdots + x_{n+1}^2 = 1, x_i > 0\}, \\ U_1^- &= \{(x_1, x_2, \dots, x_{n+1}); x_1^2 + x_2^2 + \cdots + x_{n+1}^2 = 1, x_i < 0\}, \end{aligned} \tag{1}$$

where $i = 1, 2, \dots, n+1$. Then a point on U_i^+, for example, can be written uniquely as a sequence $x_1, x_2, \dots, x_{i-1}, x_{i+1}, \dots, x_{n+1}$ of n numbers which are independent. Now let

$$\varphi_i^+: U_i^+ \to \operatorname{Int} D^n \tag{2}$$

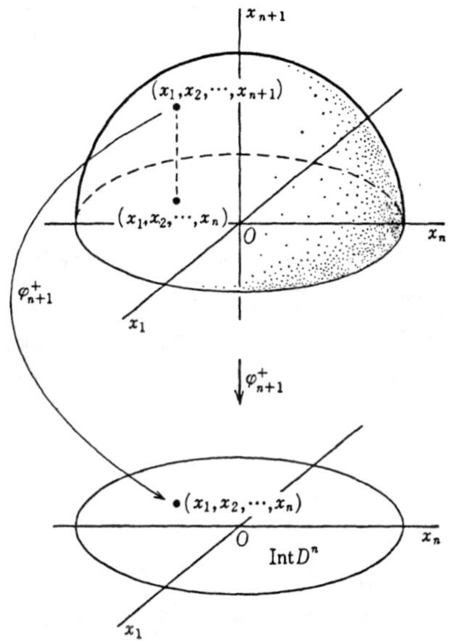

FIGURE 2.1

be the map which associates to each point $(x_1, x_2, \dots, x_{n+1})$ in U_i^+ the point $(x_1, x_2, \dots, x_{i-1}, x_{i+1}, \dots, x_{n+1})$ in $\operatorname{Int} D^n$. Then φ_i^+ is a homeomorphism (see Figure 2.1; here we have the case $i = n + 1$).

We can use cartesian coordinates to represent points in D^n and hence points in $\operatorname{Int} D^n$. On the other hand, there is no natural way we can represent the entire S^n in a cartesian coordinate system, but it is possible to represent the points in U_i^+ by cartesian coordinates via the map φ_i^+ defined in (2). In other words, φ_i^+, or $(\varphi_i^+)^{-1}$ induces 'coordinates' in U_i^+. In like manner we define $\varphi_i^- : U_i^- \to \operatorname{Int} D^n$ to give U_i^- 'coordinates'. We next investigate how two different sets of 'coordinates' representing the same point are related. Take for instance $(x_1, x_2, \dots, x_{n+1}) \in U_1^+ \cap U_2^+$, $x_1 > 0$, $x_2 > 0$, with the induced coordinates:

$$\varphi_1^+(x_1, x_2, \dots, x_{n+1}) = (x_2, x_3, \dots, x_{n+1}),$$
$$\varphi_2^+(x_1, x_2, \dots, x_{n+1}) = (x_1, x_3, \dots, x_{n+1}).$$

We then get a homeomorphism:

$$\varphi_2^+ \circ (\varphi_1^+)^{-1} : \varphi_1^+(U_1^+ \cap U_2^+) \to \varphi_2^+(U_1^+ \cap U_2^+),$$

defined by

$$\varphi_2^+ \circ (\varphi_1^+)^{-1}(y_1, y_2, \dots, y_n) = \left(\sqrt{1 - \sum_{i=1}^n y_i^2}, y_2, \dots, y_n \right), \tag{3}$$

where $(y_1, y_2, \dots, y_n) \in \operatorname{Int} D^n$.

As S^n is the union of the sets U_i^+, U_i^-, $i = 1, 2, \dots, n+1$, by breaking up S^n into U_i^+, U_i^-, we can locally induce on S^n 'coordinates' φ_i^+, φ_i^-

as given by (2) which are related by 'differentiable functions' such as $\varphi_2^+ \circ (\varphi_1^+)^{-1}$ defined by (3). This then is an example of a *manifold*, whose precise definition is to be given next. Roughly speaking a manifold is a topological space with induced local 'coordinates'.

Let r be $0, 1, \dots$, or ∞. Let M be a Hausdorff space with a countable basis—here assuming that M is a metric space with a countable basis hardly reduces the generality. We say that M is an n-dimensional *manifold*, or *n-manifold* if for every point p in M there exists a neighborhood U of p which is homeomorphic to an open subset set of n-Euclidean space $\mathbf{R}^n$.

EXAMPLE 1. $\mathbf{R}^n$ itself is an n-manifold; take the entire $\mathbf{R}^n$ as U.

EXAMPLE 2. The n-sphere S^n also is an n-manifold; take U_i^+ or U_i^- for U.

EXAMPLE 3. The torus T is a 2-manifold; take the ϵ-neighborhood $U_\epsilon(p)$ of p defined in §3 as U.

Let $\varphi: A \to B$ be a map, where A is an open subset of $\mathbf{R}^n$ and B a subset of $\mathbf{R}^m$. Then φ may be split into *component* or *coordinate* functions

$$\begin{aligned}\varphi(x_1, x_2, \dots, x_n) \\ = (\varphi_1(x_1, x_2, \dots, x_n), \varphi_2(x_1, x_2, \dots, x_n), \dots, \varphi_m(x_1, x_2, \dots, x_n)),\end{aligned}$$

where $(x_1, x_2, \dots, x_n) \in A$ and $\varphi_i(x_1, x_2, \dots, x_n)$ is the ith component of $\varphi(x_1, x_2, \dots, x_n) \in B$, $i = 1, 2, \dots, m$. Here we have real-valued functions φ_i, $i = 1, 2, \dots, m$, defined on A. We say that φ is *differentiable of class* C^r at p if each φ_i is a C^r function at p, that is all partial derivatives of φ_i up to the order r exist and are continuous at p, $i = 1, 2, \dots, m$. Throughout the rest of the book we often adopt the expression 'of class C^r' or 'C^r' in place of 'differentiable of class C^r' for simplicity. We say that φ is *of class* C^r if it is C^r at each point of A; for instance, φ is C^0 if and only if φ is a continuous map defined on A. A C^∞ function is often called *smooth*.

If $\varphi: A \to B$, $\psi: B \to C$ are both C^r then it is evident by definition that $\psi \circ \varphi: A \to C$ is C^r, where B is an open subset of $\mathbf{R}^m$ and C a subset of $\mathbf{R}^q$.

Let M be an n-manifold. Let Λ be an index set, U_λ an open set in M, and $\varphi_\lambda: U_\lambda \to V_\lambda$ a homeomorphism of U_λ onto an open set V_λ in Euclidean space $\mathbf{R}^n$, $\lambda \in \Lambda$. The family $\mathscr{S} = \{(U_\lambda, \varphi_\lambda); \lambda \in \Lambda\}$ is called a C^r *coordinate neighborhood system* on M if it satisfies the following:

($\mathrm{M_I}$) $\{U_\lambda; \lambda \in \Lambda\}$ is an open cover of M, that is $\bigcup_{\lambda\in\Lambda} U_\lambda = M$.

($\mathrm{M_{II}}$) When $U_\lambda \cap U_\mu \neq \varnothing$, $\lambda, \mu \in \Lambda$, the maps

$$\varphi_\mu \circ \varphi_\lambda^{-1}: \varphi_\lambda(U_\lambda \cap U_\mu) \to \varphi_\mu(U_\lambda \cap U_\mu),$$
$$\varphi_\lambda \circ \varphi_\mu^{-1}: \varphi_\mu(U_\lambda \cap U_\mu) \to \varphi_\lambda(U_\lambda \cap U_\mu)$$

are C^r (Figure 2.2). Here r is $0, 1, 2, \dots$, or ∞.

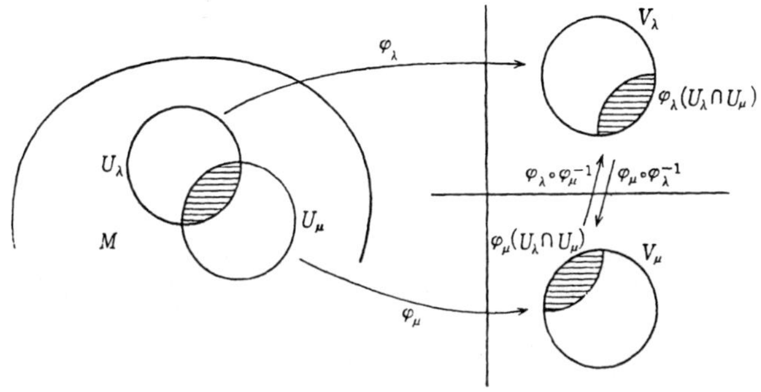

FIGURE 2.2

An n-dimensional topological manifold together with a C^r coordinate neighborhood system $\mathscr{S}$ is called an n-dimensional *differentiable manifold of class* C^r or a C^r n-manifold, which we shall denote by $(M, \mathscr{S})$ or simply by M. Sometimes we write $(M^n, \mathscr{S})$ or M^n to stress the dimension n. Elements of $\mathscr{S}$ are called *coordinate neighborhoods.*

It is evident from the definition that if M is a C^r manifold and $0 \leq r' \leq r$ then M is a $C^{r'}$ manifold. A C^0 manifold is nothing but a topological manifold.

Euclidean n-space $\mathbf{R}^n$ of Example 1 is a C^∞ manifold. The singleton family $\mathscr{S} = \{(\mathbf{R}^n, \mathrm{id})\}$, where $\mathrm{id}\colon \mathbf{R}^n \to \mathbf{R}^n$ is the identity map, is a coordinate neighborhood system for R^n: $(\mathrm{M_I})$ is trivially satisfied, and as for $(\mathrm{M_{II}})$, $\mathrm{id} \circ (\mathrm{id}^{-1}) = \mathrm{id}\colon \mathbf{R}^n \to \mathbf{R}^n$ is a C^∞ map.

The n-sphere S^n of Example 2 is a C^∞ manifold with the coordinate neighborhood system $\mathscr{S} = \{(U_i^\pm, \varphi_i^\pm); i = 1, 2, \dots, n+1\}$, where the open sets $U_i^\pm$ are defined in (1) and the coordinate functions $\varphi_i^\pm$ are defined in (2).

The torus T in Example 3 is also a C^∞ manifold. Using notation and definitions given in §3, we let $\mathscr{S}$ be the family of pairs $(U_\epsilon(p), \varphi_{p,\epsilon})$, where each $\varphi_{p,\epsilon}$ is the homeomorphism

$$\varphi_{p,\epsilon}\colon U_\epsilon(p) \to \{(\xi, \eta); \sqrt{\xi^2 + \eta^2} < \epsilon\},$$

determined by the local coordinate system $(p; \xi, \eta)$ in $U_\epsilon(p)$. The condition $(\mathrm{M_I})$ is of course satisfied, and as Figure 2.3 shows, $\varphi_{p',\epsilon'} \circ \varphi_{p,\epsilon}^{-1}$ is a parallel translate in $\mathbf{R}^2$ and hence is of class C^∞.

EXAMPLE 4. Let $(M^n, \mathscr{S})$ be a C^r manifold and let W be an open set in M. Consider $\mathscr{S}_W = \{(U_\lambda \cap W, \varphi_\lambda | U_\lambda \cap W); (U_\lambda, \varphi_\lambda) \in \mathscr{S}\}$, where $\varphi_\lambda | U_\lambda \cap W$ denotes the map φ_λ restricted to $U_\lambda \cap W$. Then $(W, \mathscr{S}_W)$ is a C^r n-manifold.

Let $(M, \mathscr{S})$ be a C^r n-manifold. Consider a pair (U, φ), where U is an open set in M and $\varphi\colon U \to V$ is a homeomorphism of U to an open

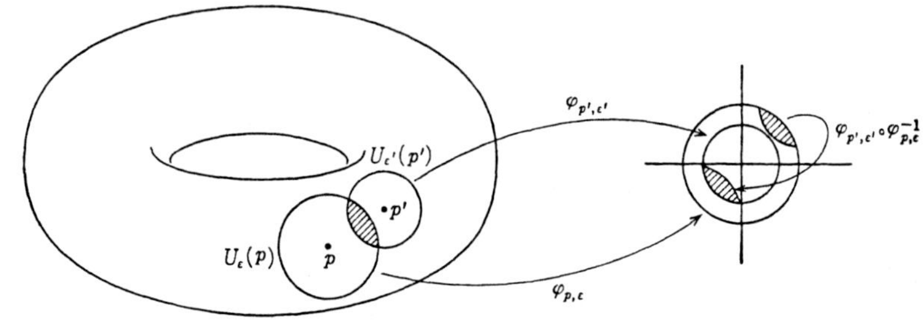

FIGURE 2.3

subset V of $\mathbf{R}^n$. We say that (U, φ) is *compatible*with $\mathcal{S}$ if the maps

$$\varphi_\lambda \circ \varphi^{-1} : \varphi(U \cap U_\lambda) \to \varphi_\lambda(U \cap U_\lambda),$$
$$\varphi \circ \varphi_\lambda^{-1} : \varphi_\lambda(U \cap U_\lambda) \to \varphi(U \cap U_\lambda),$$

are C^r for every $(U_\lambda, \varphi_\lambda) \in \mathcal{S}$ such that $U \cap U_\lambda \neq \phi$. When (U, φ) is compatible with $\mathcal{S}$ we may form a new C^r coordinate neighborhood system $\mathcal{S}'$ on M^n by adding (U, φ) to $\mathcal{S}$; $(M^n, \mathcal{S}')$ is an n-dimensional C^r manifold.

LEMMA 2.2. *Suppose both (U, φ) and (U', φ') are compatible with $\mathcal{S}$. Then the map*

$$\varphi' \circ \varphi^{-1} : \varphi(U \cap U') \to \varphi'(U \cap U')$$

is differentiable of class C^r.

PROOF. For an arbitrary point x in $U \cap U'$ let $(U_\lambda, \varphi_\lambda) \in \mathcal{S}$ be a coordinate neighborhood of x. By our assumption the functions

$$\varphi_\lambda \circ \varphi^{-1} : \varphi(U \cap U' \cap U_\lambda) \to \varphi_\lambda(U \cap U' \cap U_\lambda),$$
$$\varphi' \circ \varphi_\lambda^{-1} : \varphi_\lambda(U \cap U' \cap U_\lambda) \to \varphi'(U \cap U' \cap U_\lambda)$$

are both differentiable of class C^r. Hence,

$$\varphi' \circ \varphi^{-1} = (\varphi' \circ \varphi_\lambda^{-1}) \circ (\varphi_\lambda \circ \varphi^{-1}) : \varphi(U \cap U' \cap U_\lambda) \to \varphi'(U \cap U' \cap U_\lambda)$$

is C^r, and hence $\varphi' \circ \varphi^{-1}$ is C^r at $\varphi(x)$. Therefore $\varphi' \circ \varphi^{-1}$ is differentiable of class C^r in $\varphi(U \cap U')$.

In studying a C^r manifold it would be nice to have free access to coordinate neighborhoods suitable for particular purposes in our investigations. Thus the bigger the coordinate neighborhood system, the better. To this end, given a C^r manifold $(M, \mathcal{S})$ let $\mathcal{S}_1$ be the family of all pairs (U, φ) which are compatible with $\mathcal{S}$. By Lemma 2.2 $\mathcal{S}_1$ is a C^r coordinate neighborhood system on M. The elements of $\mathcal{S}$ are clearly compatible with $\mathcal{S}$, so $\mathcal{S} \subset \mathcal{S}_1$. The set $\mathcal{S}_1$ is maximal among C^r coordinate neighborhood systems which contain $\mathcal{S}$ as a subset; $\mathcal{S}_1$ is uniquely determined by $\mathcal{S}$. In the following we shall assume, replacing $\mathcal{S}$ by $\mathcal{S}_1$ if necessary, that a maximal coordinate neighborhood system is given, that is we add the following to the conditions $(\mathrm{M_I})$ and $(\mathrm{M_{II}})$ in the definition of a C^r coordinate neighborhoods:

$(\mathrm{M_{III}})$ $\mathscr{S}$ is maximal with respect to set inclusions.

A maximal C^r coordinate neighborhood system is called a *differentiable atlas of class* C^r or a *differential structure of class* C^r. We call elements of a C^r atlas *charts* or coordinate neighborhoods as before. Henceforth the term 'atlas' will signify a C^r atlas. We remark that there is no geometric significance attached to $(\mathrm{M_{III}})$; we use a maximal $\mathscr{S}$ because it contains ample neighborhoods and because it is uniquely determined.

Let $(M_1^n, \mathscr{S}_1)$ and $(M_2^m, \mathscr{S}_2)$ be C^r manifolds. For $(U_\lambda, \varphi_\lambda) \in \mathscr{S}_1$ such that $\varphi_\lambda \colon U_\lambda \to V_\lambda$, $V_\lambda \subset \mathbf{R}^n$, and $(U'_{\lambda'}, \varphi'_{\lambda'}) \in \mathscr{S}_2$ with $\varphi'_{\lambda'} \colon U'_{\lambda'} \to V'_{\lambda'}$, $V'_{\lambda'} \subset \mathbf{R}^n$, define a homeomorphism

$$\varphi_{\lambda,\lambda'} \colon U_\lambda \times U'_{\lambda'} \to V_\lambda \times V'_{\lambda'}$$

by $\varphi_{\lambda,\lambda'}(x, y) = (\varphi_\lambda(x), \varphi'_{\lambda'}(y))$, $x \in U_\lambda$, $y \in U'_{\lambda'}$. Then $V_\lambda \times V'_{\lambda'}$ is an open subset of $\mathbf{R}^n \times \mathbf{R}^m = \mathbf{R}^{n+m}$, and

$$\mathscr{S}_3 = \{(U_\lambda \times U'_{\lambda'}, \varphi_{\lambda,\lambda'}); (U_\lambda, \varphi_\lambda) \in \mathscr{S}_1, (U'_{\lambda'}, \varphi'_{\lambda'}) \in \mathscr{S}_2\}$$

satisfies $(\mathrm{M_I})$ and $(\mathrm{M_{II}})$ as a C^r coordinate neighborhood system on the product space $M_1^n \times M_2^m$. Denote by $\mathscr{S}_4$ the maximal coordinate neighborhood system containing $\mathscr{S}_3$. Then $(M_1^n \times M_2^m, \mathscr{S}_4)$ is an $n \times m$-dimensional C^r manifold called the *product manifold* of $(M_1^n, \mathscr{S}_1)$ and $(M_2^m, \mathscr{S}_2)$. For example, the torus T^2 is the product manifold of two copies of S^1's.

Let (M, φ) be a C^r manifold and $f \colon U \to \mathbf{R}$ a real-valued function defined on an open set U in M. For a point p in U, pick $(U_\lambda, \varphi_\lambda) \in \mathscr{S}$ with $p \in U_\lambda$. We then have the function $f \circ \varphi_\lambda^{-1} \colon \varphi_\lambda(U \cap U_\lambda) \to \mathbf{R}$, (see Figure 2.4). We say that f is *differentiable of class* C^r at p if $f \circ \varphi_\lambda^{-1}$ is differentiable of class C^r at $\varphi_\lambda(p)$. This definition does not depend on the

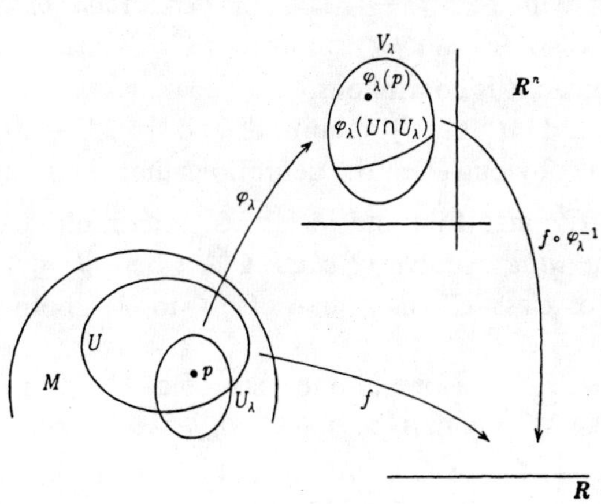

FIGURE 2.4

choice of $(U_\lambda, \varphi_\lambda)$. For consider $f \circ \varphi_\mu^{-1}: \varphi_\mu(U \cap U_\mu) \to \mathbf{R}$, where $p \in U_\mu$ and $(U_\mu, \varphi_\mu) \in \mathscr{S}$. We then have

$$(f \circ \varphi_\mu^{-1})|\varphi_\mu(U \cap U_\lambda \cap U_\mu) = (f \circ \varphi_\lambda^{-1}) \circ ((\varphi_\lambda \circ \varphi_\mu^{-1})|\varphi_\mu(U \cap U_\lambda \cap U_\mu)),$$

and since $\varphi_\lambda \circ \varphi_\mu^{-1}$ is C^r at $\varphi_\mu(p)$, $f \circ \varphi_\mu^{-1}$ is also C^r at $\varphi_\mu(p)$.

We say that f is *differentiable of class* C^r on U if it is differentiable of class C^r at every point in U.

We next consider a map $f: U \to \mathbf{R}^m$. Given $p \in U$, $f(p)$ splits into m coordinates $f(p) = (f_1(p), f_2(p), \dots, f_m(p))$, where $f_i: U \to \mathbf{R}$ are real-valued functions, $i = 1, 2, \dots, m$. We say that f is *differentiable of class* C^r at p if each f_i is differentiable of class C^r in the above definition, and that f is *differentiable of class* C^r in U if it is differentiable of class C^r at every point in U. We may generalize this definition to maps between two manifolds.

Let $(M_1^n, \mathscr{S}_1)$ and $(M_2^m, \mathscr{S}_2)$ be C^r manifolds and let $f: M_1^n \to M_2^m$ be a continuous map. Let $p \in M_1^n$. Then f is *differentiable of class* C^r at the point p if the map

$$\varphi'_{\lambda'} \circ f: f^{-1}(U'_{\lambda'}) \to \mathbf{R}^m,$$

where $f(p) \in U'_{\lambda'}$ and $(U'_{\lambda'}, \varphi'_{\lambda'}) \in \mathscr{S}_2$, is differentiable of class C^r at p in the sense as defined above. This definition is independent of the choice of a chart $(U'_{\lambda'}, \varphi'_{\lambda'})$. For consider $\varphi'_{\mu'} \circ f: f^{-1}(U'_{\mu'}) \to \mathbf{R}^m$, where $(U'_{\mu'}, \varphi'_{\mu'})$ is another chart about $f(p)$. Then we have

$$(\varphi'_{\mu'} \circ f)|f^{-1}(U'_{\lambda'} \cap U'_{\mu'}) = (\varphi'_{\mu'} \circ \varphi'^{-1}_{\lambda'}) \circ ((\varphi'_{\lambda'} \circ f)|f^{-1}(U'_{\lambda'} \cap U'_{\mu'})),$$

and as $\varphi'_{\mu'} \circ \varphi'^{-1}_{\lambda'}$ is C^r at $\varphi'_{\lambda'} \circ f(p)$ so is $\varphi'_{\mu'} \circ f$ at p.

A continuous map $f: M_1^n \to M_2^m$ is *differentiable of class* C^r if it is differentiable of class C^r at every point in M_1^n. We say that the map f is of class C^0 to mean it is continuous.

Let M_1, M_2 and M_3 be C^r manifolds and $f: M_1 \to M_2$, $g: M_2 \to M_3$ be C^r maps. It is obvious from the definitions that $g \circ f: M_1 \to M_3$ is C^r.

Let M be a C^r manifold and M' a $C^{r'}$ manifold. A map of M to M' is said to be *differentiable of class* C^k, where $k \le r$, $k \le r'$, if it is differentiable of class C^k as a map of M to M' both regarded as C^k manifolds.

Let $f: M_1 \to M_2$ be a one-to-one map of a C^r manifold M_1 onto a C^r manifold M_2. We say that f is a C^r *diffeomorphism* if both f and its inverse $f^{-1}: M_2 \to M_1$ are C^r. We say that C^r manifolds M_1 and M_2 are C^r *diffeomorphic* and write $M_1 = M_2$ when there is a C^r diffeomorphism between them. If $M_1 = M_2$, their dimensions are equal.

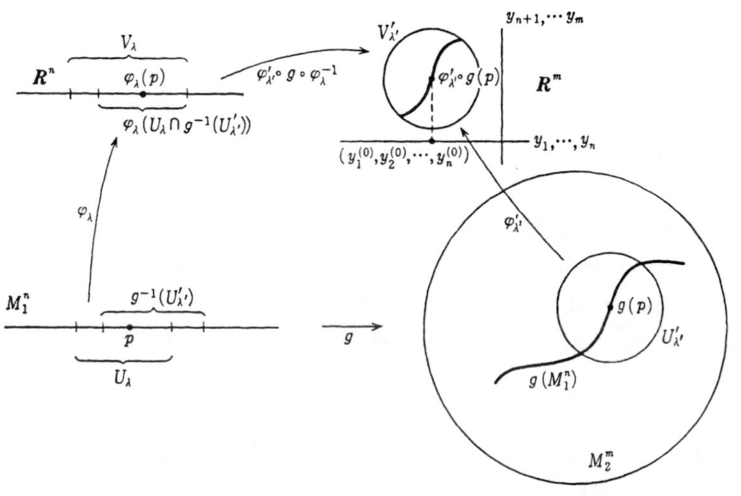

FIGURE 2.5

Let $(M_1^n, \mathscr{S}_1)$ and $(M_2^m, \mathscr{S}_2)$ be C^r manifolds, $r \geq 1$, and let $g: M_1^n \to M_2^m$ be a C^r map. For $p \in M_1^n$ choose a chart $(U_\lambda, \varphi_\lambda) \in \mathscr{S}_1$ about p and a chart $(U'_{\lambda'}, \varphi'_{\lambda'}) \in \mathscr{S}_2$ about $g(p)$. Then since g is C^r, the map

$$\varphi'_{\lambda'} \circ g \circ \varphi_\lambda^{-1}: \varphi_\lambda(U_\lambda \cap g^{-1}(U'_{\lambda'})) \to \varphi'_{\lambda'}(U'_{\lambda'}) = V'_{\lambda'} \subset \mathbf{R}^m$$

is differentiable of class C^r at $\varphi_\lambda(p)$ (Figure 2.5).

In a general setting for advanced calculus we define the Jacobian matrix of a map as follows. Let $\varphi: A \to B$ be a C^r map, $r \geq 1$, of an open subset A of $\mathbf{R}^n$ to a subset B of $\mathbf{R}^m$, which splits into m coordinate functions:

$$\varphi(x_1, x_2, \ldots, x_n) = ((\varphi_1(x_1, x_2, \ldots, x_n), \varphi_2(x_1, x_2, \ldots, x_n), \ldots, \varphi_m(x_1, x_2, \ldots, x_n)).$$

The *Jacobian matrix* $J\varphi(x)$ of φ at $x = (x_1, x_2, \ldots, x_n)$ is:

$$J\varphi(x) = \frac{\partial(\varphi_1, \varphi_2, \ldots, \varphi_m)}{\partial(x_1, x_2, \ldots, x_n)}(x) = \begin{pmatrix} \frac{\partial \varphi_1}{\partial x_1}(x) & \frac{\partial \varphi_2}{\partial x_1}(x) & \cdots & \frac{\partial \varphi_m}{\partial x_1}(x) \\ \frac{\partial \varphi_1}{\partial x_2}(x) & \frac{\partial \varphi_2}{\partial x_2}(x) & \cdots & \frac{\partial \varphi_m}{\partial x_2}(x) \\ \vdots & \vdots & & \vdots \\ \frac{\partial \varphi_1}{\partial x_n}(x) & \frac{\partial \varphi_2}{\partial x_n}(x) & \cdots & \frac{\partial \varphi_m}{\partial x_n}(x) \end{pmatrix}.$$

The rank of the Jacobian matrix $J(\varphi'_{\lambda'} \circ g \circ \varphi_\lambda^{-1})(\varphi_\lambda(p))$ is called the *rank* of g at p. This definition does not depend on the choice of charts $(U_\lambda, \varphi_\lambda)$, $(U''_\lambda, \varphi'_{\lambda'})$. For let $(U_\mu, \varphi_\mu) \in \mathscr{S}_1$ be a chart about p and take another chart $(U'_{\mu'}, \varphi'_{\mu'}) \in \mathscr{S}_2$ about $g(p)$. Consider the map

$$\varphi'_{\mu'} \circ g \circ \varphi_\mu^{-1}: \varphi_\mu(U_\mu \cap g^{-1}(U'_{\mu'})) \to \varphi'_{\mu'}(U'_{\mu'}).$$

We have

$$J(\varphi'_{\mu'} \circ g \circ \varphi_\mu^{-1}) = (J(\varphi'_{\mu'} \circ \varphi'^{-1}_{\lambda'}))(J(\varphi'_{\lambda'} \circ g \circ \varphi_\lambda^{-1}))(J(\varphi_\lambda \circ \varphi_\mu^{-1})),$$

where $\varphi'_{\mu'} \circ \varphi'^{-1}_{\lambda'}$ and $\varphi_\lambda \circ \varphi_\mu^{-1}$ are diffeomorphisms of $\varphi'_{\lambda'}(U'_{\lambda'} \cap U'_{\mu'})$ and $\varphi(U_\lambda \cap U_\mu)$ respectively. Thus the rank of $J(\varphi'_{\mu'} \circ \varphi'^{-1}_{\lambda'})$ at $\varphi'_{\lambda'}(g(p))$ is m, and the rank of $J(\varphi_\lambda \circ \varphi_\mu^{-1})$ at $\varphi_\mu(p)$ is n; hence the rank of $J(\varphi'_{\mu'} \circ g \circ \varphi_\mu^{-1})$ at $\varphi_\mu(p)$ agrees with the rank of $J(\varphi'_{\lambda'} \circ g \circ \varphi_\lambda^{-1})$ at $\varphi_\lambda(p)$.

We say that $g\colon M_1^n \to M_2^m$ is a C^r *immersion* if $n \le m$ and the rank of g is n at each point of M. If g is a C^r immersion, the map

$$\varphi'_{\lambda'} \circ g \circ \varphi_\lambda^{-1}\colon \varphi_\lambda(U_\lambda \cap g^{-1}(U'_{\lambda'})) \to \varphi'_{\lambda'}(U'_{\lambda'}) = V'_{\lambda'} \subset \mathbf{R}^m$$

is of rank n at each $\varphi_\lambda(p)$ in $\varphi_\lambda(U_\lambda \cap g^{-1}(U'_{\lambda'}))$. Let $(x_1, x_2, \ldots, x_n)$ be a point in $\varphi_\lambda(U_\lambda \cap g^{-1}(U'_{\lambda'}))$ and write

$$(\varphi'_{\lambda'} \circ g \circ \varphi_\lambda^{-1})(x_1, x_2, \ldots, x_n) = (y_1, y_2, \ldots, y_m);$$

then $y_1, y_2, \ldots, y_m$ are C^r functions of $x_1, x_2, \ldots, x_n$. The rank of the matrix $J(\varphi'_{\lambda'} \circ g \circ \varphi_\lambda^{-1})$ is n at $\varphi_\lambda(p) = (x_1^{(0)}, x_2^{(0)}, \ldots, x_n^{(0)})$, and reordering $y_1, y_2, \ldots, y_m$ if necessary, we may assume that

$$\left|\frac{\partial(y_1, y_2, \ldots, y_n)}{\partial(x_1, x_2, \ldots, x_n)}(\varphi_\lambda(p))\right| \neq 0$$

(since we took the atlas $\mathscr{S}_2$ to be maximal, any permutation of a coordinate tuple in $\mathscr{S}_2$ is again in $\mathscr{S}_2$!) Hence by the inverse function theorem (see Appendix II, Note 2), there exists a neighborhood V' of $\varphi_\lambda(p)$ such that the map $(\varphi'_{\lambda'} \circ g \circ \varphi_\lambda^{-1})|V'\colon V' \to \varphi'_{\lambda'} \circ g \circ \varphi^{-1}(V')$ is one-to-one and onto, and $x_1, x_2, \ldots, x_n$ as well as $y_{n+1}, y_{n+2}, \ldots, y_m$ are C^r functions of $y_1, y_2, \ldots, y_n$.

It then follows that the image $g(\varphi_\lambda^{-1}(V'))$ of the neighborhood $\varphi_\lambda^{-1}(V')$ of a point $p \in M_1^n$ can be expressed as

$$\begin{aligned} g(\varphi_\lambda^{-1}(V')) = \varphi'^{-1}_{\lambda'}\{ &(y_1, y_2, \ldots, y_n, \alpha_{n+1}(y_1, y_2, \ldots, y_n), \ldots, \\ &\alpha_m(y_1, y_2, \ldots, y_n)) \in \mathbf{R}^m; (y_1, y_2, \ldots, y_n) \in V''\}, \end{aligned}$$

where $(\varphi'_{\lambda'} \circ g)(p) = (y_1^{(0)}, y_2^{(0)}, \ldots, y_m^{(0)})$, V'' is some neighborhood of the point $(y_1^{(0)}, y_2^{(0)}, \ldots, y_n^{(0)}) \in \mathbf{R}^n$, and $\alpha_{n+1}, \alpha_{n+2}, \ldots, \alpha_m$ are C^r functions of $y_1, y_2, \ldots, y_n$, defined on V'' (Figure 2.5). Thus the image $g(M_1^n)$ is locally—with respect to $g(\varphi_\lambda^{-1}(V'))$ and $(U_{\lambda'}', \varphi_{\lambda'}')$—an n-dimensional surface in $\mathbf{R}^m$ defined by C^r functions $y_{n+i} = \alpha_{n+i}(y_1, y_2, \ldots, y_n)$, for $i = 1, 2, \ldots, m-n$.

In particular, we notice that $g|(\varphi^{-1}(V'))$ was one-to-one and hence that any immersion is locally one-to-one. From this observation we extend our

FIGURE 2.6

definition of an immersion to the case $r = 0$: A C^0 map

$$g: M_1^n \to M_2^m$$

is called a C^0 *immersion* if it is locally one-to-one.

EXAMPLE 5. Define a map $g: S^1 \to \mathbf{R}^2$ by

$$g((\cos\theta, \sin\theta)) = (\cos 4\theta \cos 2\theta, \cos 4\theta \sin 2\theta), \qquad 0 \leq \theta < 2\pi$$

(see Figure 2.6). Then g is a C^∞ map whose rank is everywhere one, and so it is a C^∞ immersion. Evidently g is locally one-to-one, but it is not one-to-one as a whole.

A C^r immersion $g: M_1^n \to M_2^2$ is a C^r *embedding* if $g: M_1^n \to g(M_1^n)$ is a homeomorphism in the relative topology on $g(M_1^n)$. For example, the identity map $g: S^n \to \mathbf{R}^{n+1}$ defined by

$$g(x_1, x_2, \ldots, x_{n+1}) = (x_1, x_2, \ldots, x_{n+1}), \qquad (x_1, x_2, \ldots, x_{n+1}) \in S^n,$$

is a C^∞ embedding. More generally any C^r diffeomorphism is a C^r embedding.

A C^r immersion $g: M_1^n \to M_2^m$ is not necessarily an embedding even if it is a one-to-one map. For example, the integral curve $\varphi(t, p)$ with an irrational number b/a in Theorem 1.4 is a one-to-one immersion as a map of $\mathbf{R}^1$ to T, but it is not an embedding. If M_1^n is compact it is easy to show that a one-to-one immersion of M_1^n is an embedding.

A C^r embedding of the unit circle S^1 to M^n (or its image) is called a *simple closed curve of class* C^r in M^n. In speaking of C^r immersions, embeddings and curves we sometimes omit 'C^r' when no confusion arises.

For $q < n$, the map $i: \mathbf{R}^q \to \mathbf{R}^n$ defined by

$$i(x_1, x_2, \ldots, x_q) = (x_1, x_2, \ldots, x_q, 0, \ldots, 0)$$

is a C^∞ embedding. Henceforth, we identify $\mathbf{R}^q$ with $i(\mathbf{R}^q)$ and regard $\mathbf{R}^q$ as naturally sitting in $\mathbf{R}^n$.

Let $(M^n, \mathcal{S})$ be a C^r manifold and W a subset of M^n. We say that the subset W is a q-dimensional *submanifold* or *q-submanifold* of M^n if given a point p in W there exists a chart $(U_\lambda, \varphi_\lambda) \in \mathscr{S}$ about p such that

$$\varphi_\lambda(W \cap U_\lambda) = \varphi_\lambda(U_\lambda) \cap \mathbf{R}^q.$$

EXAMPLE 6. Let $F_1, F_2, \ldots, F_{n-q}$ be C^r functions of $x_1, x_2, \ldots, x_n$ with $r \geq 1$, and

$$W = \{ (x_1, x_2, \ldots, x_n) \in \mathbf{R}^n ; F_i(x_1, x_2, \ldots, x_n) = 0, \\ i = 1, 2, \ldots, n-q \}.$$

Suppose that the rank of matrix $\partial(F_1, F_2, \ldots, F_{n-q})/\partial(x_1, x_2, \ldots, x_n)$ is $n-q$. Assume for simplicity that

$$\left| \frac{\partial(F_1, F_2, \ldots, F_{n-q})}{\partial(x_{q+1}, x_{q+2}, \ldots, x_n)} \right| \neq 0$$

at $(x_1^{(0)}, x_2^{(0)}, \ldots, x_q^{(0)}) \in W$. Then by the implicit function theorem, there exist neighborhoods V of $(x_1^{(0)}, x_2^{(0)}, \ldots, x_n^{(0)})$ in $\mathbf{R}^n$ and V' of $(x_1^{(0)}, x_2^{(0)}, \ldots, x_q^{(0)})$ in $\mathbf{R}^q$ such that $W \cap V$ can be expressed as

$$W \cap V = \{ (x_1, x_2, \ldots, x_q, h_{q+1}(x_1, x_2, \ldots, x_q), \ldots, \\ h_n(x_1, x_2, \ldots, x_q)) ; (x_1, x_2, \ldots, x_q) \in V' \},$$

where the h_i are C^r functions of $x_1, x_2, \ldots, x_q$ defined on the neighborhood V', $i = q+1, q+2, \ldots, n$. Now define a coordinate neighborhood (V, φ) containing $(x_1^{(0)}, x_2^{(0)}, \ldots, x_n^{(0)})$ in $\mathbf{R}^n$ by

$$\varphi(x_1, x_2, \ldots, x_n) = (x_1, x_2, \ldots, x_q, x_{q+1} - h_{q+1}, \ldots, x_n - h_n).$$

Then in the coordinate neighborhood (V, φ) we have $\varphi(W \cap V) = \varphi(V) \cap \mathbf{R}^n$. Hence W is a q-submanifold of $\mathbf{R}^n$.

THEOREM 2.3. (i) *If W is a q-submanifold of C^r manifold $(M^n, \mathscr{S})$, then W is a q-dimensional C^r manifold endowed with the C^r atlas naturally induced from $(M^n, \mathscr{S})$.*

(ii) *The natural inclusion map $\imath: W \to M$ is a C^r embedding when W is regarded as a q-dimensional C^r manifold by* (i).

PROOF. For each $p \in W$ choose a chart $(U_\lambda, \varphi_\lambda) \in \mathscr{S}$ with $p \in U_\lambda$ such that $\varphi_\lambda(W \cap U_\lambda) = \varphi_\lambda(U_\lambda) \cap \mathbf{R}^q$; denote by $\mathscr{S}_W = \{ (U_\lambda, \varphi_\lambda) ; \lambda \in \Lambda' \}$ the set of all such $(U_\lambda, \varphi_\lambda)$. Evidently $\{ W \cap U_\lambda ; \lambda \in \Lambda' \}$ is an open cover of the submanifold W, and we have a homeomorphism

$$\varphi_\lambda | W \cap U_\lambda : W \to \varphi_\lambda(W \cap U_\lambda),$$

from $W \cap U_\lambda$ onto the open set $\varphi_\lambda(W \cap U_\lambda)$ in $\mathbf{R}^q$. Let $(U_\mu, \varphi_\mu) \in \mathscr{S}_W$ be a chart about p. Then the map

$$(\varphi_\mu | W \cap U_\mu) \circ (\varphi_\lambda | W \cap U_\lambda)^{-1} : \varphi_\lambda(W \cap U_\lambda) \to \varphi_\mu(W \cap U_\mu)$$

is C^r since $\varphi_\mu \circ \varphi_\lambda^{-1}\colon \varphi_\lambda(U_\lambda \cap U_\mu) \to \varphi_\mu(U_\lambda \cap U_\mu)$ is C^r. Therefore the family $\{(W \cap U_\lambda, \varphi_\lambda | W \cap U_\lambda); \lambda \in \Lambda\}$ satisfies conditions $(\mathrm{M_I})$ and $(\mathrm{M_{II}})$ and the maximal set containing this family will be the desired C^r atlas for the submanifold W. Thus we have proved (i). By the definition of submanifolds, the map $\iota\colon W \to \iota(W)$ is a homeomorphism whose rank is everywhere q. This establishes (ii).

THEOREM 2.4. *Let $g\colon M_1^n \to M_2^m$ be a C^r embedding, $r \geq 1$. Then the image $g(M_1^n)$ of M_1^n by g is an n-dimensional submanifold of M_2^m.*

PROOF. We saw earlier that for $p \in M_1^n$ we can choose $(U_\lambda, \varphi_\lambda) \in \mathscr{S}_1$ about p and $(U'_{\lambda'}, \varphi'_{\lambda'}) \in \mathscr{S}_2$ about $g(p)$ such that the image of a sufficiently small neighborhood U' of p under $\varphi'_{\lambda'} \circ g$ is an n-dimensional surface in Euclidean m-space $\mathbf{R}^m$ of the form

$$\{(y_1, y_2, \dots, y_n, \alpha_{n+1}(y_1, y_2, \dots, y_n), \dots, \alpha_m(y_1, y_2, \dots, y_n))\}.$$

Take $\widehat{U}'_{\lambda'} \subset U'_{\lambda'}$ such that $p \in g^{-1}(\widehat{U}'_{\lambda'}) \subset U'$ and $\hat{\varphi}'_{\lambda'}$ a coordinate function defined on $\widehat{U}'_{\lambda'}$ which replaces $(y_1, y_2, \dots, y_n, \dots, y_{n+1}, \dots, y_m)$ by

$$(y_1, y_2, \dots, y_{n+1} - \alpha_{n+1}(y_1, y_2, \dots, y_n), \dots, y_m - \alpha_m(y_1, y_2, \dots, y_n)).$$

Then we see that $\hat{\varphi}'_{\lambda'}(g(M_1^n) \cap \widehat{U}'_{\lambda'}) = \hat{\varphi}'_{\lambda'}(\widehat{U}'_{\lambda'}) \cap \mathbf{R}^n$ with respect to the coordinate neighborhood $(\widehat{U}'_{\lambda'}, \hat{\varphi}'_{\lambda'})$.

A C^r map $g\colon M_1^n \to M_2^m$, $r \geq 1$, is called a C^r *submersion* if $n \geq m$ and the rank of g is m at every point in M_1^n. Let g be a C^r submersion and let $p \in M_1^n$. Choose $(U_\lambda, \varphi_\lambda) \in \mathscr{S}_1$ about p and $(U_{\lambda'}{}', \varphi_{\lambda'}{}') \in \mathscr{S}_2$ about $g(p)$ as above. The rank of the Jacobian matrix of the map

$$\varphi_{\lambda'}{}' \circ g \circ \varphi_\lambda^{-1}\colon \varphi_\lambda(U_\lambda \cap g^{-1}(U_{\lambda'}{}')) \to \varphi_{\lambda'}{}'(U_{\lambda'}{}')$$

at $\varphi_\lambda(p)$ is m. Write

$$(\varphi_{\lambda'}{}' \circ g \circ \varphi_\lambda^{-1})(x_1, x_2, \dots, x_n) = (y_1, y_2, \dots, y_m),$$
$$\varphi_\lambda(p) = (x_1^{(0)}, x_2^{(0)}, \dots, x_n^{(0)}).$$

By reordering $x_1, x_2, \dots, x_n$ if necessary we may assume that

$$\left|\frac{\partial(y_1, y_2, \dots, y_m)}{\partial(x_{n-m+1}, x_{n-m+2}, \dots, x_n)}\right| \neq 0$$

at $(x_1^{(0)}, x_2^{(0)}, \dots, x_n^{(0)})$. Therefore choosing a small enough $\epsilon > 0$ and defining a C^r map

$$\bar{g}\colon V_\epsilon \to \mathbf{R}^{n-m} \times \mathbf{R}^m,$$

from $V_\epsilon = \{(x_1, x_2, \dots, x_n) \in \mathbf{R}^n; |x_i - x_i^{(0)}| < \epsilon, i = 1, 2, \dots, n\}$ to the product $\mathbf{R}^{n-m} \times \mathbf{R}^m = \mathbf{R}^n$, by

$$\bar{g}(x_1, x_2, \dots, x_n) = (x_1, x_2, \dots, x_{n-m}, y_1, y_2, \dots, y_m),$$

we get

$$\left|\frac{\partial(x_1, x_2, \dots, x_{n-m}, y_1, y_2, \dots, y_m)}{\partial(x_1, x_2, \dots, x_n)}\right| \neq 0$$

at each point in V_ϵ. We take a suitable $(U'_{\lambda'}, \varphi'_{\lambda'})$ in which

$$\bar{g}(x_1^{(0)}, x_2^{(0)}, \dots, x_n^{(0)}) = (x_1^{(0)}, x_2^{(0)}, \dots, x_{n-m}^{(0)}, 0, 0, \dots, 0),$$

and for a sufficiently small ϵ' we let

$$V_{\epsilon'}' = \{(x_1, x_2, \dots, x_{n-m}) \in \mathbf{R}^{n-m}; |x_i - x_i^{(0)}| < \epsilon', i = 1, 2, \dots, n-m\},$$
$$V_{\epsilon'}'' = \{(y_1, y_2, \dots, y_m) \in \mathbf{R}^m; |y_i| < \epsilon', i = 1, 2, \dots, m\}.$$

Then by the inverse function theorem, there exists a C^r map

$$\bar{h}: V_{\epsilon'}' \times V_{\epsilon'}'' \to V_\epsilon,$$

such that

$$(\bar{g} \circ \bar{h})(x_1, x_2, \dots, x_{n-m}, y_1, y_2, \dots, y_m)$$
$$= (x_1, x_2, \dots, x_{n-m}, y_1, y_2, \dots, y_m).$$

A priori $\bar{h}(V_{\epsilon'}' \times V_{\epsilon'}'')$ is open in $\mathbf{R}^n$, and $\bar{h}: V_{\epsilon'}' \times V_{\epsilon'}'' \to \bar{h}(V_{\epsilon'}' \times V_{\epsilon'}'')$ is a C^r diffeomorphism. Now we put

$$U = (\varphi_\lambda^{-1} \circ \bar{h})(V_{\epsilon'}' \times V_{\epsilon'}''), \qquad \varphi = \bar{g} \circ (\varphi_\lambda | U).$$

Then the pair (U, φ) is a chart of $\mathscr{S}_1$ and evidently we have

$$g^{-1}(g(p)) \cap U = \varphi^{-1}(V_{\epsilon'}' \times (0, 0, \dots, 0)) = \varphi^{-1}(\varphi(U) \cap \mathbf{R}^{n-m}).$$

Thus we have the following

THEOREM 2.5. *Let $g: M_1^n \to M_2^m$ be a C^r submersion. Then for every point p' of $g(M_1^n)$, $g^{-1}(p')$ is an $(n-m)$-dimensional submanifold of M_1^n.*

Let M^n be a C^r manifold, and $f: M \to \mathbf{R}$ a real-valued function defined on M^n. The *support* of the function f is defined to be the closure of the subset $\{p \in M; f(p) \neq 0\}$ of M^n and denote it by $\operatorname{supp} f$.

Let $\mathfrak{U} = \{U_\sigma; \sigma \in \Sigma\}$ be an open cover of M^n. We say that $\mathfrak{U}$ is *locally finite* if each point p in M^n has a neighborhood U such that $U \cap U_\sigma \neq \emptyset$ for finitely many U_σ, $\sigma \in \Sigma$.

Given a locally finite open cover $\mathfrak{U} = \{U_\sigma; \sigma \in \Sigma\}$ of M^n, a *partition of unity* subordinate to $\mathfrak{U}$ is defined to be C^r functions $\mu_\sigma: M^n \to [0, 1]$, $\sigma \in \Sigma$, which satisfy the following.

(i) For each $\sigma \in \Sigma$, $\operatorname{supp} \mu_\sigma \subset U_\sigma$.
(ii) For each $p \in M^n$, $\sum_{\sigma \in \Sigma} \mu_\sigma(p) = 1$.

THEOREM 2.6. *Let* $(M^n, \mathscr{S})$ *be a compact* C^r *manifold and suppose that* $\mathfrak{U} = \{U_\sigma ; \sigma \in \Sigma\}$ *is a locally finite open cover of* M^n. *Then there exists a partition of unity subordinate to* $\mathfrak{U}$ (the theorem holds without the compactness; see Appendix II, Note 2).

PROOF. For each point p in M there is a chart $(U_{\lambda(p)}, \varphi_{\lambda(p)})$ in $\mathscr{S}$ about p such that $U_{\lambda(p)}$ is contained in U_σ for some $\sigma \in \Sigma$, and $\varphi_{\lambda(p)}(U_{\lambda(p)}) = \operatorname{Int} D^n$. Such a pair $(U_{\lambda(p)}, \varphi_{\lambda(p)})$ exists by virtue of condition $\mathrm{M}_{\mathrm{III}}$.

Set $\operatorname{Int} D^n(1/2) = \{(x_1, x_2, \ldots, x_n) \in D^n ; x_1^2 + x_2^2 + \cdots + x_n^2 < 1/4\}$, and $U_{\lambda(p)}{}' = \varphi_{\lambda(p)}^{-1}(\operatorname{Int} D^n(1/2))$. As $\{U_{\lambda(p)}{}' ; p \in M^n\}$ is obviously an open cover of M^n, we may choose finitely many $U_{\lambda(p_1)}{}', U_{\lambda(p_2)}{}', \ldots, U_{\lambda(p_s)}{}'$ from it such that $\bigcup_{i=1}^s U_{\lambda(p_i)}{}' = M^n$. Now assume that $U_{\lambda(p_i)} \subset U_{\sigma_i}$, $i = 1, 2, \ldots, s$, and using the C^∞ function Φ of Lemma 1.5, define $\bar{\mu}_{p_i} ; M \to \mathbf{R}$ by

$$\bar{\mu}_{p_i}(q) = \begin{cases} \Phi(5(x_1^2 + x_2^2 + \cdots + x_n^2)/2), & q \in U_{\lambda(p_i)}, \\ 0, & q \notin U_{\lambda(p_i)}, \end{cases}$$

where $\varphi_{\lambda(p_i)}(q) = (x_1, x_2, \ldots, x_n)$. Evidently $\operatorname{supp} \bar{\mu}_{p_i} \subset U_{\sigma_i}$, $i = 1, 2, \ldots, s$, and for any p in M^n we have

$$\sum_{i=1}^{s} \bar{\mu}_{p_i}(p) \neq 0.$$

Let $\mu_{p_i} = \bar{\mu}_{p_i} / \sum_{i=1}^s \bar{\mu}_{p_i}(p)$, and define μ_σ by

$$\mu_\sigma = \begin{cases} \sum_{\sigma_i = \sigma} \mu_{p_i}, & \sigma \in \{\sigma_1, \sigma_2, \ldots, \sigma_s\}, \\ 0, & \sigma \in \Sigma, \ \sigma \neq \sigma_i, i = 1, 2, \ldots, s. \end{cases}$$

Then $\{\mu_\sigma ; \sigma \in \Sigma\}$ is a partition of unity subordinate to $\mathfrak{U}$.

§10. Tangent spaces

Let t_1, t_2 be real numbers or $\pm\infty$ such that $t_1 < t_2$. The open interval (t_1, t_2) in $\mathbf{R}^1$ is a C^∞ manifold (§9, Example 4), and thus *a priori* it is a C^r manifold, $r = 1, 2, \ldots, \infty$.

Let $(M^n, \mathscr{S})$ be an n-dimensional C^r manifold. A C^k map

$$l : (t_1, t_2) \to M^n$$

is called a C^k *curve* in M^n. Here $k \leq r$. In Euclidean space $\mathbf{R}^n$ it was straightforward to define vectors using directed segments; however, we have no concepts of lines in manifolds, and so it is difficult to extend the definition of vectors in $\mathbf{R}^n$ directly to M^n. Nevertheless we have defined curves in the manifold M, so we shall define a tangent vector at a point p in M as a certain equivalence class of C^1 curves passing through p. This is done as

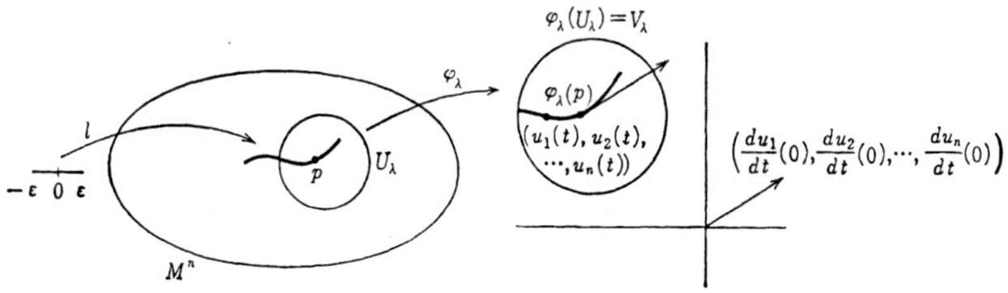

FIGURE 2.7

follows. Henceforth we shall assume that $r \geq 1$. For a point p in M we consider a C^1 curve

$$l\colon (-\epsilon, \epsilon) \to M^n, \qquad l(0) = p, \qquad \epsilon > 0,$$

and denote by $L(p)$ the set of all such curves through p. Let $l, \bar{l} \in L(p)$, and let $(U_\lambda, \varphi_\lambda) \in \mathcal{S}$ be a chart about p. Then we have C^1 maps

$$\varphi_\lambda \circ l\colon l^{-1}(U_\lambda) \to \varphi_\lambda(U_\lambda) = V_\lambda \subset \mathbf{R}^n,$$
$$\varphi_\lambda \circ \bar{l}\colon \bar{l}^{-1}(U_\lambda) \to \varphi_\lambda(U_\lambda) = V_\lambda \subset \mathbf{R}^n.$$

Here $\bar{l}\colon (-\bar{\epsilon}, \bar{\epsilon}) \to M^n$ and $\bar{l}(0) = p$. For $t \in l^{-1}(U_\lambda) \cap \bar{l}^{-1}(U_\lambda)$, we write

$$(\varphi_\lambda \circ l)(t) = (u_1(t), u_2(t), \dots, u_n(t)),$$
$$(\varphi_\lambda \circ \bar{l})(t) = (\bar{u}_1(t), \bar{u}_2(t), \dots, \bar{u}_n(t))$$

(see Figure 2.7). In this setting we say that l and $\bar{l}$ are equivalent and we shall write $l \sim \bar{l}$ if

$$\frac{du_i}{dt}(0) = \frac{d\bar{u}_i}{dt}(0), \qquad i = 1, 2, \dots, n.$$

This definition does not depend on the choice of a chart $(U_\lambda, \varphi_\lambda)$. Suppose that (U_μ, φ_μ) is another chart about p different from $(U_\lambda, \varphi_\lambda)$ and that we define C^r functions z_i of $x_1, x_2, \dots, x_n$, $i = 1, 2, \dots, n$, by

$$\begin{gathered}\varphi_\mu \circ \varphi_\lambda^{-1}(x_1, x_2, \dots, x_n) = (z_1, z_2, \dots, z_n), \\ (x_1, x_2, \dots, x_n) \in \varphi_\lambda(U_\lambda \cap U_\mu).\end{gathered} \tag{1}$$

Further we define $w_i(t)$, $\bar{w}_i(t)$, $i = 1, 2, \dots, n$, by

$$(\varphi_\mu \circ l)(t) = (w_1(t), w_2(t), \dots, w_n(t)),$$
$$(\varphi_\mu \circ \bar{l})(t) = (\bar{w}_1(t), \bar{w}_2(t), \dots, \bar{w}_n(t)),$$

for $t \in l^{-1}(U_\mu) \cap \bar{l}^{-1}(U_\mu)$. Then we have

$$\frac{dw_i}{dt} = \sum_{j=1}^n \frac{\partial z_i}{\partial x_j}\frac{du_j}{dt}, \qquad \frac{d\bar{w}_i}{dt} = \sum_{j=1}^n \frac{\partial z_i}{\partial x_j}\frac{d\bar{u}_j}{dt}. \tag{2}$$

Hence

$$\frac{dw_i}{dt}(0) = \frac{d\overline{w}_i}{dt}(0), \qquad i = 1, 2, \ldots, n.$$

The relation $l \sim \bar{l}$ in $L(p)$ is clearly an equivalence relation. An equivalence class $[l]$, the set of all curves in $L(p)$ equivalent to l, is called a *tangent vector* to M^n at the point p and is denoted by v or sometimes by $X(p)$. The set of all tangent vectors at p is denoted by $T_p(M^n)$ and called the *tangent space of* M^n *at* p. We also say that $[l]$ is the *tangent vector* to the curve l at p.

Let $p \in M^n$ and pick a chart $(U_\lambda, \varphi_\lambda) \in \mathscr{S}$ about p. Now to each element $v = [l] \in T_p(M^n)$ is associated a sequence of n numbers

$$v_i = \frac{du_i}{dt}(0), \qquad i = 1, 2, \ldots, n,$$

as shown above. Hence we can associate to each element v in $T_p(M^n)$ a vector $(v_1, v_2, \ldots, v_n)$ in $\mathbf{R}^n$. Now suppose that $(v_1', v_2', \ldots, v_n') \in \mathbf{R}^n$ is given. We choose $l'\colon (-\epsilon', \epsilon') \to M^n$, such that

$$(\varphi_\lambda \circ l')(t) = (v_i' t, v_2' t, \ldots, v_n' t),$$

then $[l']$ is an element of $T_p(M^n)$ corresponding to $(v_1', v_2', \ldots, v_n')$. In other words, for a fixed chart $(U_\lambda, \varphi_\lambda) \in \mathscr{S}$ about p, we have a one-to-one correspondence

$$\Phi_\lambda\colon T_p(M^n) \to \mathbf{R}^n,$$

defined by

$$\Phi_\lambda(v) = (v_1, v_2, \ldots, v_n). \tag{3}$$

Hence elements of $T_p(M^n)$ can be represented by elements of $\mathbf{R}^n$.

Let $\partial/\partial x_i$ be the unit vector in the direction of the x_i axis, $i = 1, 2, \ldots, n$; then the $\partial/\partial x_1, \partial/\partial x_2, \ldots, \partial/\partial x_n$ form a basis in the n-dimensional vector space $\mathbf{R}^n$. Now set

$$\Phi_\lambda(v) = v_1 \frac{\partial}{\partial x_1} + v_2 \frac{\partial}{\partial x_2} + \cdots + v_n \frac{\partial}{\partial x_n}.$$

and transfer the vector-space structure of $\mathbf{R}^n$ to $T_p(M^n)$ using map Φ_λ as follows. For $v, v' \in T_p(M^n)$ and $a, b \in \mathbf{R}$ we define $av + bv' \in T_p(M^n)$ by

$$\Phi_\lambda(av + bv') = a\Phi_\lambda(v) + b\Phi_\lambda(v').$$

Thus we have turned $T_p(M^n)$ into a vector space of dimension n.

Take another chart (U_μ, φ_μ) about p, and consider $\Phi_\mu\colon T_p(M^n) \to \mathbf{R}^n$:

$$\Phi_\mu(v) = v_1' \frac{\partial}{\partial z_1} + v_2' \frac{\partial}{\partial z_2} + \cdots + v_n' \frac{\partial}{\partial z_n}.$$

Then from (2) we obtain a relation between Φ_u and Φ_λ:

$$v'_i = \sum_{j=1}^{n} \frac{\partial z_i}{\partial x_j}(\varphi_\lambda(p))v_j, \qquad i = 1, 2, \ldots, n. \tag{4}$$

Hence the vector-space structure induced on $T_p(M^n)$ as above is independent of the choice of a chart. Here we remind the reader that $\partial z_i/\partial x_j$ is a C^{r-1} function defined on $\varphi_\lambda(U_\lambda \cap U_\mu)$, and that

$$\left|\frac{\partial(z_1, z_2, \ldots, z_n)}{\partial(x_1, x_2, \ldots, x_n)}\right| \neq 0.$$

For example, an open set U in $\mathbf{R}^n$ is a C^∞ n-manifold (§9, Example 4). Let id: $U \to U$ be the identity map. Then (U, id) is a chart on U and determines the correspondence Φ_0: $T_p(U) \to \mathbf{R}^n$, $p \in U$. Intuitively, the directed segment $\overrightarrow{pq}$ in Figure 2.8 represents an element of the tangent space $T_p(U)$, and the image of this vector under Φ_0 is $\overrightarrow{Oq'}$ which is a parallel translate of $\overrightarrow{pq}$ moving p to the origin O (Figures 2.7 and 2.8).

In particular, let $n = 1$ and let U be an open interval (t_1, t_2). We denote by $\mathbf{e}_t$ the element of the tangent space of (t_1, t_2) which is represented by the directed segment $\overrightarrow{t(t+1)}$. We call $\mathbf{e}_t$ the *unit tangent vector at* t; in other words, we define $\mathbf{e}_t$ by $\Phi_0(\mathbf{e}_t) = \partial/\partial x_1$.

Let $(M^n, \mathscr{S})$ be a C^r manifold, $r \geq 1$, and p a point of M^n. Denote by $C^r(U)$ the set of all C^r functions defined on a neighborhood U of p. If $f, g \in C^r(U)$ and $a, b \in \mathbf{R}$ then $af + bg \in C^r(U)$; so $C^r(U)$ is a vector space over $\mathbf{R}$.

For $v \in T_p(M^n)$, say $v = [l]$, we define a map

$$D_v: C^r(U) \to \mathbf{R}$$

by

$$D_v(f) = \frac{d(f \circ l)}{dt}(0), \qquad f \in C^r(U).$$

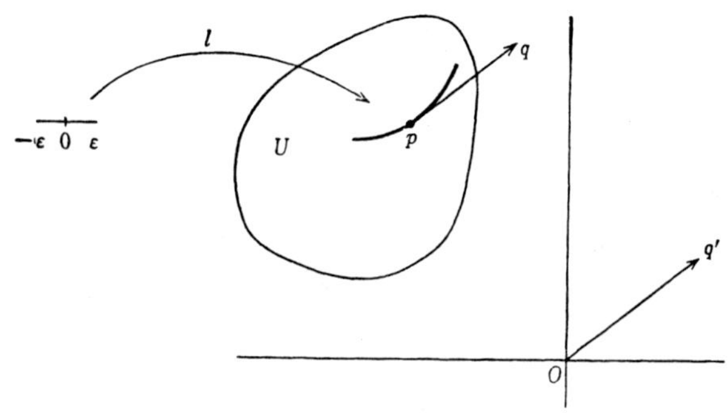

FIGURE 2.8

Now choose $(U_\lambda, \varphi_\lambda) \in \mathscr{S}$ such that $p \in U_\lambda \subset U$ and $\varphi_\lambda(p) = 0$; then a simple calculation shows that

$$D_v(f) = \sum_i v_i \frac{\partial(f \circ \varphi_\lambda^{-1})}{\partial x_i}(0),$$

where

$$\Phi_\lambda(v) = v_i \frac{\partial}{\partial x_1} + v_2 \frac{\partial}{\partial x_2} + \cdots + v_n \frac{\partial}{\partial x_n}.$$

Hence, in particular, $D_v(f)$ does not depend on the choice of a representative of $v = [l]$. By definition D_v is a linear map. It should also be evident that

$$D_{av+bv'} = aD_v + bD_{v'}.$$

THEOREM 2.7. *For $v \in T_p(M^n)$ the linear map $D_v: C^r(U) \to \mathbf{R}$ satisfies:*

$$D_v(fg) = f(p)D_v(g) + g(p)D_v(f), \qquad f, g \in C^r(U). \tag{5}$$

If $D_v = D_{v'}$ then $v = v'$, where $v, v' \in T_p(M^n)$. In case M^n is C^∞, given a linear map $\widehat{D}: C^\infty(U) \to \mathbf{R}$ satisfying

$$\widehat{D}(fg) = f(p)\widehat{D}(g) + g(p)\widehat{D}(f), \qquad f, g \in C^\infty(U),$$

there exists a unique $v \in T_p(M^n)$ such that $\widehat{D} = D_v$.

PROOF. From the definition of D_v, (5) is evident. Now take $(U_\lambda, \varphi_\lambda)$ such that $p \in U_\lambda \subset U$, $\varphi_\lambda(p) = 0$, and put $\varphi_\lambda(q) = (x_1(q), x_2(q), \ldots, x_n(q))$ for $q \in U_\lambda$. Then $x_i: U_\lambda \to \mathbf{R}$ is a C^r function defined on U_λ, $i = 1, 2, \ldots, n$.

Choose an embedding $h: \operatorname{Int} D^n \to M^n$ such that $h(0) = p$, where 0 is the center of $\operatorname{Int} D^n$, and $h(\operatorname{Int} D^n) \subset U_\lambda$. Using the C^∞ function Φ of Lemma 1.5, we define the C^r functions

$$\hat{x}_i: U \to \mathbf{R}, \qquad i = 1, 2, \ldots, n,$$

by

$$\hat{x}_i(q) = \begin{cases} \Phi(3(y_1^2 + y_2^2 + \cdots + y_n^2))x_i(h(y_1, y_2, \ldots, y_n)), \\ \qquad q \in h(\operatorname{Int} D^n), \ q = h(y_1, y_2, \ldots, y_n) \\ 0, \qquad q \in U - h(\operatorname{Int} D^n). \end{cases}$$

Then $\hat{x}_i$ is equal to x_i in some neighborhood of p, so we have

$$D_v(\hat{x}_i) = v_i, \quad \text{where } \Phi_\lambda(v) = v_1 \frac{\partial}{\partial x_1} + v_2 \frac{\partial}{\partial x_2} + \cdots + v_n \frac{\partial}{\partial x_n}.$$

Therefore it follows that $D_v = D_{v'}$ implies $v = v'$.

Next we let M^n be a C^∞ manifold. For $c \in \mathbf{R}$ we use the same c to denote the C^∞ function which assigns to each point of U the number c. Then the properties of $\widehat{D}$ clearly imply that $\widehat{D}(0) = 0$ and if $c \neq 0$, we get

$$\widehat{D}(c) = c\widehat{D}(1) = c\widehat{D}(1 \times 1) = c(\widehat{D}(1) + \widehat{D}(1)) = 2c\widehat{D}(1),$$

so $\widehat{D}(c)=0$. Let $f \in C^\infty(U)$ and write

$$f \circ \varphi_\lambda^{-1}(x_1, x_2, \dots, x_n) = f(p) + \sum_i \frac{\partial (f \circ \varphi_\lambda^{-1})}{\partial x_i}(0) x_i + \sum_{i,j} f_{ij}(x_1, x_2, \dots, x_n) x_i x_j,$$

for $(x_1, x_2, \dots, x_n) \in \varphi_\lambda(U_\lambda)$ (see Appendix II, Note 3). Here the f_{ij} are C^∞ functions defined on $\varphi_\lambda(U_\lambda)$. In the same way as we defined $\hat{x}_i$ from x_i, we define a C^∞ function $\widehat{f_{ij} \circ \varphi_\lambda}: U \to \mathbf{R}$ from $f_{ij} \circ \varphi_\lambda : U_\lambda \to \mathbf{R}$ such that $\widehat{f_{ij} \circ \varphi_\lambda} = f_{ij} \circ \varphi_\lambda$ in some neighborhood of p. Since $\widehat{D}$ is linear we have

$$\widehat{D}(f) = \widehat{D}(f(p)) + \sum_i \frac{\partial (f \circ \varphi_\lambda^{-1})}{\partial x_i}(0) \widehat{D}(\hat{x}_i) + \sum_{i,j} \widehat{D}((\widehat{f_{ij} \circ \varphi_\lambda})\, \hat{x}_i \hat{x}_j).$$

But $\widehat{D}(f(p)) = 0$ as mentioned above, and moreover

$$\begin{aligned}\widehat{D}((\widehat{f_{ij} \circ \varphi_\lambda})\, \hat{x}_i \hat{x}_j) &= \hat{x}_i(p) \hat{x}_j(p) \widehat{D}(\widehat{f_{ij} \circ \varphi_\lambda}) \\ &\quad + \widehat{f_{ij} \circ \varphi_\lambda}(p) \hat{x}_j(p) \widehat{D}(\hat{x}_i) + \widehat{f_{ij} \circ \varphi_\lambda}(p) \hat{x}_i(p) \widehat{D}(\hat{x}_j) = 0.\end{aligned}$$

Hence we have shown that

$$\widehat{D}(f) = \sum_i \frac{\partial (f \circ \varphi_\lambda^{-1})}{\partial x_i}(0) \widehat{D}(\hat{x}_i).$$

Thus if we choose v so that

$$\Phi_\lambda(v) = \widehat{D}(\hat{x}_1) \frac{\partial}{\partial x_1} + \widehat{D}(\hat{x}_2) \frac{\partial}{\partial x_2} + \dots + \widehat{D}(\hat{x}_n) \frac{\partial}{\partial x_n},$$

we will get $D_v(f) = \widehat{D}(f)$, $f \in C^\infty(U)$.

By Theorem 2.7 we identify v with D_v and adopt $\partial/\partial x_1$, $\partial/\partial x_2$, ..., $\partial/\partial x_n$ as a basis in $\Phi_\lambda(T_p(M^n))$.

The set of all tangent vectors to M^n, denoted by $T(M^n)$, is called the *tangent space* of M^n:

$$T(M^n) = \bigcup_{p \in M} T_p(M^n).$$

If $v, v' \in T(M^n)$ then $v \in T_p(M^n)$, $v' \in T_{p'}(M^n)$ for some $p, p' \in M^n$; however, $v + v'$ makes sense only when $p = p'$, and in this case we have that $v + v' \in T_p(M^n)$.

A map $\pi: T(M^n) \to M^n$ is defined by $\pi(v) = p$ if $v \in T_p(M^n)$, that is $\pi(T_p(M^n)) = p$. π is called the *projection* or *the projection map* of $T(M^n)$ onto M^n.

Up to now the tangent space $T(M^n)$ has been regarded simply as a union of infinitely many n-dimensional vector spaces. We shall now introduce a natural C^{r-1} manifold structure on $T(M^n)$.

Consider $\pi^{-1}(U_\lambda) = \bigcup_{p\in U_\lambda} T_p(M^n)$, where $(U_\lambda, \varphi_\lambda) \in \mathscr{S}$. Recall the linear map $\Phi_\lambda : T_p(M^n) \to \mathbf{R}^n$ defined by (3) for each p in U_λ. We now define a map

$$\widetilde{\Phi}_\lambda : \pi^{-1}(U_\lambda) \to \varphi_\lambda(U_\lambda) \times \mathbf{R}^n = V_\lambda \times \mathbf{R}^n, \qquad \lambda \in \Lambda$$

by $\widetilde{\Phi}_\lambda(v) = (\Phi_\lambda(p), \Phi_\lambda(v))$ for $v \in T_p(M^n)$. Then $\widetilde{\Phi}_\lambda$ is one-to-one and onto; moreover, $\widetilde{\Phi}_\lambda | T_p(M^n) = \Phi_\lambda$. Notice $V_\lambda \times \mathbf{R}^n$ is an open subset of $\mathbf{R}^{2n}$. For each $\lambda \in \Lambda$ we give $\pi^{-1}(U_\lambda)$ a topology so that $\widetilde{\Phi}_\lambda$ will be a homeomorphism.

Suppose that (U_μ, φ_μ) is a chart such that $U_\lambda \cap U_\mu \neq \varnothing$. Then by (4) the map

$$\widetilde{\Phi}_\mu \circ \widetilde{\Phi}_\lambda^{-1} : \widetilde{\Phi}_\lambda(\pi^{-1}(U_\lambda \cap U_\mu)) \to \widetilde{\Phi}_\mu(\pi^{-1}(U_\lambda \cap U_\mu))$$

satisfies:

$$(\widetilde{\Phi}_\mu \circ \widetilde{\Phi}_\lambda^{-1})(x, (v_1, v_2, \dots, v_n)) = ((\varphi_\mu \circ \varphi_\lambda^{-1})(x), (v_1', v_2', \dots, v_n')),$$

$$v_i' = \sum_{j=1}^{n} \frac{\partial z_i}{\partial x_j} v_j, \qquad i = 1, 2, \dots, n.$$

Therefore the induced topologies on $\pi^{-1}(U_\lambda)$ and $\pi^{-1}(U_\mu)$ respectively defined as above agree on $\pi^{-1}(U_\lambda \cap U_\mu)$. Thus the similarly induced topologies on $\pi^{-1}(U_\lambda)$, $\lambda \in \Lambda$, give a well-defined topology on $T(M^n)$. This construction immediately yields the fact that $T(M^n)$ is a Hausdorff space with a countable basis.

We consider $\widetilde{\mathscr{S}} = \{(\pi^{-1}(U_\lambda), \widetilde{\Phi}_\lambda); \lambda \in \Lambda\}$ for $T(M^n)$. Since $V_\lambda \times \mathbf{R}^n$ is an open set in $\mathbf{R}^{2n}$ and $\widetilde{\Phi}_\mu \circ \widetilde{\Phi}_\lambda^{-1}$ is a C^{r-1} map, $\widetilde{\mathscr{S}}$ satisfies $(\mathrm{M_I})$ and $(\mathrm{M_{II}})$; hence $\widetilde{\mathscr{S}}$ is a C^{r-1} coordinate neighborhood system on $T(M^n)$. The maximal coordinate neighborhood system containing $\widetilde{\mathscr{S}}$ is an atlas on $T(M^n)$, so the tangent space $T(M^n)$ of M^n is a C^{r-1} manifold of dimension $2n$. The projection $\pi : T(M^n) \to M^n$ is clearly a C^{r-1} map.

Let $(M^n, \mathscr{S})$ be a C^r manifold, $r \geq 1$. We say that M^n is *orientable* if there exists a subset $\mathscr{S}'$ of $\mathscr{S}$ which satisfies the following.

(i) $\{U_\lambda; (U_\lambda, \varphi_\lambda) \in \mathscr{S}'\}$ is an open cover of M^n.

(ii) For $(U_\lambda, \varphi_\lambda)$, $(U_\mu, \varphi_\mu) \in \mathscr{S}'$ with $U_\lambda \cap U_\mu \neq 0$ the Jacobian determinant satisfies

$$\left|\frac{\partial(z_1, z_2, \dots, z_n)}{\partial(x_1, x_2, \dots, x_n)}\right| > 0, \qquad (x_1, x_2, \dots, x_n) \in \varphi_\lambda(U_\lambda \cap U_\mu),$$

where $\varphi_\mu \circ \varphi_\lambda^{-1}(x_1, x_2, \dots, x_n) = (z_1, z_2, \dots, z_n)$ as in (1).

For example, the n-dimensional sphere S^n is orientable, for we may choose the set $\{(U_i^\pm, \varphi_i^\pm; i = 1, 2, \dots, n+1\}$ of §9 with a sign adjustment for the subatlas $\mathscr{S}'$. It is also routine to see that the torus is orientable (Example 3, §9).

We *orient* an orientable manifold M^n by specifying an orientation for the tangent space $T_p(M^n)$ at each p—fixing an ordered basis of each vector space $T_p(M^n)$—in such a way that the orientations for the tangent spaces $T_{p_1}(M^n)$ and $T_{p_2}(M^n)$ are mapped by Φ_λ to an identical orientation for $\mathbf{R}^n$ whenever p_1 and p_2 belong to a common chart $(U_\lambda, \varphi_\lambda) \in \mathscr{S}'$. Thus if M^n is connected we can orient M^n in two distinct ways since there are exactly two ways to order a basis in $T_p(M^n)$. If M has a given orientation then $-M$ is the manifold M with the other orientation, that is the orientation obtained by reversing the given orientation for each $T_p(M^n)$.

Let $(M_1^n, \mathscr{S}_1)$, $(M_2^m, \mathscr{S}_2)$ be C^r manifolds, $r \geq 1$, and $g: M_1^n \to M_2^m$ a C^r map. Let $v = [l]$, $l: (-\epsilon, \epsilon) \to M_1^n$ with $l(0) = p$, be an element of the tangent space $T_p(M_1^n)$ at $p \in M_1^n$. The C^1 map

$$g \circ l: (-\epsilon, \epsilon) \to M_2^m$$

determines a tangent vector $[g \circ l]$ to M_2^m at $g(p)$, since $g \circ l(0) = g(p)$.

Choose charts $(U_\lambda, \varphi_\lambda)$ about p and $(U_{\lambda'}', \varphi_{\lambda'}')$ about $g(p)$ and let

$$\varphi_{\lambda'}' \circ g \circ \varphi_\lambda^{-1}(x_1, x_2, \dots, x_n) = (y_1, y_2, \dots, y_m),$$

for $(x_1, x_2, \dots, x_n) \in \varphi_\lambda(U_\lambda \cap g^{-1}(U_{\lambda'}'))$. Now if we write

$$(\varphi_\lambda \circ l)(t) = (u_1(t), u_2(t), \dots, u_n(t)), \qquad t \in l^{-1}(U_\lambda),$$
$$(\varphi_{\lambda'}' \circ g \circ l)(t) = (\hat{u}_1(t), \hat{u}_2(t), \dots, \hat{u}_m(t)), \qquad t \in (g \circ l)^{-1}(U_{\lambda'}'),$$

we get

$$\frac{d\hat{u}_i}{dt}(0) = \sum_{j=1}^{n} \frac{\partial y_i}{\partial x_j}(\varphi_\lambda(p)) \frac{du_j}{dt}(0), \qquad i = 1, 2, \dots, m. \tag{6}$$

Hence it follows that $l \sim \bar{l}$ implies $g \circ l \sim g \circ \bar{l}$, and so $[g \circ l]$ is well defined by $v = [l]$. Thus we have obtained a map

$$g_*: T_p(M_1^n) \to T_{g(p)}(M_2^m), \qquad p \in M_1^n,$$

defined by $g_*(v) = [g \circ l]$. Putting

$$\Phi_\lambda(v) = (v_1, v_2, \dots, v_n), \qquad \Phi'_{\lambda'}(g_*(v)) = (v_1', v_2', \dots, v_m'),$$

we get from (6) the identity

$$v_i' = \sum_{j=1}^{n} \frac{\partial y_i}{\partial x_j}(\varphi_\lambda(p)) v_j, \qquad i = 1, 2, \dots, m, \tag{7}$$

and we see that the map g_* is a homomorphism of the n-dimensional vector space $T_p(M_1^n)$ to the m-dimensional vector space $T_p(M_2^m)$.

We obtain a map

$$g_*: T(M_1^n) \to T(M_2^m)$$

by assigning $g_*(v)$ to each element $v \in T(M_1^n)$. We call g_* the *differential* of g and often write dg for g_*.

We saw earlier that $T(M_1^n)$ and $T(M_2^m)$ are C^{r-1} manifolds; now it immediately follows from (7) that g_* is a C^{r-1} map.

THEOREM 2.8. *Let $g: M_1^n \to M_2^m$ be a C^r map, $r \geq 1$. Then g is a C^r immersion if and only if at each point p in M_1^n we have $g_*^{-1}(0) = 0$, where $g_*: T_p(M_1^n) \to T_{g(p)}(M_2^n)$. A necessary and sufficient condition for the map g to be a C^r submersion is that $g_*(T_p(M_1^n)) = T_{g(p)}(M_2^m)$ for each p in M_1^n.*

PROOF. Equation (7) says that $g_*^{-1}(0) = 0$ if and only if the rank of Jacobian matrix $\partial(y_1, y_2, \ldots, y_m)/\partial(x_1, x_2, \ldots, x_n)$ is n at $\varphi_\lambda(p)$. Also $g_*(T_p(M_1^n)) = T_{g(p)}(M_2^m)$ if and only if the rank of this Jacobian matrix is m at $\varphi_\lambda(p)$. Hence the theorem follows from the definitions of immersions and submersions.

Let $g: M_1^n \to M_2^n$ be a C^r diffeomorphism between oriented C^r manifolds M_1^n and M_2^n, $r \geq 1$. Let $e_1, e_2, \ldots, e_n$ be a basis in $T_p(M_1^n)$ which defines the given orientation at each p in M_1^n. We say that g is *orientation preserving* if the induced orientation, $g_*(e_1), g_*(e_2), \ldots, g_*(e_n)$, agrees with the given orientation for $T_{g(p)}(M_2^n)$ at each p in M_1^n.

Let $(M^n, \mathscr{S})$ be a C^r manifold and W^q a q-submanifold of M^n, $r \geq 1$. Let $p \in W^q$ and consider a C^1 curve

$$l^W : (-\epsilon, \epsilon) \to M^n$$

such that

$$l^W(0) = p, \qquad l^W((-\epsilon, \epsilon)) \subset W^q.$$

Denote by $T_p(W^q)$ the set of all tangent vectors $[l^W]$ at p determined by all such curves l^W. Then $T_p(W^q) \subset T_p(M^n)$. By considering a chart $(U_\lambda, \varphi_\lambda)$ about p such that $\varphi_\lambda(W^q \cap U_\lambda) = \varphi_\lambda(U_\lambda) \cap \mathbf{R}^q$ we see that $T_p(W^q)$ is a q-subspace of the n-dimensional vector space $T_p(M^n)$. We can think of $T_p(W^q)$ also as the tangent space to W^q at p which is a C^r q-manifold in the sense of Theorem 2.3(i).

Let W_1^q and W_2^{n-q} be submanifolds of M^n. We say that W_1^q and W_2^{n-q} *intersect transversely* at $p \in W_1^n \cap W_2^{n-q}$ if

$$T_p(M^n) = T_p(W_1^q) \oplus T_p(W_2^{n-q}).$$

Let W, W', M^n be C^r manifolds, and $g: W \to M^n$, $g': W' \to M^n$ be C^r immersions, $r \geq 1$. We say that immersions g and g' are *transverse* if we have

$$g_*(T_x(W)) + g'_*(T_{x'}(W')) = T_p(M^n),$$

for each $p \in g(W) \cap g'(W')$, $p = g(x) = g'(x')$ where $p \in M^n$, $x \in W$, and $x' \in W'$.

Next let $(M^n, \mathcal{S})$ be a C^s manifold, $s \geq 1$. A map

$$X: M^n \to T(M^n)$$

is called a *vector field* on M^n if the composite

$$\pi \circ X: M^n \to M^n$$

is the identity map. Here $\pi: T(M^n) \to M^n$ is the projection. Suppose that $r \leq s-1$. If X is C^r as a map between C^r manifolds M^n and $T(M^n)$, we say that X is a C^r *vector field*. A vector field X assigns to each point of M^n a tangent vector $X(p) \in T_p(M^n)$; in this context, we often adopt set notation $\{X(p); p \in M^n\}$ for X. For example, in §3 we defined vector fields X on the torus T, which in our current discussion can also be regarded as vector fields on the torus T, viewed as a two-dimensional smooth manifold.

Let X be a vector field on M^n. If $X(p) = 0$ we say that p is a *singular point* or *zero* of X. We say that X is *nonsingular* If $X(p) \neq 0$ at each point p in M^n.

Let W^q be a submanifold of M^n. We define a C^r *vector field*

$$Y = \{Y(p); p \in W^q\}$$

of $T(M^n)$ defined on W^q to be a C^r map $Y: W^q \to T(M^n)$ such that $(\pi \circ Y)(p) = p$, $p \in W^q$.

A vector field Y of $T(M^n)$ on a submanifold W^{n-1} is said to be *transverse* to W^{n-1} if $Y(p) \notin T_p(W^{n-1})$ for each point p of W^{n-1}. Let X be a vector field on M^n and $X|W^{n-1} = \{X(p); p \in W^{n-1}\}$ the restriction of X to W^{n-1}; then we say that X is *transverse* to W^{n-1} if $X|W^{n-1}$ is transverse to W^{n-1}.

So far we have considered only a submanifold W; however more generally we can take W'^{n-1} to be a C^s manifold and $f: W'^{n-1} \to M^n$ to be a one-to-one C^s immersion, and define a vector field of $T(M^n)$ on $f(W'^{n-1})$, a vector field of $T(M^n)$ transverse to $f(W'^{n-1})$, etc., verbatim as above.

Let M^n be a C^r manifold and $T(M^n)$ the tangent space of M^n, $r \geq 1$. Suppose for each $p \in M^n$ the tangent space $T_p(M^n)$ at p has an inner product $\langle v, v' \rangle_p$, $v, v' \in T_p(M^n)$. Define a real-valued function $g: T(M^n) \to \mathbf{R}$ by

$$g(v) = \langle v, v \rangle_p, \qquad v \in T_p(M^n);$$

we say that the set $\{\langle\ ,\ \rangle_p; p \in M^n\}$ is a *Riemannian metric* on M^n if g is differentiable of class C^{r-1}.

The length $\| v \|$ of v, $v \in T_p(M^n)$, is defined by $\| v \| = \sqrt{\langle v, v \rangle_p}$. We say that v and v' in $T_p(M^n)$ are *orthogonal* when $\langle v, v' \rangle_p = 0$.

Let W be a q-submanifold of M^n. Suppose M^n has a Riemannian metric $\{\langle\ ,\ \rangle_p ; p \in M^n\}$. By Theorem 2.3 (i), W is a C^r q-manifold; the restriction of $\langle v , v'\rangle_p$ for $p \in W$ and $v , v' \in T_p(M^n)$ defines a Riemannian metric on W.

A familiar Riemannian metric on $\mathbf{R}^n$ is given by the standard inner product

$$\langle v , v'\rangle_p = \sum_{i=1}^{n} v_i v'_i , \qquad p \in \mathbf{R}^n , \qquad v , v' \in T_p(\mathbf{R}^n), \tag{8}$$

where $v = \sum_{i=1}^n v_i \partial/\partial x_i$, $v' = \sum_{i=1}^n v'_i \partial/\partial x_i$.

Theorem 2.9. *A C^r manifold M^n admits a Riemannian metric, $r \geq 1$.*

Proof. We shall prove the theorem for a compact M^n. Consider an atlas $\mathcal{S} = \{(U_\lambda , \varphi_\lambda); \lambda \in \Lambda\}$ for M^n, and choose $\lambda_1 , \lambda_2 , \dots , \lambda_m \in \Lambda$ such that $\{U_{\lambda_i} ; i = 1 , 2 , \dots , m\}$ is an open cover of M^n. Let μ_i, $i = 1, 2, \dots , m$, be a partition of unity subordinate to this open cover. For $p \in M^n$ and $v , v' \in T_p(M^n)$ define $\langle v , v'\rangle_p$ by

$$\langle v , v'\rangle_p = \sum_i \mu_i(p)\langle(\varphi_{\lambda_i})_*(v), (\varphi_{\lambda_i})_*(v')\rangle_{\varphi_{\lambda_i}(p)} .$$

Here $\langle\ ,\ \rangle_{\varphi_{\lambda_i}}$ is the inner product on $\mathbf{R}^n$ defined in (8), and $\sum_i$ sums over i with $p \in U_{\lambda_i}$. It is routine to check that $\langle v , v'\rangle_p$ defines a Riemannian metric on M^n.

A similar proof, which we omit, can be given for a noncompact M^n (see Appendix II, Note 4).

Let M^n be a C^r manifold endowed with a Riemannian metric. Let α, β be real numbers and let $\varphi : [\alpha , \beta] \to M^n$ be a C^r curve, $r \geq 1$. We define the *length* φ to be the integral

$$\int_\alpha^\beta \| \varphi_*(\mathbf{e}_t) \| \, dt,$$

where $\alpha \leq t \leq \beta$, and $\mathbf{e}_t$ is the unit tangent vector at t.

Suppose M^n is arcwise connected. Then it is easy to show that for any points p, p' in M^n there exist a sequence $\alpha_0 < \alpha_1 < \cdots < \alpha_m$ of real numbers and C^r curves

$$\varphi_i : [\alpha_{i-1} , \alpha_i] \to M^n , \qquad i = 1, 2, \dots , m,$$

such that $\varphi_1(\alpha_0) = p$, $\varphi_m(\alpha_m) = p'$, $\varphi_i(\alpha_i) = \varphi_{i+1}(\alpha_i)$, $i = 1, 2, \dots , m-1$. The set $\{\varphi_i\}$ is called a C^r *broken line*joining p and p'. The *length*of a C^r broken line is defined to be

$$L(\{\varphi_i\}) = \sum_{i=1}^{m} \int_{\alpha_{i-1}}^{\alpha_i} \| (\varphi_i)_*(\mathbf{e}_t) \| \, dt.$$

For $p, p' \in M^n$ define $\rho(p, p')$ to be the lower bound of the lengths of all C^r broken lines joining p and p', that is

$$\rho(p, p') = \inf L(\{\varphi_i\}).$$

Then ρ defines a metric on M^n, and thus M^n becomes a metric space; further the metric topology agrees with the original topology of M^n. We omit the proof (see Appendix II, Note 5).

§11. C^r manifolds with boundaries

Let p be a point of the n-ball D^n. If $p \in \operatorname{Int} D^n$ there is a neighborhood of p in D^n which is homeomorphic to an open set in $\mathbf{R}^n$, and hence $\operatorname{Int} D^n$ is a C^∞ manifold. If, however, p is on the boundary S^{n-1} of D^n no neighborhood of p is homeomorphic to an open set of $\mathbf{R}^n$ (Figure 2.9). In order to include sets such as D^n in our discussion we extend the definition of C^r manifolds defined in §9 as follows.

We call $\mathbf{R}^n_+ = \{(x_1, x_2, \dots, x_n) \in \mathbf{R}^n; x_n \geq 0\}$ the upper half plane. Euclidean space $\mathbf{R}^{n-1}$ is a subset of $\mathbf{R}^n_+$:

$$\mathbf{R}^{n-1} = \{(x_1, x_2, \dots, x_n) \in \mathbf{R}^n_+; x_n = 0\}.$$

Let M be a Hausdorff space with a countable basis. We say that M is a topological manifold of dimension n if each point p in M has a neighborhood U homeomorphic to an open subset of $\mathbf{R}^n_+$. The n-ball D^n is an n-dimensional topological manifold (Figure 2.9).

Let A be a subset of $\mathbf{R}^n$, B a subset of $\mathbf{R}^m$ and let $\varphi: A \to B$ be a map. We say that φ is a C^r map if there exists an open set U in $\mathbf{R}^n$ with $A \subset U$ and a C^r map $\hat{\varphi}$ from U to $\mathbf{R}^m$ such that $\hat{\varphi}|A = \varphi$.

Let M^n be an n-dimensional topological manifold. Let Λ be an index set, U_λ an open set in M^n, and $\varphi_\lambda: U_\lambda \to V_\lambda$ a homeomorphism of U_λ to an open set V_λ in $\mathbf{R}^n_+$, $\lambda \in \Lambda$. Consider $\mathscr{S} = \{(U_\lambda, \varphi_\lambda); \lambda \in \Lambda\}$. We say that $\mathscr{S}$ is a C^r *atlas* for M^n, $r = 0, 1, 2, \dots$, or ∞, if it satisfies the following.

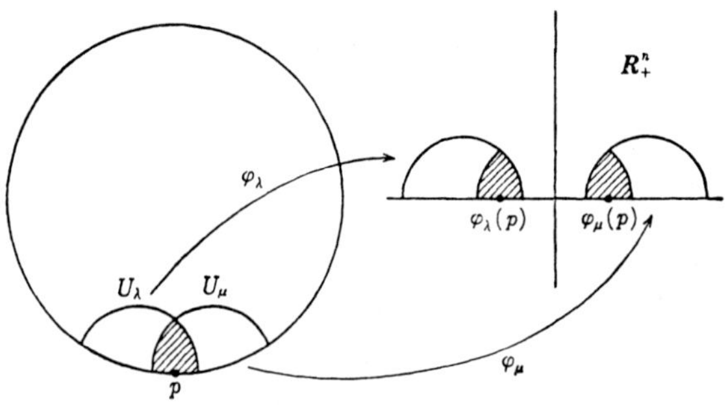

FIGURE 2.9

(M'_{I}) $\{U_\lambda ; \lambda \in \Lambda\}$ is an open cover of M^n.

($\mathrm{M}'_{\mathrm{II}}$) If $U_\lambda \cap U_\mu \neq \varnothing$, $\lambda, \mu \in \Lambda$, both maps

$$\varphi_\mu \circ \varphi_\lambda^{-1} : \varphi_\lambda(U_\lambda \cap U_\mu) \to \varphi_\mu(U_\lambda \cap U_\mu), \quad \varphi_\lambda \circ \varphi_\mu^{-1} : \varphi_\mu(U_\lambda \cap U_\mu) \to \varphi_\lambda(U_\lambda \cap U_\mu)$$

are C^r in the sense defined above (Figure 2.9).

($\mathrm{M}'_{\mathrm{III}}$) $\mathscr{S}$ is maximal with respect to set inclusions. The pair $(M^n, \mathscr{S})$ is called an n-dimensional *differentiable manifold of class* C^r. We might simply write M^n or M instead of $(M^n, \mathscr{S})$. An element $(U_\lambda, \varphi_\lambda)$ is called a *chart* or a *coordinate neighborhood.*

A point p in M^n is called a *boundary point* of M^n if $\varphi_\lambda(p) \in \mathbf{R}^{n-1}$ where $(U_\lambda, \varphi_\lambda) \in \mathscr{S}$ is a chart about p. We denote by ∂M^n the set of all boundary points of M^n; ∂M^n is called the *boundary* of M^n. For example, we have $\partial D^n = S^{n-1}$.

We call M^n a *manifold without boundary* if $\partial M^n = \varnothing$; in this case, the $V_\lambda = \varphi_\lambda(U_\lambda)$, $\lambda \in \Lambda$, are open sets in $\mathbf{R}^n_+ - \mathbf{R}^{n-1}$ hence in $\mathbf{R}^n$, and so $(M^n, \mathscr{S})$ is a C^r manifold defined in §9. It should be evident to the reader that a C^r manifold defined in §9 is a C^r manifold without boundary. A compact manifold without boundary is called *closed.*

When $\partial M^n \neq \varnothing$ we say that M^n is a *manifold with boundary*; here it is not hard to see that ∂M^n is an $(n-1)$-dimensional topological manifold defined in §9 and the set $\mathfrak{U} = \{(U_\lambda \cap \partial M^n, \varphi_\lambda | U_\lambda \cap \partial M^n); \lambda \in \Lambda\}$ satisfies (M_{I}) and (M_{II}) of §9 with respect to ∂M^n. Hence taking the maximal set containing $\mathfrak{U}$, which we shall denote by $\partial\mathscr{S}$, we have obtained an $(n-1)$-dimensional C^r manifold $(\partial M^n, \partial\mathscr{S})$.

The definitions of C^r *maps,* C^r *diffeomorphisms,* C^r *immersions,* C^r *embeddings, submanifolds,* and so on are identical with those given in §9. Let $(M^n, \mathscr{S})$ be a C^r manifold and t_1, t_2 be real numbers with $t_1 < t_2$. We call a C^r map $l: L \to M^n$ a C^r curve in M^n, where L is one of one-dimensional manifolds $[t_1, t_2)$, $[t_1, t_2]$, $(t_1, t_2]$, or (t_1, t_2).

Let M^n be a C^r manifold with boundary. If $p \in M^n - \partial M^n$, the definition of the tangent space $T_p(M^n)$ of M^n at p in §10 is adopted in our new definition. Suppose $p \in \partial M^n$. We then consider maps

$$l: [0, \epsilon) \to M^n, \qquad l(0) = p,$$

and define the equivalence relation $\sim$ on the set of all such l in the manner adopted in §10. An equivalence class $v = [l]$ is called a tangent vector to M^n. For a chart $(U_\lambda, \varphi_\lambda)$ about p, we can define $\Phi_\lambda(v)$ by (3) in §10, but in our present case $v_n \geq 0$ when we set $\Phi_\lambda(v) = (v_1, v_2, \dots, v_n)$. To modify this situation, whenever the nth coordinate of $\Phi_\lambda(v)$ is positive we introduce $-v$ artificially and define $\Phi_\lambda(-v) = (-v_1, -v_2, \dots, -v_n)$. Denote by $T_p(M^n)$ the set consisting of all tangent vectors $v = [l]$ and their negatives $-v$. Then $\Phi_\lambda: T_p(M^n) \to \mathbf{R}^n$ becomes a one-to-one onto map; hence we can give $T_p(M^n)$ the structure of an n-dimensional vector space defined in §10. Analogous to our discussion in §10, $T(M^n) = \bigcup_{p\in M^n} T_p(M)$ is an $2n$-dimensional C^{r-1} manifold with boundary, which we shall call the tangent manifold of M^n, and the projection $\pi\colon T(M^n) \to M^n$ is a C^{r-1} map. Similarly we define C^r vector fields, vector fields without singularities, etc.

Likewise the notion of the orientability and the Riemannian metrics defined in §10 carries over to manifolds with boundaries.

Suppose now that M^n is oriented and $\partial M^n \neq \varnothing$. For a chart $(U_\lambda, \varphi_\lambda)$ about p let the orientation for $T_p(M^n)$ be given by

$$\Phi_\lambda^{-1}\left(\frac{\partial}{\partial x_1}\right), \Phi_\lambda^{-1}\left(\frac{\partial}{\partial x_2}\right), \dots, \Phi_\lambda^{-1}\left(\frac{\partial}{\partial x_n}\right).$$

If $p \in \partial M^n$ we can orient $\partial T_p(M^n)$ by

$$\Phi_\lambda^{-1}\left(\frac{\partial}{\partial x_1}\right), \Phi_\lambda^{-1}\left(\frac{\partial}{\partial x_2}\right), \dots, \Phi_\lambda^{-1}\left(\frac{\partial}{\partial x_{n-1}}\right), \tag{1}$$

and so we can orient the boundary ∂M^n. This is a 'natural' way to orient the boundary of a manifold with a given orientation, at least according to the definitions of manifolds; however we define the *induced orientation* for the boundary ∂M^n, to be the orientation (1) times $(-1)^n$, that is the orientation (1) if n is even and the opposite of the orientation (1) if n is odd, in order to preserve consistency with homology theory. The adjustment $(-1)^n$ will play a role in Stokes' theorem in §27.

CHAPTER III

Dynamical Systems and Limit Sets

§12. Dynamical systems

Let $(M^n, \mathscr{S})$ be an n-dimensional C^s manifold and consider a C^r vector field on M^n, $X = \{X(p); p \in M\}$. Here we take $1 \leq r \leq s-1$. Such a vector field may also be called a C^r *dynamical system* .

Let t_1, t_2 be real numbers or $\pm\infty$ with $t_1 < t_2$, and let

$$\varphi : (t_1, t_2) \to M$$

be a C^{r+1} curve in M^n.

For $t_1 < t < t_2$ choose ϵ with $t_1 < t - \epsilon < t + \epsilon < t_2$, and consider the restriction of the curve φ to the open interval $(t-\epsilon, t+\epsilon)$:

$$\varphi|(t-\epsilon, t+\epsilon) : (t-\epsilon, t+\epsilon) \to M^n.$$

Evidently the tangent vector $[\varphi|(t-\epsilon, t+\epsilon)]$ to M^n at point $\varphi(t)$ is independent of the choice of ϵ. This $[\varphi|(t-\epsilon, t+\epsilon)]$ is called the tangent vector to the curve φ at $\varphi(t)$ and will be denoted by $v(\varphi, t)$ (Figure 3.1).

Applying the differential of the curve φ, $\varphi_* : T((t_1, t_2)) \to T(M^n)$, to the unit tangent vector $\mathbf{e}_t$ of the interval (t_1, t_2) at t (see §10), we get

$$\varphi_*(\mathbf{e}_t) = v(\varphi, t).$$

If φ satisfies $v(\varphi, t) = X(\varphi(t))$ at each point t in (t_1, t_2), φ is called an *integral curve* or a *solution curve* of the vector field X.

In particular when $t_1 < 0 < t_2$ and $\varphi(0) = p$ we say that φ is an *integral curve of X starting at the point p, through the point p*, or *with the initial point p*.

Let $\bar{t}$ be a point on the open interval (t_1, t_2) and $(U_\lambda, \varphi_\lambda) \in \mathscr{S}$ be a chart about the point $\varphi(\bar{t})$ in M^n. Then the vector $X(q)$ at q in U_λ has the expression

$$\Phi_\lambda(X(q)) = (v_1(q), v_2(q), \dots, v_n(q)),$$

where the map Φ_λ was defined by (3) in §10.

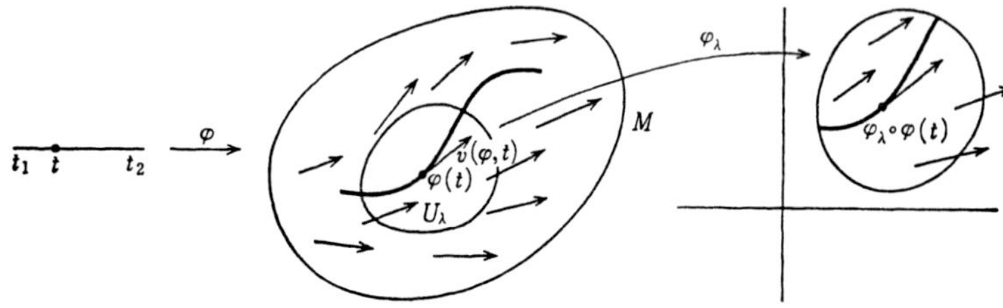

FIGURE 3.1

Let φ be an integral curve of X and write

$$(\varphi_\lambda \circ \varphi)(t) = (u_1(t), u_2(t), \ldots, u_n(t)), \qquad i = 1, 2, \ldots, n,$$

where $t \in \varphi^{-1}(U_\lambda)$. By the definition of Φ_λ we have the following:

$$\frac{du_i}{dt}(t) = v_i(\varphi(t)), \qquad i = 1, 2, \ldots, n. \tag{1}$$

Conversely if a C^{r+1} curve φ in M satisfies (1) for every $\bar{t} \in (t_1, t_2)$ and a chart $(U_\lambda, \varphi_\lambda)$ about $\varphi_\lambda(\bar{t})$, then φ is an integral curve of X.

In the following we shall assume that M^n is a closed manifold. Let p be a point of M^n and choose a chart $(U_\lambda, \varphi_\lambda)$ about p. In complete analogy with the argument in §3 for the vector fields on the torus, the theorem on the existence and uniqueness for differential equations implies a unique solution of (1):

$$\psi\colon (-\tau, \tau) \to \varphi_\lambda(U_\lambda), \qquad \psi(0) = \varphi_\lambda(p),$$

defined on some open interval $(-\tau, \tau)$ through $\varphi_\lambda(p)$. The curve ψ is of class C^{r+1}. Define $\varphi_1\colon (-\tau, \tau) \to M^n$ by $\varphi_\lambda \circ \varphi_1(t) = \psi(t)$; then φ_1 is an integral curve of X through p. We shall show that φ_1 can be extended to an integral curve defined on the entire line $(-\infty, \infty)$.

Given an integral curve $\varphi_1\colon (-\tau, \tau) \to M^n$ of X, choose a sequence of real numbers $x_1, x_2, \ldots, x_m, \ldots$ such that

$$-\tau < x_1 < x_2 < \cdots < x_m < \cdots < \tau, \qquad \lim_{m\to\infty} x_m = \tau.$$

By Theorem 2.1 the sequence $\varphi_1(x_1), \varphi_1(x_2), \ldots, \varphi_1(x_m), \ldots$ in M^n has a convergent subsequence

$$\varphi_1(x_{i_1}), \varphi_1(x_{i_2}), \ldots, \varphi_1(x_{i_m}), \ldots, \qquad \lim_{m\to\infty} \varphi_1(x_{i_m}) = \bar{p}.$$

Here we recall that M^n is a metric space (§10).

Choose a chart (U_μ, φ_μ) about $\bar{p}$; then the vector $X(q')$ of X at a point q' in U_μ can be characterized as

$$\Phi_\mu(X(q')) = (\bar{v}_1(q'), \bar{v}_2(q'), \ldots, \bar{v}_n(q')).$$

As before the integral curve φ of X satisfies

$$\frac{d\bar{u}_i}{dt}(t) = \bar{v}_i(\varphi(t)), \qquad i = 1, 2, \ldots, n, \tag{2}$$

where $(\varphi_\mu \circ \varphi)(t) = (\overline{u}_1(t), \overline{u}_2(t), \ldots, \overline{u}_n(t))$, $t \in \varphi^{-1}(U_\mu)$. By the existence and uniqueness theorem for ordinary differential equations again—as in §2 and §3—we can choose an appropriate neighborhood $V \subset \varphi_\mu(U_\mu)$ of the point $\varphi_\mu(\overline{p})$ and a positive number δ such that for any $\overline{q} \in V$, there is a solution of the differential equations in (2) through $\overline{q}$ and defined on an open interval $(-\delta, \delta)$:

$$\overline{\psi}: (-\delta, \delta) \to \varphi_\mu(U_\mu), \qquad \overline{\psi}(0) = \overline{q}.$$

Select $x_{i_m} \in (-\tau, \tau)$ such that

$$|x_{i_m} - \tau| < \delta/2, \quad \varphi_1(x_{i_m}) \in U_\mu, \quad \varphi_\mu(\varphi_1(x_{i_m})) \in V,$$

and let $\overline{\psi}_0 : (-\delta, \delta) \to \varphi_\mu(U_\mu)$ be a solution of (2) starting at $\varphi_\mu(\varphi_1(x_{i_m}))$ as above.

Using $\overline{\psi}_0$, we define a curve $\varphi_2 : (-\tau, x_{i_m} + \delta) \to M$ by

$$\varphi_2(t) = \begin{cases} \varphi_1(t), & -\tau < t \le x_{i_m}, \\ \varphi_\mu^{-1}(\overline{\psi}_0(t - x_{i_m})), & x_{i_m} \le t < x_{i_m} + \delta. \end{cases}$$

Then by the uniqueness of the solutions of differential equations φ_2 is an integral curve of X starting at p. Since $x_{i_m} + \delta > \tau + \delta/2$ we may apply this procedure successively in both the positive and negative directions of t; thus we have constructed an integral curve of X starting at p defined on the entire line $(-\infty, \infty)$. This solution curve will be denoted by

$$\varphi_{\{p\}}: (-\infty, \infty) \to M^n, \qquad p \in M^n.$$

By the uniqueness of solutions, the curve $\varphi_{\{p\}}$ is uniquely determined by its initial point p.

In Chapter I we wrote $\varphi(t, p)$ for $\varphi_{\{p\}}$. We remind the reader that the curve $\varphi(t, p)$, $-\infty < t < \infty$, defined in §3 is an example of an integral curve where the torus is M^n in our current discussion.

THEOREM 3.1. *Let X be a C^r dynamical system on a closed C^s manifold M^n of dimension n, $1 \le r \le s - 1$. Then we have the following.*

(i) *For every point p of M^n, there exists through p a unique integral curve $\varphi_{\{p\}}: (-\infty, \infty) \to M^n$ of class C^{r+1}.*

(ii) *For $t \in \mathbf{R}$, define $\Psi_t: M^n \to M^n$ by $\Psi_t(p) = \varphi_{\{p\}}(t)$. Then Ψ_0 is the identity map, and $\Psi_{t_1} \circ \Psi_{t_2} = \Psi_{t_1+t_2}$; $\varphi_{\{\varphi_{\{p\}}(t_2)\}}(t_1) = \varphi_{\{p\}}(t_1 + t_2)$.*

(iii) *The map $\Psi: \mathbf{R} \times M^n \to M^n$ defined by $\Psi(t, p) = \varphi_{\{p\}}(t)$ is differentiable of class C^r.*

PROOF. (i) was shown above. (ii) and (iii) follow directly from the uniqueness of solutions and the differentiability in initial values of solutions for differential equations.

The set $\{\varphi_{\{p\}}(t);\ -\infty < t < \infty\}$ which we shall denote by $C(p)$ is called the *trajectory* or the *orbit* of X through p. By Theorem 3.1(ii), $p' \in C(p)$ implies $C(p') = C(p)$.

Let $\varphi_{\{p\}}: (-\infty, \infty) \to M$ be the integral curve of X through p. Suppose that $X(p) \neq 0$ and that $\varphi_{\{p\}}(t) = p$ for some $t > 0$. Then the smallest among such t denoted by t_0 satisfies $\varphi_{\{p\}}(t_0) = p$, and $\varphi_{\{p\}}(t) \neq p$ for each number t, $0 < t < t_0$. Hence by Theorem 3.1(ii) we have

$$C(p) = \{\varphi_{\{p\}}(t);\ 0 \le t \le t_0\},$$

and $C(p)$ is compact. Such an integral curve is said to be *periodic* or *closed* and its trajectory $C(p)$ *periodic* or *closed*. If p is a singular point of X, in other words $X(p) = 0$, then $\varphi_{\{p\}}(t) = p$ for all numbers t, so the trajectory of X through p is a singleton set: $C(p) = \{p\}$.

The set

$$L^+(p) = \bigcap_{0 \le s < \infty} \overline{\{\varphi_{\{p\}}(t);\ s \le t < \infty\}}$$

is called the *ω-limit set* of $\varphi_{\{p\}}$, and the set

$$L^-(p) = \bigcap_{-\infty < s \le 0} \overline{\{\varphi_{\{p\}}(t);\ -\infty < t \le s\}}$$

is called the *α-limit set* of $\varphi_{\{p\}}$. These sets are often called simply *limit sets*. Since M^n is compact the intersection of closed sets of a family with the finite intersection property is nonempty; hence both $L^-(p)$ and $L^+(p)$ are nonempty closed sets.

A set $A \subset M^n$ is an *invariant set* of a dynamical system X if $\varphi_{\{p\}}(t)$ belongs to A for each $p \in A$ and for all t, $-\infty < t < \infty$. For example a closed orbit is an invariant set. An α-limit set $L^-(p)$ or an ω-limit set $L^+(p)$ is also invariant; the method of proof used for Theorem 1.3 works here. We say that A is a *minimal set* if A is a nonempty closed invariant set whose only closed invariant subsets are A and the empty set.

THEOREM 3.2. *Under the same hypothesis as for Theorem* 3.1, *the limit sets* $L^{\pm}(p)$ *are connected.*

PROOF. We give a proof for $L^+(p)$. Since $L^+(p)$ is closed in M^n, it is compact. Suppose that $L^+(p)$ is not connected. Set $L^+(p) = L_1 \cup L_2$, where L_1, L_2 are closed nonempty subsets of $L^+(p)$ with $L_1 \cap L_2 = \varnothing$. As the manifold M^n is Hausdorff and L_1, L_2 are compact, there exist open sets U_1, U_2 in M^n such that

$$L_1 \subset U_1, \qquad L_2 \subset U_2, \qquad U_1 \cap U_2 = \varnothing.$$

Since both L_1 and L_2 are subsets of the ω-limit set $L^+(p)$, there exist sequences $x_1, x_2, \ldots, x_m, \ldots$, and $y_1, y_2, \ldots, y_m, \ldots$ such that

$$\lim_{m\to\infty} x_m = \infty, \qquad \lim_{m\to\infty} \varphi_{\{p\}}(x_m) = x \in L_1,$$
$$\lim_{m\to\infty} y_m = \infty, \qquad \lim_{m\to\infty} \varphi_{\{p\}}(y_m) = y \in L_2.$$

Taking subsequences if necessary we may assume

$$x_1 < y_1 < x_2 < y_2 < \cdots < x_m < y_m < \cdots,$$
$$\varphi_{\{p\}}(x_m) \in U_1, \qquad \varphi_{\{p\}}(y_m) \in U_2, \qquad m = 1, 2, \ldots.$$

Since $U_1 \cap U_2 = \varnothing$, for $x_m < y_m$ there exist z_m with $x_m < z_m < y_m$, $m = 1, 2, \ldots$, such that $\varphi_{\{p\}}(z_m) \in M^n - U_1 - U_2$. Since $M^n - U_1 - U_2$ is compact, the sequence $\varphi_{\{p\}}(z_1), \varphi_{\{p\}}(z_2), \ldots, \varphi_{\{p\}}(z_m), \ldots$ has, by Theorem 2.1, a subsequence which converges to a point a of the set $M - U_1 - U_2$. Hence $a \in L^+(p)$. But this is a contradiction. Hence we conclude that the ω-limit set $L^+(p)$ is connected. A similar proof applies to the limit set $L^-(p)$.

§13. Dynamical systems on S^2 and the Poincaré-Bendixson theorem

Let X be a C^r dynamical system on the two-sphere S^2, $r \geq 1$. A limit set $L^{\pm}(p)$ of the trajectory $C(p)$ through p is $\{p\}$ if p is a singular point and it is $C(p)$ if its integral curve $\varphi_{\{p\}}$ is periodic. For limit sets $L^{\pm}(p)$ we have the following

THEOREM 3.3 (Poincaré-Bendixson). *For any point p on the two-dimensional sphere S^2, a limit set $L^+(p)$ or $L^-(p)$ containing no singular point is a closed orbit.*

PROOF. The theorem is trivial when $\varphi_{\{p\}}$ is periodic, so let us assume otherwise. Let $p' \in L^+(p)$; then $C(p') \subset L^+(p)$ as $L^+(p)$ is an invariant set. Here the periodicity of $C(p')$ can be proven in the same way as adopted for the proof of Lemma 1.16. Let us assume $L^+(p) - C(p') \neq \varnothing$. Suppose that $L^+(p) - C(p')$ is closed in $L^+(p)$. Then $C(p')$ splits into a disjoint union of two nonempty closed sets

$$L^+(p) = (L^+(p) - C(p')) \cup C(p'), \qquad (L^+(p) - C(p')) \cap C(p') = \varnothing.$$

This contradicts Theorem 3.2. Hence $L^+(p) - C(p')$ is not closed in $L^+(p)$, and so the closure $\overline{L^+(p) - C(p')}$ of $L^+(p) - C(p')$ in $L^+(p)$ contains a point of $C(p')$, say

$$a \in \overline{L^+(p) - C(p')} \cap C(p').$$

Let U be a sufficiently small neighborhood of a. We may think of U as a subset of $\mathbf{R}^2$. By assumption a is not a singular point, so we can draw a line segment l passing through a and orthogonal to the vector $X(a)$ (Figure 3.2). Since U is sufficiently small the change of $X(q)$, $q \in U$, in U is so

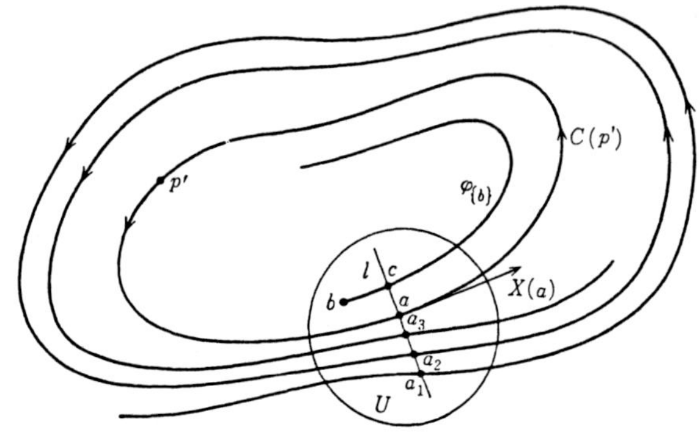

FIGURE 3.2

small that the integral curve $\varphi_{\{p\}}$ of X traverses l in a constant direction as t increases.

Since $a \in \overline{L^+(p) - C(p')}$, there exists a point $b \in L^+(p) - C(p')$ close to a. We may assume that $b \in U$. The integral curve $\varphi_{\{b\}}$ through b intersects l at some point c. As $L^+(p)$ is an invariant set, $c \in L^+(p)$. Also if $c \in C(p')$ then $b \in C(p')$, and so we must have $c \notin C(p')$. Hence we have found two distinct points a and c of $L^+(p)$ on the line segment l.

On the other hand, since $a \in L^+(p)$, there exists a sequence $x_1, x_2, \ldots, x_m, \ldots$ such that

$$0 < x_1 < x_2 < \cdots < x_m < \cdots, \qquad \lim_{m\to\infty} x_m = \infty,$$
$$\lim_{m\to\infty} \varphi_{\{p\}}(x_m) = a, \qquad \varphi_{\{p\}}(x_m) \in U.$$

Hence $\{\varphi_{\{p\}}(t); t > 0\}$ intersects l at infinitely many points. Let t_1 be the smallest t for which $\{\varphi_{\{p\}}(t)\}$ intersects l and write $a_1 = \varphi_{\{p\}}(t_1)$. As t increases from t_1 let $a_2 = \varphi_{\{p\}}(t_2)$ be the next point of intersection of $\{\varphi_{\{p\}}(t); t > 0\}$ and l (Figure 3.2). Similarly we choose points $a_i = \varphi_{\{p\}}(t_i)$, $i = 1, 2, \ldots$, on l (Figure 3.2). Since the union of the trajectory $\{\varphi_{\{p\}}(t); t_1 < t < t_2\}$ and the segment $\overline{a_1 a_2}$ on l is a simple closed curve on S^2, by the Jordan Curve Theorem it separates S^2 into two domains. The set $\{\varphi_{\{p\}}(t); t_2 < t < \infty\}$ is contained in one of these domains, and so the point $a_3 = \varphi_{\{p\}}(t_3)$ sits on the opposite side of a_1 with respect to a_2 on l. In general, a_{i+1} sits on the opposite side of a_{i-1} with respect to a_i on l, $i = 2, 3, \ldots$. Therefore, the points $a_1, a_2, \ldots, a_m, \ldots$, are located on l in this order, and

$$\lim_{m\to\infty} a_m = a.$$

Hence we must have

$$l \cap L^+(p) = \{a\}.$$

But this contradicts the fact that $a, c \in l \cap L^+(p)$. Hence $L^+(p) = C(p')$. The case of $L^-(p)$ is similar.

§14. Schweitzer's dynamical system on S^3

Let M be a closed n-dimensional C^s manifold. It is well known that M has a nonsingular vector field if and only if the Euler number $\chi(M)$ is zero (see Appendix III, Note 1).

The torus is the only closed orientable two-dimensional C^s manifold whose Euler number is zero. In Chapter I we discussed topological properties of trajectories and limit sets of a nonsingular vector field on the torus. The Euler number of a closed three-dimensional C^s manifold is always zero; every closed three-manifold admits a nonsingular vector field. In this section we shall consider the three-dimensional sphere S^3, the simplest closed three-manifold.

A standard example of a nonsingular vector field on S^3 might be the following C^∞ vector field X_H constructed from the Hopf map.

Identifying a point (x_1, x_2, x_3, x_4) of S^3 with the pair (z_1, z_2) of complex numbers $z_1 = x_1 + x_2 i$, $z_2 = x_3 + x_4 i$, we have

$$S^3 = \{(z_1, z_2); |z_1|^2 + |z_2|^2 = 1\}.$$

We define a relation on S^3 by $(z_1, z_2) \sim (z_1', z_2')$ if

$$z_1' = z_1 e^{i\theta}, \qquad z_2' = z_2 e^{i\theta}, \qquad 0 \le \theta < 2\pi,$$

where (z_1, z_2), $(z_1', z_2') \in S^3$. The relation $\sim$ is an equivalence relation. We denote by $[z_1, z_2]$ the equivalence class of a point (z_1, z_2). Consider a correspondence which assigns to an equivalence class $[z_1, z_2]$ the point $(2\mathscr{R}(z_1\bar{z}_2), 2\mathscr{I}(z_1\bar{z}_2), |z_1|^2 - |z_2|^2) \in S^2$. This correspondence does not depend on the choice of representatives of equivalence classes. The correspondence is one-to-one from the set of all equivalence classes onto S^2; so we identify these two sets.

Now define a map $\pi : S^3 \to S^2$ by

$$\pi((z_1, z_2)) = [z_1, z_2].$$

Then π, called the *Hopf map*, is of class C^∞. It is immediate from the definition that the inverse image of a point $[z_1, z_2]$ on S^2 is

$$\pi^{-1}([z_1, z_2]) = \{(z_1 e^{i\theta}, z_2 e^{i\theta}); 0 \le \theta < 2\pi\},$$

which is C^∞ diffeomorphic to S^1.

For a point (z_1, z_2) on S^3, we define a C^∞ curve

$$l^{(z_1, z_2)} : (-\epsilon, \epsilon) \to S^3$$

by

$$l^{(z_1, z_2)}(t) = (z_1 e^{it}, z_2 e^{it}), \qquad -\epsilon < t < \epsilon,$$

and consider the tangent vector $[l^{(z_1, z_2)}]$ to S^3 at (z_1, z_2). Define a vector field on S^3:

$$X_H = \{X_H((z_1, z_2)); (z_1, z_2) \in S^3\} \tag{1}$$

by

$$X_H((z_1, z_2)) = [l^{(z_1, z_2)}].$$

Then X_H is a nonsingular C^∞ vector field.

The integral curve $\varphi_{\{(z_1, z_2)\}}$ of X_H through a point (z_1, z_2) on S^3 is contained in $\pi^{-1}([z_1, z_2])$ and its trajectory $C((z_1, z_2))$ satisfies:

$$C((z_1, z_2)) = \pi^{-1}([z_1, z_2]).$$

Hence every trajectory of X_H is periodic.

In 1950, Seifert proved that any C^r vector field, $r \geq 1$, obtained through a small perturbation of X_H, has at least one periodic trajectory. From this result came the *Seifert Conjecture* which states that every nonsingular C^r vector field on S^3 has a periodic trajectory, $r \geq 1$. However, in 1972, Schweitzer proved the following theorem, which shows that the conjecture is false for C^1 vector fields (Appendix III).

THEOREM 3.4 (Schweitzer). *There exists a nonsingular C^1 vector field on the three-dimensional sphere S^3 which has no closed integral curves.*

In order to prove this theorem we first construct a C^∞ vector field on S^3 with exactly one periodic integral curve.

Consider $S^3 - \pi^{-1}([z_1, 0])$, where $[z_1, 0]$ is a point on S^2. Identifying the complex plane with $\mathbf{R}^2$ define a map

$$\eta: S^3 - \pi^{-1}([z_1, 0]) \to \mathbf{R}^2 \times S^1$$

by

$$\eta((z_1, z_2)) = (z_1/z_2, z_2/|z_2|).$$

Then η is a C^∞ diffeomorphism. For $(z_1, z_2) \in S^3 - \pi^{-1}([z_1, 0])$, we shall write $\eta((z_1, z_2)) = ((x, y), e^{i\theta}) \in \mathbf{R}^2 \times S^1$, and define a C^∞ curve

$$\tilde{l}^{(z_1, z_2)}: (-\epsilon, \epsilon) \to S^3 - \pi^{-1}([z_1, 0])$$

by

$$\tilde{l}^{(z_1, z_2)}(t) = \eta^{-1}((x + t, y), e^{i\theta}), \qquad -\epsilon < t < \epsilon.$$

Define a vector field X_1' on S^3 by

$$X_1'((z_1, z_2)) = \begin{cases} [\tilde{l}^{(z_1, z_2)}], & (z_1, z_2) \in S^3 - \pi^{-1}([z_1, 0]), \\ 0, & (z_1, z_2) \in \pi^{-1}([z_1, 0]). \end{cases}$$

Then X_1' is a C^∞ vector field.

We go on to define a nonsingular C^∞ vector field X_1 on S^3 by

$$X_1((z_1, z_2)) = X_H((z_1, z_2)) + X_1'((z_1, z_2)). \tag{2}$$

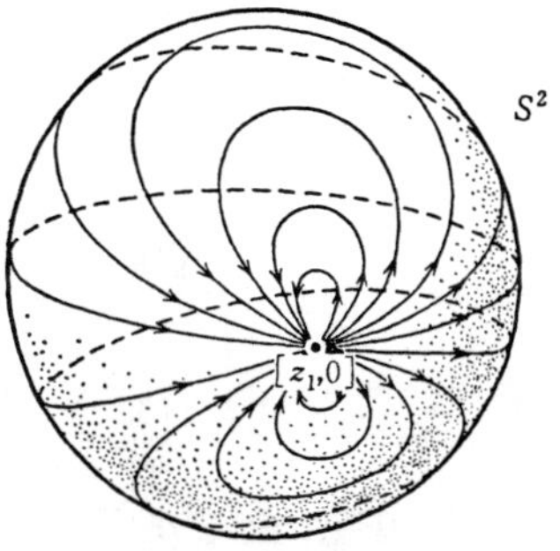

FIGURE 3.3

Evidently we have $\pi_*(X_H((z_1, z_2))) = 0$, where $\pi_*\colon T(S^3) \to T(S^2)$ is the differential of the map π. Further if $(z_1, z_2) \sim (z_1', z_2')$, from the definition of $\tilde{l}^{(z_1, z_2)}$ we have

$$\pi_*(X_1'((z_1, z_2))) = \pi_*(X_1'((z_1', z_2'))),$$

and hence

$$\pi_*(X_1((z_1, z_2))) = \pi_*(X_1((z_1', z_2'))).$$

Thus the correspondence $Y_1([z_1, z_2]) = \pi_*(X_1((z_1, z_2)))$, $[z_1, z_2] \in S^2$, is independent of the choice of a representative (z_1, z_2); Y_1 is a C^∞ vector field on S^2 with exactly one singular point $[z_1, 0]$. The trajectories of the vector field Y_1 are shown in Figure 3.3. The limit sets are all equal to the singleton set $\{[z_1, 0]\}$. There are no periodic integral curves.

It follows from the definition of Y_1 that $\pi \circ \varphi_{\{(z_1, z_2)\}}$ is the integral curve of Y_1 through $\pi((z_1, z_2))$, where $\varphi_{\{(z_1, z_2)\}}$ is the integral curve of X_1 through $(z_1, z_2) \in S^3$. If $(z_1 e^{i\theta}, 0) \in \pi^{-1}([z_1, 0])$, then the trajectory $C((z_1 e^{i\theta}, 0))$ of X_1 through $(z_1 e^{i\theta}, 0)$ is $\pi^{-1}([z_1, 0])$ and hence is closed; whereas, if $(z_1', z_2') \notin \pi^{-1}([z_1, 0])$, the limit sets of $\varphi_{\{(z_1', z_2')\}}$ are each $\pi^{-1}([z_1, 0])$, so $C((z_1', z_2'))$ is not a closed trajectory. Therefore, the set $\pi^{-1}([z_1, 0])$ is the only closed trajectory of X_1.

One next constructs a C^1 vector field of Theorem 3.4 by modifying a part of the vector field X_1; here Denjoy's vector field of §6 plays an essential role.

Let X_D be Denjoy's C^1 vector field defined on the torus T in the proof of Theorem 1.12. Let C be a trajectory of X_D. Recall $\overline{C} \neq T$. Thus there exists a C^∞ embedding $j\colon \operatorname{Int} D^2 \to T$ such that $j(\operatorname{Int} D^2) \subset T - \overline{C}$.

Let $N = T - j(\operatorname{Int} D^2)$ (Figure 3.6 see page 75), and construct a C^1 vector field

$$Y = \{Y((x, t));\, x \in N,\, t \in [-2, 2]\}$$

on $N \times [-2, 2]$ as follows:

First, we choose a C^∞ function on N, $\gamma\colon N \to [0, 1]$, satisfying

(i) $\gamma = 0$ in some neighborhood of ∂N,

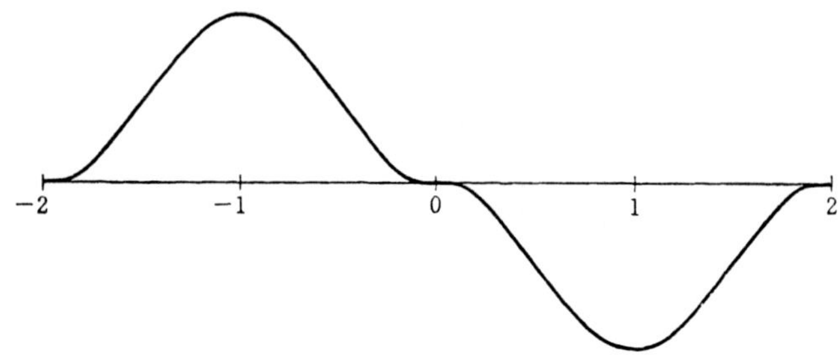

FIGURE 3.4

(ii) $\gamma = 1$ in $\overline{C}$.
We next construct a C^∞ function $\rho: [-2, 2] \to [-1, 1]$ satisfying the following:

(i) $\rho(-t) = -\rho(t)$,
(ii) $\rho(t) = 0$ near 0, ± 2,
(iii) $\rho(-1) = 1$, $\rho(1) = -1$,
(iv) $|\rho(t)| < 1$ if $t \neq \pm 1$,

(see Figure 3.4). We note that the functions γ and ρ can be routinely constructed via the C^∞ map Φ of Lemma 1.5.

Now for $(x, t) \in N \times [-2, 2]$, let $Y((x, t))$ be defined by

$$Y((x, t)) = \rho(t)\gamma(x)X_D(x) + (1 - |\rho(t)|\gamma(x))\mathbf{e}_t,$$

where $\mathbf{e}_t$ is the unit tangent vector of interval $[-2, 2]$ at t. Then

$$Y = \{ Y((x, t)); x \in N, t \in [-2, 2]\}$$

is a nonsingular C^1 vector field on $N \times [-2, 2]$, and $Y((x, t)) = \mathbf{e}_t$ in some neighborhood of $(\partial N \times [-2, 2]) \cup (N \times (\{-2\} \cup \{2\}))$ (see Appendix III, Note 2).

Since $Y((x, -2))$ points inwards on $N \times [-2, 2]$, we can define the integral curve $\varphi_{\{(x, -2)\}}$ of Y for $t \geq 0$. The component $(1 - |\rho(t)|\gamma(x))\mathbf{e}_t$ of Y in the direction of $[-2, 2]$ is never negative; it is zero for $(x, \pm 1) \in \overline{C} \times \{\pm 1\}$ and positive for $(x, t) \in N \times ([-2, 2] - \{\pm 1\})$. Moreover we have

$$Y((x, t)) = -\rho(t)\gamma(x)X_D(x) + (1 - |\rho(t)|\gamma(x))\mathbf{e}_{-t}.$$

Hence trajectories of Y must be symmetric with respect to $N \times \{0\}$ (Figure 3.5). From these facts we conclude that as t increases from zero the situation of the integral curve $\varphi_{\{(x, -2)\}}$ must be one of the following (Figure 3.5).

(i) $\varphi_{\{(x, -2)\}}$ reaches $(x, 2) \in N \times \{2\}$. Here because of the symmetry discussed above, the point of departure $(x, -2)$ and the point of arrival $(x, 2)$ share the same x.

(ii) $\varphi_{\{(x, -2)\}}$ approaches $\gamma^{-1}(1) \times \{-1\}$ and does not appear in the set $N \times (-1, 2]$. In particular this is the case when $x \in \overline{C}$.

In case (ii) the ω-set of $\varphi_{\{(x, -2)\}}$ is contained in $\gamma^{-1}(1) \times \{-1\}$. Similarly we define the integral curve $\varphi_{\{(x, 2)\}}$ through $(x, 2) \in N \times \{2\}$ for $t \leq 0$, and

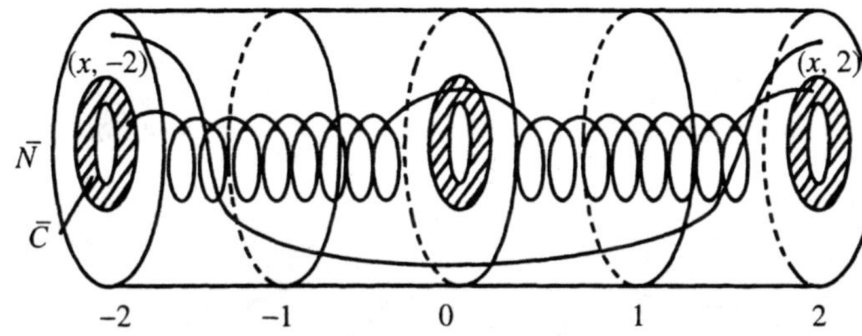

FIGURE 3.5

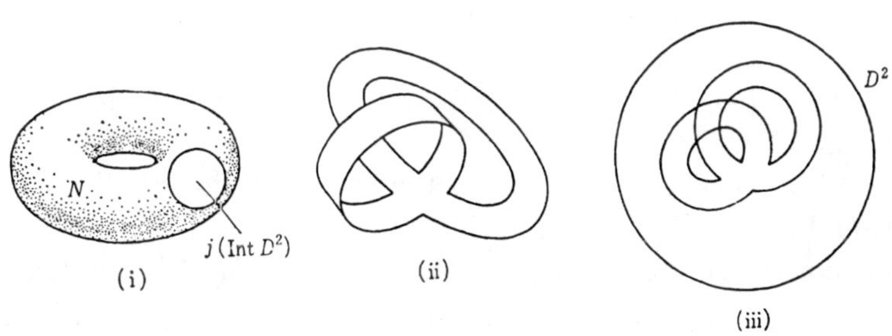

FIGURE 3.6

one of the situations similar to (i) or (ii) above happens. Here the α-limit set of the integral curve $\varphi_{\{(x,2)\}}$ in the case corresponding to (ii) is contained in $\gamma^{-1}(1) \times \{1\}$. There also is an integral curve of Y in $N \times (-1, 1)$ whose α-limit set is contained in $\gamma^{-1}(1) \times \{-1\}$ and ω-limit set in $\gamma^{-1}(1) \times \{1\}$ (Figure 3.5).

Now make the following observation. The N in Figure 3.6 (i) can also be depicted as in Figure (ii); we consider an immersion of N in D^2 which can be visualized as in Figure (iii). Thicken D^2 to $D^2 \times [0, \delta]$, and then by undoing the self-intersections as seen in (iii) in the direction of $[0, \delta]$, obtain an embedding of N in $D^2 \times [0, \delta]$. Further we thicken the image of this embedding in the $[0, \delta]$-direction to define a C^∞ embedding

$$g: N \times [-\epsilon, \epsilon] \to D^2 \times [0, \delta].$$

We can make g in such a way that the image $g_*(\mathbf{e}_t)$ of the tangent vector $\mathbf{e}_t$ to $N \times [-\epsilon, \epsilon]$ at (x, t) is the tangent vector $\mathbf{e}_t$ to $D^2 \times [0, \delta]$ at $g(x, t)$, where both unit tangent vectors at $t \in [-\epsilon, \epsilon]$ and $t \in [-\delta, \delta]$ respectively are denoted by $\mathbf{e}_t$. We may also assume that $g(\overline{C} \times \{-\epsilon\}) \cap \{O\} \times [0, \delta] \neq \varnothing$. Here O is the center of D^2.

We define a C^1 vector field X_0 on $D^2 \times [0, \delta]$ as follows.

(i) $X_0((x, t)) = \mathbf{e}_t$ for $(x, t) \in D^2 \times [0, \delta] - g(N \times (-\epsilon, \epsilon))$.
(ii) $X_0(g(x, t)) = (2/\epsilon) g_*(\kappa_*^{-1}(Y(\kappa(x, t))))$ in $g(N \times (-\epsilon, \epsilon))$.

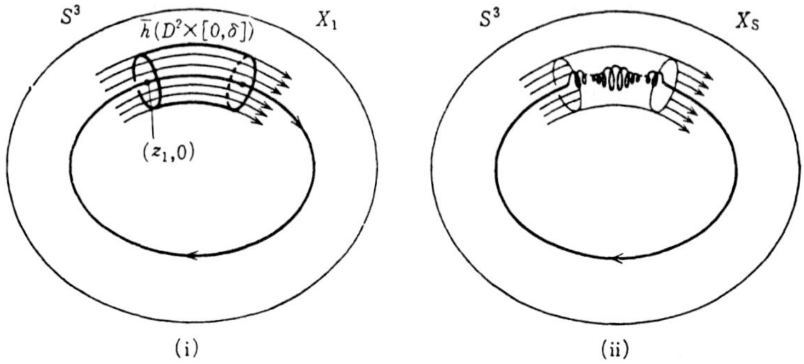

FIGURE 3.7

Here $\kappa\colon N\times(-\epsilon\,,\,\epsilon)\to N\times(-2\,,\,2)$ is a C^∞ diffeomorphism defined by $\kappa(x\,,\,t)=(x\,,\,2t/\epsilon)$.

The integral curves of X_0 also exhibit the behavior of Y as given in cases (i) and (ii).

Now choose an embedding $h\colon D^2\to S^3$ such that $h(D^2)$ intersects orthogonally the closed trajectory $\pi^{-1}([z_1\,,\,0])$ of the C^∞ vector field X_1 on S^3 defined by (2); say $h(D^2)$ is orthogonal to the vector $X_1(z_1\,,\,0)$ at $(z_1\,,\,0)$. Further, by thickening $h(D^2)$ obtain an embedding

$$\bar{h}\colon D^2\times[0\,,\,\delta]\to S^3,$$

such that $\bar{h}(\{x\}\times[0\,,\,\delta])$ is on the trajectory of X_1 through the point $h(x)$ for each $x\in D^2$ (Figure 3.7). We may choose $\bar{h}$ in such a way that

$$\bar{h}(0\,,\,0)=(z_1\,,\,0),\qquad X_1(\bar{h}(x\,,\,t))=\bar{h}_*(\mathbf{e}_t).$$

We next modify X_1 in $\bar{h}(D^2\times[0\,,\,\delta])$ and define a nonsingular vector field X_S on S^3 as follows (Figure 3.7).

(i) X_S agrees with X_1 on $S^3-\bar{h}(D^2\times[0\,,\,\delta])$.

(ii) $X_S(\bar{h}(x\,,\,t))=\bar{h}_*(X_0((x\,,\,t)))$ at $\bar{h}(x\,,\,t)$, $x\in D^2$, $t\in[0\,,\,\delta]$.

In the new vector field X_S the closed trajectory of X_1 was erased in the thickened disk $\bar{h}(D^2\times[0\,,\,\delta])$; moreover there is no emergence of new closed trajectories in the process of our modification (Figure 3.7(ii)). Hence we have established Theorem 3.4.

The method applied in the proof uses the result of Denjoy and so it cannot upgrade the theorem for C^2 vector fields. The Seifert Conjecture therefore remains open for C^r vector fields, with $r\geq 2$ ([1]).

([1])Translator's note. Recently J. Harrison solved the problem for the case $r=2$, *Topology* **27** (1988), pp. 249–278.

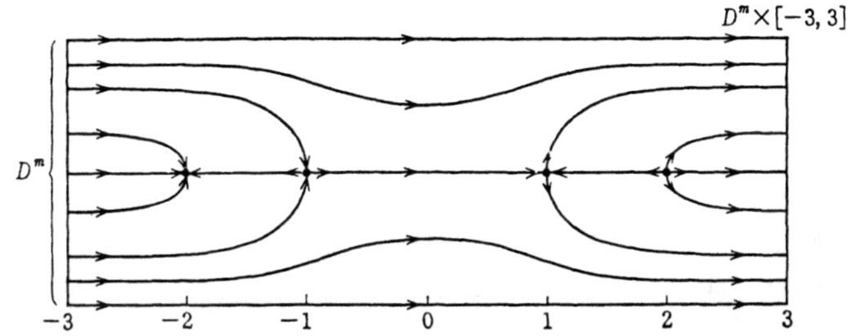

FIGURE 3.8

§15. Wilson's dynamical system

In the proof of Schweitzer's theorem in the previous section one replaced the standard vector field on $D^2 \times [0, \delta]$ by a new vector field, thereby removing the existing closed trajectory. This technique had been used by Wilson as early as 1966 when he proved the following

THEOREM 3.5 (Wilson). *Let X be a nonsingular C^r vector field on an n-dimensional closed C^s manifold, $1 \leq r \leq s-1$. Then for each integer k, $0 < k \leq n-2$, it is possible to modify X to a nonsingular C^r vector field X_W whose limit sets are a finite collection of k-tori, $S^1 \times \cdots \times S^1$ (the product manifold of k copies of S^1).*

Wilson first considered a C^∞ vector field Y_1 defined on $D^m \times [-3, 3]$, $m \geq 1$, as follows (Figure 3.8). Let $x \in D^m$, $t \in [-3, 3]$.

(i) $Y_1((x, t)) = \mathbf{e}_t$ (the unit tangent vector at t) in some neighborhood of $(\partial D^m \times [-3, 3]) \cup (D^m \times (\{-3\} \cup \{3\}))$.

(ii) $\pi_{1*}(Y_1((x, -t))) = -\pi_{1*}(Y_1((x, t)))$, and $\pi_{2*}(Y_1((x, -t))) = \pi_{2*}(Y_1((x, t)))$, where $\pi_1 \colon D^m \times [-3, 3] \to D^m$, $\pi_2 \colon D^m \times [-3, 3] \to [-3, 3]$ are projections.

(iii) Y_1 has four singular points, $\{O\} \times \{\pm 2\}$, $\{O\} \times \{\pm 1\}$. Here O is the center of D^m. Y_1 has no closed orbits.

Using Y_1 he then constructed a C^∞ vector field Y_1' on $S^1 \times D^m \times [-3, 3]$. Regard S^1 as $[0, 2\pi]$ with the endpoints $\{0\}$ and $\{2\pi\}$ identified, and denote by $\mathbf{e}_\theta'$ the unit vector at a point θ of S^1. Choose a C^∞ function on the thickened m-ball $D^m \times [0, 3]$:

$$\nu \colon D^m \times [0, 3] \to [0, 1] \tag{1}$$

which satisfies the following.

(i) $\nu = 0$ in some neighborhood of $(\partial D^m \times [0, 3]) \cup (D^m \times (\{0\} \cup \{3\}))$.

(ii) $\nu = 1$ in some neighborhoods of the points $(O, 1)$, $(O, 2) \in D^m \times [0, 3]$. Here O is the center of D^m.

Define a C^∞ vector field Y_1' on $S^1 \times D^m \times [-3, 3]$:

$$Y_1' = \{\, Y_1'((\theta, x, t)) ;\, \theta \in S^1, x \in D^m, -3 \leq t \leq 3 \,\}$$

by

$$Y_1'((\theta, x, t)) = \begin{cases} -c_1\nu(x, t)\mathbf{e}_\theta' + Y_1((x, t)), & t \geq 0, \\ c_1\nu(x, |t|)\mathbf{e}_\theta' + Y_1((x, t)), & t \leq 0, \end{cases} \tag{2}$$

where c_1 is a nonzero real number. Then Y_1' is a nonsingular C^∞ vector field with the following properties.

(i) $Y_1'((\theta, x, t)) = \mathbf{e}_t$ in some neighborhood of

$$(\partial(S^1 \times D^m) \times [-3, 3]) \cup (S^1 \times D^m \times (\{-3\} \cup \{3\})).$$

(ii) Y_1' has four closed orbits $S^1 \times \{O\} \times \{\pm 2\}$, $S^1 \times \{O\} \times \{\pm 1\}$.

(iii) The integral curve $\varphi_{\{(\theta, x, -3)\}}$ of Y_1' through $(\theta, x, -3)$ is defined for $t \geq 0$, and one of the following occurs:

(a) $\varphi_{\{(\theta, x, -3)\}}$ reaches $(\theta, x, 3)$.

(b) $\varphi_{\{(\theta, x, -3)\}}$ spirals towards a closed orbit: the ω-limit set of $\varphi_{\{(\theta, x, -3)\}}$ is a closed orbit.

With a suitable embedding $S^1 \times D^m \to D^{m+1}$ one regards $S^1 \times D^m$ as a subset of D^{m+1}, making sure that the center of D^{m+1} is in $S^1 \times O$, where O is the center of D^m.

Define a nonsingular C^∞ vector field Y_2 on $D^{m+1} \times [-3, 3]$ by

$$\begin{aligned} Y_2((y, t)) &= \mathbf{e}_t, & y &\in D^{m+1} - (S^1 \times D^m), \\ Y_2((y, t)) &= Y_1'((\theta, x, t)), & y &= (\theta, x) \in S^1 \times D^m. \end{aligned} \tag{3}$$

Construct a nonsingular C^∞ vector field Y_2' on $S^1 \times D^{m+1} \times [-3, 3]$ as one defined Y_1' in (2); except this time instead of c_1 one takes a nonzero real number c_2 with c_1/c_2 irrational. Then one goes on to define a nonsingular vector field Y_3 on $D^{m+2} \times [-3, 3]$ from Y_2' as in (3). The limit sets of Y_3 are a set of four two-dimensional tori.

By repeating the above construction, one obtains a C^∞ vector field Y_{k+1} on $D^{m+k} \times [-3, 3]$ whose only limit sets are a set of four k-tori.

For the vector field Y_1 there exists a neighborhood U' of O in D^m such that none of the integral curves $\varphi_{\{(x, -3)\}}$ of Y_1 through the point $(x, -3)$, $x \in U'$, reaches $D^m \times \{3\}$. Hence it follows that there exists an open set U in D^{m+k} such that none of the integral curves $\varphi_{\{(x, -3)\}}$ of Y_{k+1} through the point $(x, -3)$, $x \in U$, reaches $D^{m+k} \times \{3\}$.

Now given a C^∞ vector field X on M^n, let $m = n - k - 1$, and following the construction of $\overline{h}$ in §14, select a C^r map

$$\overline{h}\colon D^{n-1} \times [-3, 3] \to M^n$$

such that

$$X(\overline{h}(x, t)) = c\overline{h}_*(\mathbf{e}_t), \qquad c \text{ a constant.}$$

Modify X in $\overline{h}(D^{n-1} \times [-3, 3])$ to define a C^r vector field X'_W on M^n as follows.

(i) X'_W agrees with X on $M^n - \overline{h}(D^{n-1} \times [-3, 3])$.

(ii) $X'_W(\overline{h}(x, t)) = c\overline{h}_*(Y_{k+1}((x, t)))$, $x \in D^{n-1}$, $t \in [-3, 3]$, (see Figure 3.7).

No integral curve of X'_W passing trough a point in $\overline{h}(U \times \{-3\})$ reaches $\overline{h}(D^{n-1} \times \{3\})$ as t increases, so the ω-sets of the integral curves through $\overline{h}(U \times \{-3\})$ are contained in $\overline{h}(D^{n-1} \times [-3, 3])$.

Since M^n is compact, by performing finitely many adaptations of this type one can construct a vector field X_W with the desired properties of Theorem 3.5.

As a corollary to Theorem 3.5 one has the following

THEOREM 3.6 (Wilson). *Let M^n be a closed n-dimensional C^∞ manifold with Euler number zero. Assume $n \geq 4$. Then there exists a nonsingular C^∞ vector field with no closed orbits.*

PROOF. Since $\chi(M)$ is zero, there is a nonsingular C^∞ vector field on it (see Appendix III, Note 1). Apply Theorem 3.5 to this vector field with $k \geq 2$.

Applying Theorem 3.5 to a closed C^∞ manifold M^3 one obtains a nonsingular C^∞ vector field X_W with a finite number of closed trajectories. One then applies Schweitzer's modification technique in §14 to each of the closed trajectories to get the following

THEOREM 3.7 (Schweitzer). *Every closed n-dimensional C^∞ manifold admits a nonsingular C^1 vector field with no closed orbits.*

CHAPTER IV

Foliations

§16. Definitions and examples of foliations

In Chapter I we discussed integral curves of nonsingular vector fields on the torus; we saw that the torus was covered by trajectories comprising a family of lines locally—in a suitable coordinate neighborhood (see Figures 1.10, 1.12, 4.2). We noted that these trajectories trace 'striped patterns' on the torus, and we made topological observations of these patterns. As one can see in Theorem 1.4, although families of trajectories look alike locally, they are by no means identical as a whole on the torus. A *foliation* is an extension of the striped pattern on the torus to a general manifold.

First consider a most basic example. If we think of n-dimensional Euclidean space $\mathbf{R}^n$ as $\mathbf{R}^n = \mathbf{R}^k \times \mathbf{R}^{n-k}$, $0 \le k \le n$, we have

$$\mathbf{R}^n = \bigcup_{(x_{k+1}, x_{k+2}, \ldots, x_n)} \mathbf{R}^k \times (x_{k+1}, x_{k+2}, \ldots, x_n),$$

that is, $\mathbf{R}^n$ decomposes into a union of $\mathbf{R}^k \times (x_{k+1}, x_{k+2}, \ldots, x_n)$'s each of which is C^∞ diffeomorphic to $\mathbf{R}^k$. This decomposition of $\mathbf{R}^n$ is called a trivial k-dimensional foliation or a trivial codimension $n-k$ foliation of Euclidean space $\mathbf{R}^n$, and $\mathbf{R}^k \times (x_{k+1}, x_{k+2}, \ldots, x_n)$ is called a leaf. The case $n = 3$, $k = 2$ is shown in Figure 4.1.

Let $(M^n, \mathscr{S})$ be an n-dimensional C^s manifold with or without boundary: we take $\varphi_\lambda \colon U_\lambda \to V_\lambda \subset \mathbf{R}^n_+$, $\mathbf{R}^n_+ = \{(x_1, x_2, \ldots, x_n) \in \mathbf{R}^n; x_n \ge 0\}$ for a chart $(U_\lambda, \varphi_\lambda) \in \mathscr{S}$. If $r \le s$, M^n is *a priori* of class C^r, and we shall denote by $(M^n, \mathscr{S}^{(r)})$ the manifold M^n with a C^r atlas $\mathscr{S}^{(r)}$. Let k be an integer, $0 \le k \le n$.

Let $\mathscr{F} = \{L_\alpha; \alpha \in A\}$ be a family of arcwise connected subsets L_α of the manifold M^n. We say that $\mathscr{F}$ is a *k-dimensional C^r foliation* of M^n if $\mathscr{F}$ satisfies the following.

($\mathscr{F}_{\mathrm{I}}$) $L_\alpha \cap L_\beta = \varnothing$ for $\alpha, \beta \in A$ with $\alpha \ne \beta$.

($\mathscr{F}_{\mathrm{II}}$) $\bigcup_{\alpha \in A}(L_\alpha) = M^n$.

($\mathscr{F}_{\mathrm{III}}$) Given a point p in M^n, there exists a chart $(U_\lambda, \varphi_\lambda) \in \mathscr{S}^{(r)}$ about p such that for L_α with $U_\lambda \cap L_\alpha \ne \varnothing$, $\alpha \in A$, each (arcwise) connected component of $\varphi_\lambda(U_\lambda \cap L_\alpha)$ is of the form

$$\{(x_1, x_2, \dots, x_n) \in \varphi_\lambda(U_\lambda);\, x_{k+1} = c_{k+1},\, x_{k+2} = c_{k+2}, \dots, x_n = c_n\},$$

where $c_{k+1}, c_{k+2}, \dots, c_n$ are constants determined by the (arcwise) connected component (Figure 4.2).

We call L_α a *leaf* of the foliation $\mathcal{F}$. A k-dimensional C^r foliation is also referred as a C^r *codimension* $n-k$ *foliation* or a C^r *foliation of codimension* $n-k$. This is a more conventional term and so we shall call $\mathcal{F}$ a C^r *codimension* q *foliation*, where $n-k=q$. A manifold M^n coupled with a foliation $\mathcal{F}$ is called *foliated* and often denoted by $(M^n, \mathcal{F})$. By the conditions $(\mathcal{F}_{\text{I}})$ and $(\mathcal{F}_{\text{II}})$, M^n is covered by pairwise disjoint leaves, and by $(\mathcal{F}_{\text{III}})$ the leaves locally form a trivial k-dimensional foliation of $\mathbf{R}^n$. A chart $(U_\lambda, \varphi_\lambda) \in \mathcal{S}^{(r)}$ which satisfies $(\mathcal{F}_{\text{III}})$ is called a *foliated chart* or *foliated neighborhood.* The set of all foliated charts is called a *foliated atlas* and denoted by $\mathcal{S}^{(r)}_{\mathcal{F}}$.

If the boundary ∂M^n of M^n is nonempty and $p \in \partial M^n \cap L_\alpha$ for some leaf L_α, by $(\mathcal{F}_{\text{III}})$ $L_\alpha \subset \partial M^n$. Now denote

$$\partial\mathcal{F} = \{L_\alpha;\, \alpha \in A,\, L_\alpha \subset \partial M^n\}.$$

Then by $(\mathcal{F}_{\text{I}})$, $(\mathcal{F}_{\text{II}})$, $(\mathcal{F}_{\text{III}})$, $\partial\mathcal{F}$ is a C^r codimension $n-k-1$ foliation of an $(n-1)$-dimensional C^r manifold ∂M^n—notice that $\partial\mathcal{F}$ is of dimension k. In particular if $k = n-1$, $\mathcal{F}$ is a codimension one foliation, and each connected component of ∂M^n is a leaf of $\mathcal{F}$ (see Example B).

EXAMPLE A. Let $X_{a,b}$ be a vector field on the torus T given in Theorem 1.4 (Figure 1.10), and let $\mathcal{F}_{a,b}$ be the family of all trajectories $C(p)$ of

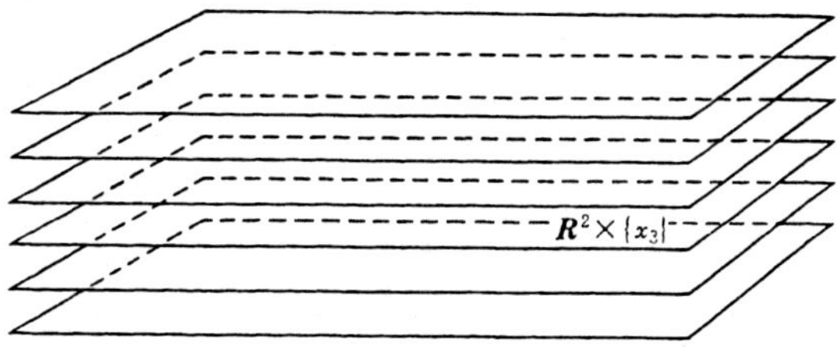

FIGURE 4.1

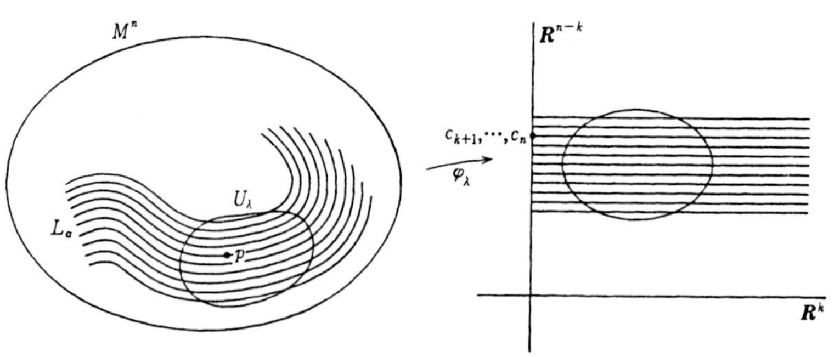

FIGURE 4.2

the vector field $X_{a,b}$. Then $\mathcal{F}_{a,b}$ is a C^∞ codimension one foliation of the torus T which is a two-dimensional C^∞ manifold.

More generally, suppose that X is a nonsingular C^r vector field on a closed C^s manifold M^n, $1 \le r \le s-1$; then the family $\mathcal{F}$ of the trajectories of X is a C^r codimension one foliation. From Theorem 3.2, $\mathcal{F}$ satisfies $(\mathcal{F}_{\mathrm{I}})$ and $(\mathcal{F}_{\mathrm{II}})$, and we get a foliated chart of $(\mathcal{F}_{\mathrm{III}})$ as follows. For $p \in M^n$, choose a C^r embedding $g\colon D^{n-1} \to M^n$ of the $(n-1)$-ball D^{n-1} in M^n such that $g(O) = p$, where O is the origin of D^{n-1}, and at each point $q \in g(D^{n-1})$ the integral curve $\varphi_{\{q\}}$ through q intersects $g(D^{n-1})$ transversely (Figure 4.3). We can easily construct such a map g using local coordinates about p. Now define

$$\tilde{g}\colon (-\epsilon, \epsilon) \times \operatorname{Int} D^{n-1} \to M^n$$

by $\tilde{g}(t, x) = \varphi(t, g(x))$, $-\epsilon < t < \epsilon$, $x \in \operatorname{Int} D^{n-1}$; for $\epsilon > 0$ sufficiently small, $\tilde{g}$ is an embedding. Define U_λ, φ_λ by

$$U_\lambda = \tilde{g}\big((-\epsilon, \epsilon) \times \operatorname{Int} D^{n-1}\big), \qquad \varphi_\lambda\big(\tilde{g}(t, x)\big) = (t, x).$$

Then $(U_\lambda, \varphi_\lambda)$ satisfies $(\mathcal{F}_{\mathrm{III}})$—we assumed M^n to be closed so that the results in Chapter 3 are applicable; otherwise the closedness is not essential here.

It follows from the above discussion that the family of the trajectories of the Denjoy C^1 vector field on the torus T of §6 is a C^1 codimension one foliation of T. Further, the family of the trajectories of the Schweitzer dynamical system on S^3 in §14 is a C^1 codimension two foliation of S^3.

EXAMPLE B. Let $f\colon (-1, 1) \to \mathbf{R}$ be a C^∞ function defined by

$$\begin{aligned} &f(0) = 0, \qquad f(t) \ge 0, \qquad f(t) = f(-t), \qquad -1 < t < 1, \\ &\lim_{t\to\pm 1} \frac{d^k}{dt^k} f(t) = \infty, \qquad \lim_{t\to\pm 1} \frac{d^k}{dt^k}\Big(1/\tfrac{df(t)}{dt}\Big) = 0, \qquad k = 0, 1, 2, \ldots, \end{aligned} \tag{1}$$

(see Figure 4.4). For example, $f(t) = e^{t^2/(1-t^2)} - 1$ will work.

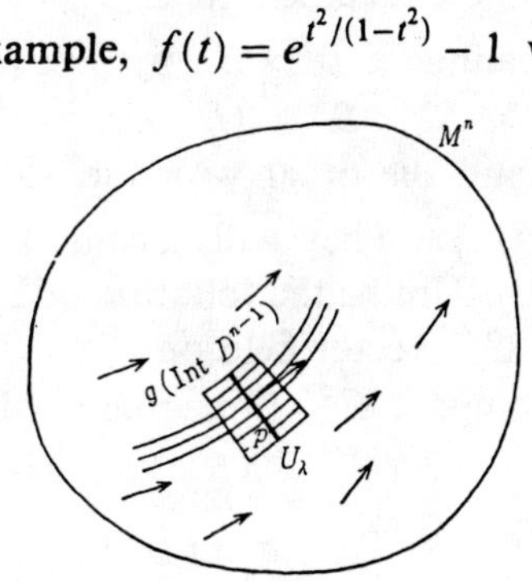

FIGURE 4.3

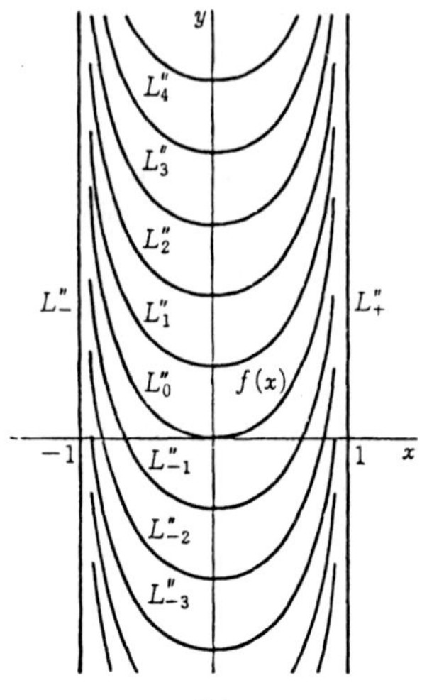

(i)

(ii)

FIGURE 4.4

Define subsets L''_α, $-\infty < \alpha < \infty$, and $L''_\pm$ of $D^1 \times \mathbf{R}^1$ by

$$L''_\alpha = \{ (t, \alpha + f(t)) ; -1 < t < 1 \}, \qquad L''_\pm = \{ (\pm 1, y) ; -\infty < y < \infty \}$$

(Figure 4.4 (i)). Denote by $\mathcal{F}'''_R$ the family whose members are the sets $L''_\pm$ and L''_α, $-\infty < \alpha < \infty$; then $\mathcal{F}'''_R$ is a C^∞ codimension one foliation of $D^1 \times \mathbf{R}^1$ (Figure 4.4 (i)). (To verify that $\mathcal{F}'''_R$ is indeed a C^∞ foliation, it is best to express $\mathcal{F}'''_R$ in terms of *differential forms* as in Example 2, §30 and show that it is of class C^∞ and *completely integrable*).

Define subsets L'_α, $0 \le \alpha < 1$, and $L'_\pm$ of the annulus $D^1 \times S^1$ by

$$L'_\alpha = \{ (t, e^{2\pi(\alpha + f(t))i}) ; -1 < t < 1 \}, \qquad L'_\pm = \{ (\pm 1, e^{2\pi\theta i}) ; 0 \le \theta < 1 \}$$

(Figure 4.4 (ii)). The family $\mathcal{F}''_R$ of sets $L'_\pm$ and L'_α, $-\infty < \alpha < \infty$, is a smooth codimension one foliation of the annulus $D^1 \times S^1$ (Figure 4.4 (ii)). The foliation $\mathcal{F}''_R$ has two compact leaves $L'_\pm$.

Join two copies of the annulus $D^1 \times S^1$ by pasting along the boundaries; the result is the torus $T = S^1 \times S^1 = (D^1 \times S^1) \cup (D^1 \times S^1)$. Foliate one of the annuli with $\mathcal{F}''_R$ and the other with the C^∞ codimension one foliation $\{ \{x\} \times S^1 ; x \in D^1 \}$. They will become a C^∞ codimension one foliation of T (Figure 4.5). Unlike the foliation of Example A, this foliation does not come from any C^∞ vector field on T.

Consider $D^2 \times S^1$, a compact C^∞ three-manifold, and define subsets L_α of $D^2 \times S^1$ by

$$L_\alpha = \{ (x, e^{2\pi(\alpha + f(|x|))i}) ; x \in \operatorname{Int} D^2 \}, \qquad 0 \le \alpha < 1,$$

where f was defined in (1) and $|x|$ is the distance between the origin O and x (Figure 4.6). The family $\mathcal{F}'_R$ of sets $\partial(D^2 \times S^1) = S^1 \times S^1$ and the

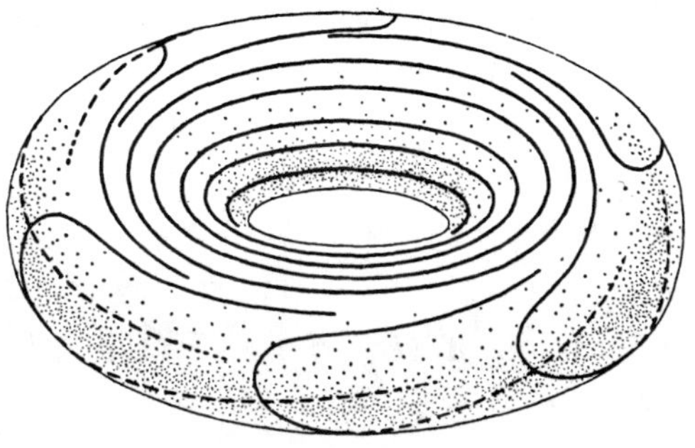

FIGURE 4.5

FIGURE 4.6

L_α, $0 \le \alpha < 1$, is a C^∞ codimension one foliation of $D^2 \times S^1$ (Figure 4.6). The only compact leaf of the foliation $\mathcal{F}'_R$ is $S^1 \times S^1$ and the rest of the leaves are all C^∞ diffeomorphic to $\mathbf{R}^2$. We call $\mathcal{F}'_R$ the *Reeb foliation* of the solid torus $D^2 \times S^1$.

Take two copies of the solid torus $D^2 \times S^1$ one labeled by $D_1^2 \times S_1^1$ and the other by $D_2^2 \times S_2^1$, each endowed with the Reeb foliation. Note that each of their boundaries $\partial D_1^2 \times S_1^1$ and $\partial D_2^2 \times S_2^1$ is the compact leaf of the Reeb foliation. We obtain the three-sphere S^3 as follows

$$S^3 = (D_1^2 \times S_1^1) \cup (D_2^2 \times S_2^1),$$

pasting the boundaries $\partial D_1^2 \times S_1^1$ and $\partial D_2^2 \times S_2^1$ by the identification of a point $(x, y) \in \partial D_1^2 \times S_1^1$ with $(y, x) \in \partial D_2^2 \times S_2^1$. The resulting S^3 has a smooth foliation also called the *Reeb foliation*, which will be denoted by $\mathcal{F}_R$. The only compact leaf of $\mathcal{F}_R$ is $\partial D_1^2 \times S_1^1 = \partial D_2^2 \times S_2^1$ and every other leaf is C^∞ diffeomorphic to $\mathbf{R}^2$.

Let $\mathcal{F}$ be a C^r codimension q foliation of a C^s manifold M^n and let $\mathcal{S}_{\mathcal{F}}^{(r)}$ be a foliated atlas. We now consider the standard topology on $\mathbf{R}^{n-q}$, the discrete topology on $\mathbf{R}^q$ and the product topology of these two on $\mathbf{R}^n = \mathbf{R}^{n-q} \times \mathbf{R}^q$. For each $(U_\lambda, \varphi_\lambda) \in \mathcal{S}_{\mathcal{F}}^{(r)}$, introduce a topology on U_λ such that $\varphi_\lambda \colon U_\lambda \to \varphi_\lambda(U_\lambda) \subset \mathbf{R}_+^n \subset R^n$ is a homeomorphism with respect to the product topology. Then by $(\mathcal{F}_{\rm III})$ the topologies on the U_λ's are coherent with each other and together they define a new topology on M^n called the *leaf topology* $\mathscr{O}_{\mathscr{F}}$. In the leaf topology $\mathscr{O}_{\mathscr{F}}$ each connected component of M^n is a leaf of $\mathscr{F}$. Each leaf inherits the *relative leaf topology* from $\mathscr{O}_{\mathscr{F}}$. We call a leaf L_α *compact* if it is compact in the relative leaf topology. For a leaf L_α, choose $(U_\lambda, \varphi_\lambda) \in \mathscr{S}_{\mathscr{F}}^{(r)}$ with $L_\alpha \cap U_\lambda \neq \varnothing$, and let $L_{\alpha,\lambda,\kappa}$ be a connected component of $L_\alpha \cap U_\lambda$. Then by $(\mathscr{F}_{\text{III}})$, $\varphi_\lambda(L_{\alpha,\lambda,\kappa})$ is an open subset of $\mathbf{R}^{n-q} \times (c_{n-q+1}, c_{n-q+2}, \dots, c_n)$; we may think of it as an open subset of $\mathbf{R}^{n-q}$. In addition, if $(U_\mu, \varphi_\mu) \in \mathscr{S}_{\mathscr{F}}^{(r)}$ is a foliated chart with $L_\alpha \cap U_\lambda \cap U_\mu \neq \varnothing$, then for $L_{\alpha,\lambda,\kappa}$ and $L_{\alpha,\mu,\kappa'}$ with

$$L_{\alpha,\lambda,\kappa} \cap L_{\alpha,\mu,\kappa'} \neq \varnothing,$$

the map

$$\varphi_\mu \circ \varphi_\lambda^{-1}: \varphi_\lambda(L_{\alpha,\lambda,\kappa} \cap L_{\alpha,\mu,\kappa'}) \to \varphi_\mu(L_{\alpha,\lambda,\kappa} \cap L_{\alpha,\mu,\kappa'})$$

is a C^r diffeomorphism of an open subset of $\mathbf{R}^{n-q}$ onto an open subset of $\mathbf{R}^{n-q}$. From the above observation, if we define a family $\mathscr{S}_\alpha$ by

$$\begin{aligned}\mathscr{S}_\alpha = \{ &(\text{connected component of}(L_\alpha \cap U_\lambda), \\ &\varphi_\lambda | \text{connected component of } (L_\alpha \cap U_\lambda)); \\ &(U_\lambda, \varphi_\lambda) \in \mathscr{S}_F^{(r)}, L_\alpha \cap U_\lambda \neq \varnothing \},\end{aligned}$$

$(L_\alpha, \mathscr{S}_\alpha)$ is a C^r $(n-q)$-manifold.

Let $\iota_\alpha: L_\alpha \to M^n$ be the inclusion map of $(L_\alpha, \mathscr{S}_\alpha)$ in M^n. Then ι_α is a one-to-one map and by $(\mathscr{F}_{\text{III}})$ it is a C^r immersion. However, ι_α is not always an embedding; for example, $\mathscr{F}_{a,b}$ on the torus with b/a irrational in Example A is not embedded in T. We say that L_α is a *proper leaf* when ι_α is an embedding. It is routine to check that a compact leaf is proper. A leaf L_β is said to be *locally dense* if $\operatorname{Int} \bar{L}_\beta \neq \varnothing$. A leaf which is neither proper nor locally dense is called *exceptional*.

If either $a = 0$, or b/a is rational, every leaf of the codimension one foliation $\mathscr{F}_{a,b}$ of the torus T in Example A is compact and hence proper. On the other hand if b/a is irrational, every leaf of $\mathscr{F}_{a,b}$ is locally dense, and no leaf is proper.

Recall the C^1 codimension one foliation of the torus T in Example A consisting of the trajectories of Denjoy's C^1 vector field which was investigated in §6. Let $S^1 = [0, 1] \cup (\bigcup_{m \in \mathbf{Z}} I_m)$ be as given in §6 (see Fig. 1.17). A trajectory $C((x, 1))$ through $(x, 1) \in S^1 \times S^1$ is a proper leaf if $x \in \operatorname{Int} I_m$, and it is an exceptional leaf if $x \in [0, 1]$.

Every leaf of the Reeb foliation in Example B is proper.

Suppose that $\mathscr{F}$ is a C^r codimension q foliation of a C^s manifold M^n. For $(U_\lambda, \varphi_\lambda)$ and $(U_\mu, \varphi_\mu) \in \mathscr{S}_{\mathscr{F}}^{(r)}$ with $U_\lambda \cap U_\mu \neq \varnothing$, the map

$$\varphi_\mu \circ \varphi_\lambda^{-1}: \varphi_\lambda(U_\lambda \cap U_\mu) \to \varphi_\mu(U_\lambda \cap U_\mu)$$

decomposes into component functions $y_1, y_2, \dots, y_n$:

$$(\varphi_\mu \circ \varphi_\lambda^{-1})(x_1, x_2, \dots, x_n) = (y_1, y_2, \dots, y_n),$$

which by $(\mathscr{F}_{\mathrm{III}})$ satisfy

$$\begin{cases} y_i = y_i(x_1, x_2 \dots, x_n), & i = 1, 2, \dots, n-q, \\ y_{n-q+j} = y_{n-q+j}(x_{n-q+1}, x_{n-q+2}, \dots, x_n), & j = 1, 2, \dots, q. \end{cases} \tag{2}$$

Suppose conversely that for a C^s manifold $(M^n, \mathscr{S})$ there exists a subfamily $\hat{\mathscr{S}}^{(r)}$ of $\mathscr{S}^{(r)}$ satisfying the following

$(\mathscr{F}_{\mathrm{I}}')$ $M^n = \bigcup_{(U_\gamma, \varphi_\gamma) \in \hat{\mathscr{S}}^{(r)}} U_\lambda$,

$(\mathscr{F}_{\mathrm{II}}')$ For $(U_\lambda, \varphi_\lambda)$, $(U_\mu, \varphi_\mu) \in \hat{\mathscr{S}}^{(r)}$ with $U_\lambda \cap U_\mu \neq \varnothing$, the component functions of $\varphi_\mu \circ \varphi_\lambda^{-1}$ are of the form given in (2).

For each $(U_\lambda, \varphi_\lambda) \in \hat{\mathscr{S}}^{(r)}$ consider a foliation of U_λ whose leaves are the arcwise connected components of

$$\begin{gathered} \varphi_\lambda^{-1}(\varphi_\lambda(U_\lambda) \cap (\mathbf{R}^{n-q} \times (x_{n-q+1}, x_{n-q+2}, \dots, x_n))), \\ (x_{n-q+1}, x_{n-q+2}, \dots, x_n) \in \mathbf{R}^q. \end{gathered} \tag{3}$$

Then by $(\mathscr{F}_{\mathrm{II}}')$ whenever $U_\lambda \cap U_\mu \neq \varnothing$, the foliations of U_λ and U_μ agree on $U_\lambda \cap U_\mu$. Hence there is a foliation of M^n whose restriction to each U_λ is the original foliation of U_λ given by (3). In other words, we give $\mathbf{R}^{n-q}$ the standard topology, $\mathbf{R}^q$ the discrete topology, and $\mathbf{R}^n = \mathbf{R}^{n-q} \times \mathbf{R}^q$ the product topology of the two; we give M^n a new topology such that for each $(U_\lambda, \varphi_\lambda) \in \hat{\mathscr{S}}^{(r)}$, φ_λ is a homeomorphism in the product topology. We then define leaves to be connected components in the new topology; that is to say, we may use $\hat{\mathscr{S}}^{(r)}$ which satisfies $(\mathscr{F}_{\mathrm{I}}')$, $(\mathscr{F}_{\mathrm{II}}')$ as an alternative definition of a foliation.

When $r \geq 1$ we say that a foliation $\hat{\mathscr{S}}^{(r)}$ is *transversely orientable* if it satisfies in addition to $(\mathscr{F}_{\mathrm{I}}')$, $(\mathscr{F}_{\mathrm{II}}')$ the following.

$(\mathscr{F}_{\mathrm{III}}')$ For $(U_\lambda, \varphi_\lambda)$ and $(U_\mu, \varphi_\mu) \in \hat{\mathscr{S}}^{(r)}$, if the component decomposition of $\varphi_\mu \circ \varphi_\lambda^{-1}$ is given by (2), we have

$$\left| \frac{\partial(y_{n-q+1}, y_{n-q+2}, \dots, y_n)}{\partial(x_{n-q+1}, x_{n-q+2}, \dots, x_n)} \right| > 0,$$

at $y_{n-q+j} = y_{n-q+j}(x_{n-q+1}, x_{n-q+2}, \dots, x_n)$, $j = 1, 2, \dots, q$.

The codimension one foliation of T in Example B corresponding to Figure 4.5 is not transversely orientable.

The rest of this section is mainly devoted to the case $q = 1$—the codimension one case. Let $\mathscr{F}$ be a C^r codimension one foliation of M^n, $r \geq 1$. A C^{r-1} vector field $X = \{X(p); p \in M^n\}$ on M^n is said to be *transverse* to the foliation $\mathscr{F}$ if the restriction of X to each leaf L_α is transverse to L_α.

THEOREM 4.1. *A C^r codimension one foliation $\mathscr{F}$ of M^n is transversely orientable if and only if there exists a C^{r-1} vector field transverse to $\mathscr{F}$.*

PROOF. Let $\mathscr{F}$ be transversely orientable. By $(\mathscr{F}'_{\mathrm{III}})$, $\varphi_\mu \circ \varphi_\lambda^{-1}$ decomposed as in (2) satisfies

$$\frac{\partial y_n}{\partial x_n} > 0.$$

Introduce a Riemannian metric on M^n: $\langle\ ,\ \rangle_p$, $p \in M^n$, and define a vector field Y as follows. For each p in M^n let $Y(p)$ be an element of $T_p(M^n)$ satisfying

(i) $\| Y(p) \| = 1$;
(ii) if L_α is a leaf containing p then $\langle Y(p) v\rangle_p = 0$ for each element v of $T_p(L_\alpha)$;
(iii) if $(U_\lambda, \varphi_\lambda) \in \hat{\mathscr{S}}^{(r)}$ is a foliated chart about p and

$$\Phi_\lambda(Y(p)) = \hat{v}_1 \frac{\partial}{\partial x_1} + \hat{v}_2 \frac{\partial}{\partial x_2} + \cdots + \hat{v}_n \frac{\partial}{\partial x_n},$$

then $\hat{v}_n > 0$, where Φ_λ is given by (2) in §10.

We claim that $Y(p)$ is well defined. Since $\partial Y_n / \partial x_n > 0$, the condition $\hat{v}_n > 0$ in (iii) does not depend on the choice of $(U_\lambda, \varphi_\lambda)$. Now the vector field Y defined by $Y = \{ Y(p); p \in M^n \}$ is C^{r-1} transverse to $\mathscr{F}$.

Conversely assume the existence of a C^{r-1} vector field X transverse to the foliation $\mathscr{F}$. Let $\hat{\mathscr{S}}^{(r)}$ satisfy $(\mathscr{F}'_{\mathrm{I}})$, $(\mathscr{F}'_{\mathrm{II}})$ and consider a subfamily of $\hat{\mathscr{S}}^{(r)}$, whose elements are the foliated charts $(U_\lambda, \varphi_\lambda)$ in $\hat{\mathscr{S}}^{(r)}$ such that $v_n > 0$, where

$$\Phi_\lambda(X(p)) = v_1 \frac{\partial}{\partial x_1} + v_2 \frac{\partial}{\partial x_2} + \cdots + v_n \frac{\partial}{\partial x_n}.$$

This subfamily satisfies $(\mathscr{F}'_{\mathrm{I}})$, $(\mathscr{F}'_{\mathrm{II}})$, and $(\mathscr{F}'_{\mathrm{III}})$.

Let M^n be a C^s manifold without boundary and $\mathscr{F}$ a C^r codimension one foliation of M^n, $r \geq 2$. Give M^n a suitable Riemannian metric. For each point $p \in M^n$ let $\mathrm{Z}(p)$ be an element of $T_p(M^n)$ which satisfies (i) and (ii) of Theorem 4.1. Then $\mathrm{Z}(p)$ is well defined up to its sign. Each $p \in M^n$ has a neighborhood U_p such that

$$\mathrm{Z} = \{ \epsilon_{p'} \mathrm{Z}(p'); p' \in U_p \}$$

is a C^{r-1} vector field on U_p; here $\epsilon_{p'} = \pm 1$. The trajectories of Z determine a C^{r-1} codimension $n-1$ foliation of U_p. This foliation depends only on $\mathscr{F}$ and the particular Riemannian metric defined on M^n, and it is independent of the sign of $\mathrm{Z}(p)$. The union of the leaves taken over the foliations of the U_p, $p \in M^n$ defines a C^{r-1} codimension $n-1$ foliation of M^n.

More generally let $\mathscr{F}$ be a C^r codimension q foliation and let $\mathscr{F}'$ be a $C^{r'}$ codimension $n-q$ foliation of a C^s manifold M^n, $r, r' \geq 1$. We say that the foliations $\mathscr{F}$ and $\mathscr{F}'$ are *transverse* if any two leaves L_α of $\mathscr{F}$ and $L'_{\alpha'}$ of $\mathscr{F}'$ with $L_\alpha \cap L'_{\alpha'} \neq \varnothing$, intersect transversely at each intersection point. From the preceding considerations we have the following

THEOREM 4.2. *Let F be a C^r codimension one foliation of a C^s manifold M^n without boundary, $r \geq 2$. Then there exists a C^{r-1} codimension $n-1$ foliation $\mathscr{F}'$ of M^n which is transverse to $\mathscr{F}$.*

The construction of the vector field Z in Theorem 4.2 can be carried over to a manifold with boundary; the only trouble we encounter is that the union of trajectories of Z does not quite give a foliation as defined in the present section, but rather they form a *foliation transverse to the boundary* which we shall define in §29. Therefore the assumption on ∂M^n can be removed if we extend our definition of foliations to include foliations transverse to the boundary.

§17. C^r bundles

Let E be an n-dimensional C^r manifold, $r \geq 1$. Let B be a q-dimensional C^r manifold, $n \geq q$, and assume that there exists a C^r submersion

$$\pi: E \to B.$$

If $\partial E \neq \varnothing$, we assume that $\partial B \neq \varnothing$ and that $\pi(\partial E) \subset \partial B$. Define a subset L_b of E by

$$L_b = \pi^{-1}(b), \qquad b \in B.$$

Then the family $\mathscr{F} = \{L_b ; b \in B\}$ is a C^r codimension q foliation of the manifold E. The conditions $(\mathscr{F}_{\mathrm{I}})$ and $(\mathscr{F}_{\mathrm{II}})$ are evidently satisfied. As for choosing foliated charts $(U_\lambda, \varphi_\lambda)$ for $(\mathscr{F}_{\mathrm{III}})$, we refer the reader to the proof of Theorem 2.5.

We next define a C^r bundle as a refinement of a C^r submersion.

Let E and B be as above and $\pi: E \to B$ a C^r map. We say that $\pi: E \to B$ is a C^r *fiber bundle*, or simply a C^r *bundle*, if there exist an $(n-q)$-dimensional C^r manifold F and an open cover $\{U_\xi ; \xi \in \Xi\}$ of B satisfying the following.

(i) For each point b of B, $\pi^{-1}(b)$ is an $(n-q)$-dimensional submanifold of E which is C^r diffeomorphic to F.

(ii) For each U_ξ there is a C^r diffeomorphism

$$\psi_\xi: \pi^{-1}(U_\xi) \to U_\xi \times F$$

such that

$$\psi_\xi(\pi^{-1}(b)) = \{b\} \times F, \qquad b \in U_\xi.$$

We call E the *total space*, B the *base space*, F the *fiber*, π the *projection map* and $\{U_\xi ; \xi \in \Xi\}$ the *coordinate system* of the C^r bundle $\pi : E \to B$. By (ii), π is a C^r submersion.

Now denote by $\mathrm{Diff}(F)$ the set of all C^r diffeomorphisms of F. Whenever we have $U_\xi \cap U_\eta \neq \varnothing$ with $\xi, \eta \in \Xi$, we can write the map

$$\psi_\xi \circ \psi_\eta^{-1} : (U_\xi \cap U_\eta) \times F \to (U_\xi \cap U_\eta) \times F$$

as

$$\psi_\xi \circ \psi_\eta^{-1}(b, y) = (b, g_{\xi\eta}(b)(y)), \tag{1}$$

where $g_{\xi\eta}: U_\xi \cap U_\eta \to \mathrm{Diff}(F)$. With a suitable topology, the C^r topology for example, $g_{\xi\eta}$ is a continuous map (see Appendix IV, Note 1). In discussing C^r bundles one usually includes some conditions on $\{g_{\xi\eta}; \xi, \eta \in \Xi\}$ as the structural group of a bundle in the definition; however, as we have no occasion to consider structural groups in this book we have restricted our definition to include only (i) and (ii). Sometimes $\pi: E \to B$ is called a *locally trivial fiber space* when it satisfies conditions (i) and (ii) with no restrictions on the structural group. We give some examples of C^r bundles.

EXAMPLE 1. Let $E = B \times F$, and define $\pi: B \times F \to B$ by $\pi(b, y) = b$, $b \in B$, $y \in F$. Clearly $\pi: B \times F \to B$ is a C^r bundle. We may take B itself for U_ξ. This bundle is called a *product bundle.*

EXAMPLE 2. Let M^n be a C^r manifold, $T(M^n)$ the tangent space of M^n and $\pi: T(M^n) \to M^n$ the projection map (§10); $\pi: T(M^n) \to M^n$ defines a C^{r-1} fiber bundle with the fiber $\mathbf{R}^n$, which is called the *tangent bundle* of M^n. Here we have regarded M^n as a C^{r-1} manifold. Further introduce a Riemannian metric on M^n. Denote

$$T_S(M^n) = \{v \in T(M^n); \| v \| = 1\},$$

and define a map

$$\pi_S: T_S(M^n) \to M^n$$

to be the restriction of π to $T_S(M^n)$. Then we have a C^{r-1} bundle whose fiber is S^{n-1}; this bundle is called the *tangent sphere bundle* of M^n.

EXAMPLE 3. The Hopf map $\pi: S^3 \to S^2$ (§14) gives rise to a C^∞ bundle whose fiber is S^1.

Given a C^r bundle $\pi: E \to B$, $\mathscr{F} = \{\pi^{-1}(b); b \in B\}$ is a C^r codimension q foliation of E whose leaves are all proper and C^r diffeomorphic to the fiber F. Here q is the dimension of the base B. A C^r foliation defined by a C^r bundle is called a *bundle foliation* or a *simple foliation.* The foliation $\mathscr{F}_{a,b}$ in Example A is a bundle foliation when $a = 0$ or $b = 0$.

Let $\pi: E \to B$ be a C^r bundle; the dimensions of E and B are assumed to be n and q respectively. Consider $g_{\xi\eta}: U_\xi \cap U_\eta \to \mathrm{Diff}(F)$ defined by (1). The fiber bundle $\pi: E \to B$ is *locally constant* if each $g_{\xi\eta}(b)$ is

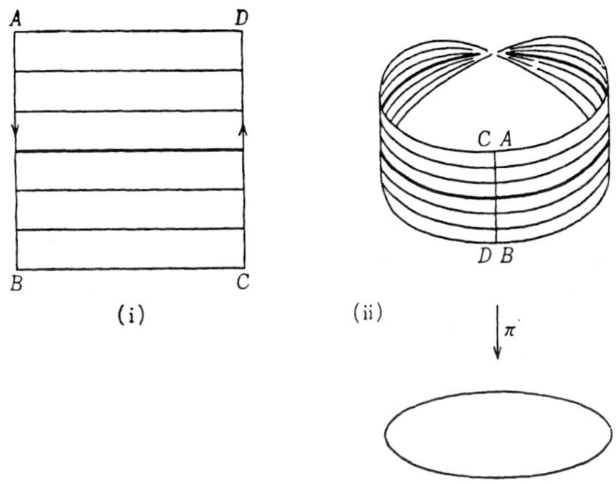

FIGURE 4.7

constant, independent of b in $U_\xi \cap U_\eta$. In this case we may define a C^r codimension $n-q$ foliation as follows: First give each $\pi^{-1}(U_\xi)$ a C^r codimension $n-q$ foliation, $\{\psi^{-1}(U_\xi \times \{y\}); y \in F\}$, $\xi \in \Xi$. By the local constancy of the bundle, the foliations of $\pi^{-1}(U_\xi)$ and $\pi^{-1}(U_\eta)$ restrict to the same foliation of $\pi^{-1}(U_\xi \cap U_\eta)$; putting together the respective bundle foliations of the $\pi^{-1}(U_\xi)$ s, we get a C^r codimension $n-q$ foliation of E.

EXAMPLE 4. Denote by I the interval $[0, 1]$. Identify $\overrightarrow{AB}$ and $\overrightarrow{CD}$ on the square $I \times I$ as indicated in Figure 4.7. The resulting space is the Möbius strip, which we denote by E. Let $\pi: E \to S^1$ be defined as illustrated in Figure 4.7 (ii). Then π is a locally constant C^∞ bundle. We also see in Figure 4.7 (ii), a C^∞ codimension 1 foliation of the Möbius strip when we decompose E into $I \times \{t\}$, $0 \le t \le 1$ as in Figure 4.7 (i).

The total space of a locally constant fiber bundle $\pi: E \to B$ is called a *covering space* of B if the fiber F is a discrete set.

Lastly in the following theorem we give a sufficient condition for a C^r submersion to be a C^r bundle.

THEOREM 4.3. *Let E be a C^r manifold of dimension n, B a C^r manifold of dimension q, and $\pi: E \to B$ a C^r submersion. If $r \ge 2$, E is compact, and B is connected, then $\pi: E \to B$ is a C^r bundle.*

PROOF. It is immediate from the assumption that π is onto. We shall first prove the theorem for $q = 1$, in which case B is either the unit circle S^1 or the closed interval $[0, 1]$. Let $Y = \{Y(b); b \in B\}$ be a nonsingular C^{r-1} vector field on B. Let x be a point of E; then we see in the proof of Theorem 2.5 that there exist a neighborhood U_x and a C^r vector field $X_x = \{X_x(y); y \in U_x\}$ on U_x such that

$$\pi_*(X_x(y)) = Y(\pi(y)), \qquad y \in U_x,$$

where $\pi_*: T(E) \to T(B)$. Since E is compact we may choose finitely many U_{x_i}, X_{x_i}, $i = 1, 2, \ldots, m$, with this property such that $\bigcup_{i=1}^m U_{x_i} = E$. Take a partition of unity

$$\mu_i, \qquad i = 1, 2, \ldots, m, \qquad \operatorname{supp} \mu_i \subset U_{x_i},$$

subordinate to the open cover $\{ U_{x_i} ; i = 1, 2, \ldots, m \}$, and define a vector field X_i on E by

$$X_i(y) = \begin{cases} \mu_i(y) X_{x_i}(y), & y \in U_{x_i}, \\ 0, & y \notin U_{x_i}, \end{cases}$$

$i = 1, 2, \ldots, m$. Notice that each X_i is of class C^{r-1}. Write

$$X(y) = \sum_{i=1}^m X_i(y), \qquad y \in E.$$

Then $X = \{X(y) ; y \in E\}$ is a C^{r-1} vector field on E, and we have

$$\pi_*(X(y)) = Y(\pi(y)), \qquad y \in E.$$

Let $b \in B$; by Theorem 2.5, $\pi^{-1}(b) = F_b$ is an $(n-1)$-dimensional compact submanifold of E. For each x in F_b we consider the integral curve $\varphi_{\{x\}}(t)$ of X through x. By the choice of X we see that $\pi \circ \varphi_{\{x\}}(t)$ is the integral curve $\overline{\varphi}_{\{b\}}(t)$ of Y through b. Taking $\epsilon > 0$ sufficiently small, we define

$$\overline{\psi}_b : F_b \times (-\epsilon, \epsilon) \to E$$

by

$$\overline{\psi}_b(x, t) = \varphi_{\{x\}}(t), \qquad -\epsilon < t < \epsilon.$$

Then we have

$$\pi \circ \overline{\psi}_b(F_b \times \{t\}) = \overline{\varphi}_{\{b\}}(t), \qquad -\epsilon < t < \epsilon,$$

and $\overline{\psi}_b | F \times \{t\}$ is a C^r diffeomorphism between $F_b \times \{t\}$ and $\pi^{-1}(\overline{\varphi}_{\{b\}}(t))$. Hence it follows that the $F_b = \pi^{-1}(b)$, $b \in B$, are all C^r diffeomorphic. We denote F_b by F. Set $U_b = \pi(\overline{\psi}_b(F \times (-\epsilon, \epsilon)))$, and define

$$\psi_b : \pi^{-1}(U_b) \to U_b \times F$$

by $\psi_b = \overline{\psi}_b^{-1}$. Then $\{ (U_b, \psi_b) ; b \in B \}$ becomes a C^r bundle structure for $\pi: E \to B$.

More generally, let $q > 1$. For each b in B, F_b is an $(n-q)$-dimensional compact submanifold of E (Theorem 2.5). Let $e_1, e_2, \ldots, e_q$ be a basis in $T_b(B)$; then there exist C^{r-1} vector fields $X_i = \{ X_i(x) \in T_x(E) ; x \in F_b \}$ of E defined on F_b such that $\pi_*(X_i(x)) = e_i$, $i = 1, 2, \ldots, q$. We can construct the X_i locally as in §10, and assemble them by a partition of unity over the entire E. We now give E and B appropriate Riemannian metrics, and using the exponential map Exp define a C^r diffeomorphism

$$\psi\colon \pi^{-1}(U) \to U \times F_b$$

by

$$\psi^{-1}(\operatorname{Exp}(\sum_i \alpha_i e_i), x) = \operatorname{Exp}(\sum_i \alpha_i X_i(x)),$$

where U is a small enough neighborhood of b in B, and $|\alpha_i|$ is a sufficiently small number. This is a C^r bundle structure for E (see Appendix IV, Note 2).

§18. Topological properties of leaves

Denote by $(a, b)^m$ the m-fold product of the open interval (a, b). The product $(a, b)^m$ is an open subset of Euclidean n-space $\mathbf{R}^m$.

Let M^n be an n-dimensional C^s manifold. Let $\mathscr{F} = \{L_\alpha ; \alpha \in A\}$ be a C^r codimension q foliation of M^n and $\mathscr{S}_{\mathscr{F}}^{(r)}$ be its foliated atlas, $r \geq 1$. Wc say that $(U_\lambda, \varphi_\lambda)$ in $\mathscr{S}_{\mathscr{F}}^{(r)}$ is a *distinguished chart* if it satisfies the following.

(i) $\varphi_\lambda(U_\lambda) = (-1, 1)^n$.

(ii) There exists $(U_{\hat\lambda}, \varphi_{\hat\lambda}) \in \mathscr{S}_{\mathscr{F}}^{(r)}$ such that $\overline{U}_\lambda \subset U_{\hat\lambda}$, and $\varphi_{\hat\lambda}|U_\lambda = \varphi_\lambda$.

It is evident that each point p of $\operatorname{Int} M^n$ belongs to some distinguished chart $(U_\lambda, \varphi_\lambda)$ with $\varphi_\lambda(p) = (0, 0, \dots, 0)$.

Define a projection map $\hat\pi\colon (-1, 1)^n \to (-1, 1)^q$ by

$$\hat\pi((x_1, x_2, \dots, x_n)) = (x_{n-q+1}, x_{n-q+2}, \dots, x_n).$$

If $(U_\lambda, \varphi_\lambda)$ is a distinguished chart and x is a point of $(-1, 1)^q$, we see that $\varphi_\lambda^{-1}(\hat\pi^{-1}(x))$ is contained in a leaf of $\mathscr{F}$. The set $\varphi_\lambda^{-1}(\hat\pi^{-1}(x))$ is called a *plaque* of U_λ and is often denoted by Q_λ (Figure 4.8). Each plaque of U_λ is C^r diffeomorphic to $(-1, 1)^{n-q}$, and we have

$$U_\lambda = \bigcup_{x \in (-1, 1)^q} \varphi_\lambda^{-1}(\hat\pi^{-1}(x)).$$

LEMMA 4.4. *Let $(U_\lambda, \varphi_\lambda)$ and (U_ξ, φ_ξ) be distinguished charts such that $U_\lambda \subset U_\xi$. If Q_λ is a plaque of U_λ, there exists a distinguished chart $(U_{\lambda'}, \varphi_{\lambda'})$ which satisfies the following.*

(i) *$Q_\lambda \subset U_{\lambda'} \subset U_\lambda$ and Q_λ is a plaque of $U_{\lambda'}$.*

(ii) *If Q_ξ is a plaque of U_ξ with $Q_\xi \cap U_{\lambda'} \neq \varnothing$, then $Q_\xi \cap U_{\lambda'}$ is a plaque of $U_{\lambda'}$.*

PROOF. By assumption a plaque of U_λ is contained in a plaque of U_ξ. For each $x \in (-1, 1)^q$ there is a unique x' such that $\varphi_\lambda^{-1}(\hat\pi^{-1}(x)) \subset \varphi_\xi^{-1}(\hat\pi^{-1}(x'))$. Let

$$\zeta\colon (-1, 1)^q \to (-1, 1)^q$$

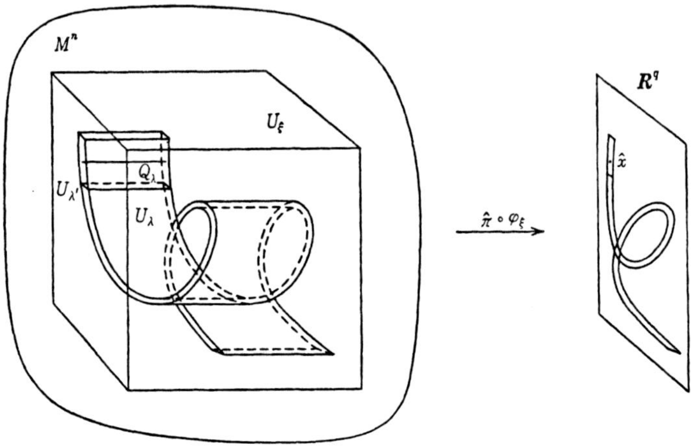

FIGURE 4.8

be the map which sends x to x'. It is immediate that ζ is a C^r immersion. Suppose now that we take a plaque Q_λ:

$$Q_\lambda = \varphi_\lambda^{-1}(\hat{\pi}^{-1}(\hat{x})), \qquad \hat{x} = (\hat{x}_1, \hat{x}_2, \ldots, \hat{x}_q) \in (-1, 1)^q.$$

Take a sufficiently small $\epsilon > 0$ to define $\hat{U}$ by

$$\hat{U} = (\hat{x}_1 - \epsilon, \hat{x}_1 + \epsilon) \times (\hat{x}_2 - \epsilon, \hat{x}_2 + \epsilon) \times \cdots \times (\hat{x}_q - \epsilon, \hat{x}_q + \epsilon) \subset (-1, 1)^q.$$

Then $\zeta|\hat{U}$ is a C^r embedding. Define $U_{\lambda'}$ by $U_{\lambda'} = \varphi_\lambda^{-1}(\hat{\pi}^{-1}(\hat{U}))$ and $\varphi_{\lambda'}$ by $\varphi_{\lambda'} = \varphi_\lambda|U_{\lambda'}$ with suitable adjustment on $\varphi_\lambda|U_{\lambda'}$. We now have the desired $(U_{\lambda'}, \varphi_{\lambda'})$ (see Figure 4.8).

Let $\mathscr{C}$ be a sequence

$$\mathscr{C} = \{U_{\lambda_1}, U_{\lambda_2}, \ldots, U_{\lambda_m}\}$$

of distinguished charts $(U_{\lambda_i}, \varphi_{\lambda_i})$, $i = 1, 2, \ldots, m$. Let x belong to a plaque Q_1 of U_{λ_1}, and let L_α be a leaf containing x. If there exist plaques Q_i of U_{λ_i}, $i = 2, 3, \ldots, m$, such that

$$Q_i \cap Q_{i+1} \neq \varnothing, \qquad i = 1, 2, \ldots m-1,$$

we call $\mathscr{C}$ a *chain at* x (Figure 4.9) and the Q_i, $i = 1, 2, \ldots, m$, a *plaque chain associated with* $\mathscr{C}$. From the definition it is evident that

$$Q_i \subset L_\alpha, \qquad i = 1, 2, \ldots, m.$$

We say that m is the *length* of the chain $\mathscr{C}$.

LEMMA 4.5. *Let $\mathscr{C} = \{U_{\lambda_1}, U_{\lambda_2}, \ldots, U_{\lambda_m}\}$ be a chain at x. Denote by O_1 the set of all points z such that $\mathscr{C}$ forms a chain at each z. Then the set O_1 is open in U_{λ_1} and it is a union of plaques of U_{λ_1}.*

PROOF. Let $O_m = U_{\lambda_m}$, and inductively define open sets O_i by

$$O_i = \varphi_{\lambda_i}^{-1}(\hat{\pi}^{-1} \circ \hat{\pi}(\varphi_{\lambda_i}(U_{\lambda_i} \cap O_{i+1}))), \qquad i = m-1, m-2, \ldots, 1.$$

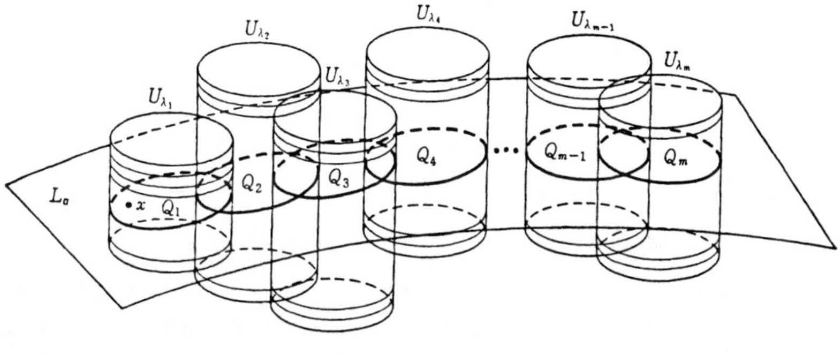

FIGURE 4.9

Then O_1 is the desired open set.

LEMMA 4.6. *Let X be a metric space, K a compact subset of X and U_i' open sets in X, $i = 1, 2, \ldots, v'$, such that $K \subset \bigcup_{i=1}^{v'} U_i'$. Then there exist open sets U_i, $i = 1, 2, \ldots, v$, satisfying the following.*

(i) $\bigcup_{i=1}^{v} U_i \supset K$.

(ii) *If $U_i \cap U_j \neq \emptyset$, then $U_i \cup U_j \subset U_k'$ for some U_k', $1 \leq k \leq v'$.*

PROOF (outline). Given U_i', $i = 1, 2, \ldots, m'$, we can find a Lebesgue number $\delta > 0$ such that for any subset G of X with its diameter $d(G) < \delta$ and $G \cap K \neq \emptyset$, there exists U_k' with $G \subset U_k'$ (see Appendix IV, Note 3). Hence it is enough to choose U_i, $i = 1, 2, \ldots, v$, satisfying (i) with $d(U_i) < \delta/2$.

LEMMA 4.7. *Let L_α be a leaf and let x and y be points on L_α. Then there exists a chain $\mathscr{C} = \{U_{\lambda_1}, U_{\lambda_2}, \ldots U_{\lambda_m}\}$ at x which has the following properties.*

(i) $x \in U_{\lambda_1}$, $y \in U_{\lambda_m}$.

(ii) *Suppose that $\mathscr{C}$ also forms a chain at a point z of U_{λ_1} and that z is on a leaf L_β. Let Q_i' be a plaque chain associated with $\mathscr{C}$ at z such that*

$$Q_i' \subset U_{\lambda_i} \cap L_\beta, \qquad z \in Q_1', \qquad i = 1, 2, \ldots, m.$$

Then Q_{i+1}' is the only plaque of $U_{\lambda_{i+1}}$ which intersects Q_i', and Q_i' is the only plaque of U_{λ_i} which intersects Q_{i+1}', $i = 1, 2 \ldots, m-1$.

PROOF. Let $l\colon [0, 1] \to L_\alpha$ be a C^0 curve with $l(0) = x$ and $l(1) = y$. Cover $l([0, 1])$ by distinguished charts and choose finitely many among them to construct a chain $\mathscr{C}'' = \{U_{\xi_1}, U_{\xi_2}, \ldots, U_{\xi_{m'}}\}$ at x such that $x \in U_{\xi_1}$, $y \in U_{\xi_{m'}}$. We shall think of $\bigcup_{i=1}^{m'} U_{\xi_i}$ as a metric space (§10), and apply Lemma 4.6 with $U_i' = U_{\xi_i}$, $v' = m'$, $K = l([0, 1])$ to obtain a chain $\mathscr{C}' = \{U_{\lambda_1'}, U_{\lambda_2'}, \ldots, U_{\lambda_m'}\}$ at x with $y \in U_{\lambda_m'}$ such that if $U_{\lambda_i'} \cap U_{\lambda_j'} \neq \emptyset$, we have

$U_{\lambda'_i} \cup U_{\lambda'_j} \subset U_{\xi_k}$, for some U_{ξ_k} in $\mathscr{C}''$. Let Q_i, $i = 1, 2, \dots, m$, be a plaque chain associated with $\mathscr{C}'$ at x such that $x \in Q_1$, $y \in Q_m$.

Since $\varnothing \neq Q_1 \cap Q_2 \subset U_{\lambda'_1} \cap U_{\lambda'_2}$, there exists U_{ξ_k} with $U_{\lambda'_1} \cup U_{\lambda'_2} \subset U_{\xi_k}$. We apply Lemma 4.4 to $U_{\lambda'_1} \subset U_{\xi_k}$ and $U_{\lambda'_2} \subset U_{\xi_k}$, and choose distinguished charts $(U_{\lambda''_1}, \varphi_{\lambda''_1})$ and $(U_{\lambda''_2}, \varphi_{\lambda''_2})$ such that

$$Q_1 \subset U_{\lambda''_1} \subset U_{\lambda'_1}, \qquad Q_2 \subset U_{\lambda''_2} \subset U_{\lambda'_2},$$

each satisfying condition (ii) of Lemma 4.4. Now we may assume that $U_{\lambda''_2} \cup U_{\lambda'_3} \subset U_{\xi'_k}$ because $\varnothing \neq Q_2 \cap Q_3 \subset U_{\lambda''_2} \cap U_{\lambda'_3}$, and applying Lemma 4.4 to $U_{\lambda''_2} \subset U_{\xi'_k}$ and $U_{\lambda'_3} \subset U_{\xi'_k}$, we choose $(U_{\lambda'''_2}, \varphi_{\lambda'''_2})$ and $(U_{\lambda''_3}, \varphi_{\lambda''_3})$ to satisfy (i) and (ii) of Lemma 4.4. Continuing this procedure we select $(U_{\lambda''_1}, \varphi_{\lambda''_1})$, $(U_{\lambda'''_i}, \varphi_{\lambda'''_i})$, $i = 2, 3, \dots, m-1$, and $(U_{\lambda''_m}, \varphi_{\lambda''_m})$. Finally define

$$(U_{\lambda_i}, \varphi_{\lambda_i}), \qquad i = 1, 2, \dots, m$$

by

$$\begin{aligned} &(U_{\lambda_1}, \varphi_{\lambda_1}) = (U_{\lambda''_1}, \varphi_{\lambda''_1}), \qquad (U_{\lambda_m}, \varphi_{\lambda_m}) = (U_{\lambda''_m}, \varphi_{\lambda''_m}), \\ &(U_{\lambda_i}, \varphi_{\lambda_i}) = (U_{\lambda'''_i}, \varphi_{\lambda'''_i}), \qquad i = 2, 3, \dots, m-1. \end{aligned} \tag{1}$$

Evidently the pairs $(U_{\lambda_i}, \varphi_{\lambda_i})$ in (1) satisfy condition (i) of the lemma. Next we show that they satisfy condition (ii). Since $U_{\lambda_1} \cup U_{\lambda_2} \subset U_{\xi_k}$, there exists a unique plaque Q''_k of U_{ξ_k} with $Q'_1 \subset Q''_k$. Now any plaque of U_{λ_2} which intersects Q'_1 is contained in $U_{\lambda_2} \cap Q''_k$. But by Lemma 4.4, $U_{\lambda_2} \cap Q''_k$ contains only one plaque of U_{λ_2}. Hence, no plaque of U_{λ_2} other than Q'_2 meets Q'_1. The proof for the rest of U_{λ_i}, $U_{\lambda_{i+1}}$ is similar.

THEOREM 4.8. *Let x and y be points on a leaf L_α. Given a distinguished chart (U_ν, φ_ν) about y, there exists a distinguished chart (U_μ, φ_μ) about x such that for any leaf L_β with $L_\beta \cap U_\mu \neq \varnothing$ we have $L_\beta \cap U_\nu \neq \varnothing$. Further, U_μ can be chosen in such a way that if U_μ has a plaque Q'_1 with $Q'_1 \subset L_\beta$ and $x \notin Q'_1$, then L_β contains a plaque of U_ν not containing the y.*

PROOF. Consider the chain at x, $\mathscr{C} = \{U_{\lambda_1}, U_{\lambda_2}, \dots, U_{\lambda_m}\}$, given in Lemma 4.7. We may assume that $U_{\lambda_m} \subset U_\nu$. Apply Lemma 4.5 to $\mathscr{C}$, and let (U_μ, φ_μ) be a distinguished chart where U_μ is an open subset of $O_1 \subset U_{\lambda_1}$ and $\varphi_\mu = \varphi_{\lambda_1}|U_\mu$ readjusted appropriately. We now have the desired distinguished chart. Choose a plaque chain Q'_i, $i = 1, 2, \dots, m$, which satisfies condition (ii) of Lemma 4.7; then plaques Q'_i and Q'_{i+1} are uniquely determined by each other. Hence if $x \notin Q'_1$ then $y \notin Q'_m$.

In the following, we shall state some consequences of Theorem 4.8 on the topology of leaves. These results have their parallels in §12 where the topology of the trajectories of dynamical systems were investigated.

THEOREM 4.9. *Let $\{L_\alpha; \alpha \in A'\}$ be a set of leaves. Suppose a point x belonging to $\overline{\bigcup_{\alpha\in A'} L_\alpha}$ is on some leaf L_β. Then $L_\beta \subset \overline{\bigcup_{\alpha\in A'} L_\alpha}$. In particular, therefore, the closure $\bar{L}_\alpha$ of a leaf L_α is the union of leaves*

$$\bar{L}_\alpha = \bigcup_{L_\beta \subset \bar{L}_\alpha} L_\beta.$$

PROOF. Let y be a point of L_β and let (U_ν, φ_ν) be a distinguished chart about y. By the first half of Theorem 4.8, there exists a distinguished chart (U_μ, φ_μ) about x such that every leaf intersecting U_μ also intersects U_ν. Therefore, we have

$$U_\nu \cap \Big(\bigcup_{\alpha\in A'} L_\alpha\Big) \neq \varnothing,$$

and so

$$y \in \overline{\bigcup_{\alpha\in A'} L_\alpha},$$

since we can make U_ν arbitrarily small.

THEOREM 4.10. *Let O be an open set in M^n. Then the union of all leaves which intersect O:*

$$Y = \bigcup_{L_\alpha \cap O \neq \varnothing} L_\alpha,$$

is open in M^n.

PROOF. Given $x \in Y$, there exists a leaf L_α containing x with $L_\alpha \cap O \neq \varnothing$. For $y \in L_\alpha \cap O$ choose a distinguished chart (U_ν, φ_ν) about y such that $U_\nu \subset O$. Then by the first part of Theorem 4.8, there exists a distinguished chart (U_μ, φ_μ) about x with the property that every leaf intersecting U_μ intersects U_ν, and hence O. Consequently $U_\mu \subset Y$.

THEOREM 4.11. *A leaf L_α is proper if and only if there exists a distinguished foliated chart $(U_\lambda, \varphi_\lambda)$ such that $L_\alpha \cap U_\lambda$ is the only plaque of U_λ.*

PROOF. Suppose that L_α is a proper leaf. Let $(U_{\lambda'}, \varphi_{\lambda'})$ be a distinguished chart such that $U_{\lambda'}$ has a plaque Q' which is contained in L_α. Since L_α is proper, there exists a neighborhood V of Q' in $U_{\lambda'}$ such that $L_\alpha \cap V = Q'$. Now the construction of a $(U_\lambda, \varphi_\lambda)$ satisfying the necessary condition of the theorem is routine.

Conversely assume the existence of a $(U_\lambda, \varphi_\lambda)$ which satisfies the necessary condition. Let x be a point of L_α. Then by Theorem 4.8 there is a distinguished chart (U_μ, φ_μ) about x such that for any leaf L_β, $L_\beta \cap U_\mu \neq \varnothing$ implies $L_\beta \cap U_\lambda \neq \varnothing$, and if L_β contains a plaque Q_1' of U_μ with $x \notin Q_1'$ then L_β contains a plaque in U_λ which is different from $L_\alpha \cap U_\lambda$. But the only plaque of U_λ contained in L_α is $L_\alpha \cap U_\lambda$, so if $x \notin L_\beta$ then $L_\beta \neq L_\alpha$. Hence the only plaque of U_μ contained in L_α is the plaque containing x. Hence L_α is a proper leaf.

THEOREM 4.12. *A necessary condition that a leaf L_α be compact is that for an arbitrary distinguished chart $(U_\lambda, \varphi_\lambda)$, $L_\alpha \cap U_\lambda$ consists of finitely many plaques of U_λ. The condition is also sufficient for a leaf L_α to be compact if there exists a compact subset E such that $L_\alpha \subset E \subset M^n$.*

PROOF. First let L_α be compact. Suppose that for some distinguished chart $(U_\lambda, \varphi_\lambda)$, $L_\alpha \cap U_\lambda$ contains infinitely many plaques. By the definition of distinguished charts there exists $(U_{\hat\lambda}, \varphi_{\hat\lambda}) \in \mathscr{S}_{\mathscr{F}}^{(r)}$ such that $\overline{U}_\lambda \subset U_{\hat\lambda}$. Hence there exists a point z in $L_\alpha \cap U_{\hat\lambda}$ whose neighborhoods all meet infinitely many plaques of U_λ; this contradicts the compactness of L_α.

Conversely suppose that there is a compact set E, $L_\alpha \subset E \subset M^n$, and that for any $(U_\lambda, \varphi_\lambda)$, $L_\alpha \cap U_\lambda$ is a finite family of plaques. Since E is compact it is covered by a finite number of distinguished charts. Hence L_α is a finite union of finite families of plaques, each of which is closed. Therefore the leaf L_α is compact.

LEMMA 4.13. *Let L_α be a noncompact proper leaf. If $\bar{L}_\alpha$ is compact the union*

$$G = \bigcup_{L_\beta \subset \bar{L}_\alpha,\ L_\beta \neq L_\alpha} L_\beta = \bar{L}_\alpha - L_\alpha$$

is a nonempty closed set.

PROOF. By Theorem 4.9, $G \neq \varnothing$. Let x be a point of $\overline{G}$. Since $G \subset \bar{L}_\alpha$ we also have $x \in \bar{L}_\alpha$. Now if $x \in L_\gamma$ by Theorem 4.9 we must have $L_\gamma \subset \overline{L}_\alpha$. Claim $L_\gamma \neq L_\alpha$. For suppose $x \in L_\alpha$, and let V be an arbitrary neighborhood of x in M^n. Then V contains a point y which is an element of G, and since each neighborhood of y contains infinitely many points of L_α which are not close to each other in the relative leaf topology of L_α, V also includes infinitely many such points. This contradicts our assumption that L_α is proper. Hence x must be in G.

THEOREM 4.14. *Let M^n be compact. Suppose that L_α is a proper leaf and that $\bar{L}_\alpha$ is a union of proper leaves. Then $\bar{L}_\alpha$ contains at least one compact leaf.*

PROOF. Let $\mathscr{K}$ be a family of all nonempty compact sets contained in $\bar{L}_\alpha$ each of which is a union of leaves. Then $\mathscr{K} \neq \varnothing$ since $\bar{L}_\alpha \in \mathscr{K}$. We order $\mathscr{K}$ by $\prec$ according to set inclusions. We characterize the family $\{K_\sigma \in \mathscr{K}; \sigma \in \Sigma\}$ by:

$$\text{if } \sigma, \sigma' \in \Sigma, \text{ then } K_{\sigma'} \prec K_\sigma, \text{ or } K_\sigma \prec K_{\sigma'}.$$

This is a family of closed sets with the finite intersection property; therefore, $\bigcap_{\sigma \in \Sigma} K_\sigma \neq \varnothing$. Hence $\bigcap_{\sigma\in\Sigma} K_\sigma \in \mathscr{K}$. Hence, by Zorn's lemma, there exists a minimal element K in $\mathscr{K}$ with respect to $\prec$. Suppose K contains a noncompact leaf L_β. By Lemma 4.13, the union of all L_γ different from L_β

with $L_\gamma \subset \bar{L}_\beta$ is nonempty and closed, so it is compact, which contradicts the minimality of K. Therefore K is a union of compact leaves, and further the minimality of K implies that K is indeed a compact set.

The reader should find noncompact leaves of the Reeb foliation on S^3 to be an enlightening example of L_α in Theorem 4.14.

CHAPTER V

Stability Theorems on Foliations

§19. Coherent systems of charts

Let M^n be an n-dimensional C^s manifold, and let $\mathscr{F} = \{L_\alpha; \alpha \in A\}$ be a C^r codimension q foliation of M^n, $r \geq 1$. In order to describe the behavior of the foliation in a neighborhood of one of its leaves, it turns out that we must construct a system of distinguished charts with certain properties. The following theorem gives one way to do it.

THEOREM 5.1. *Let K be a compact subset of a leaf $\widehat{L}_\alpha$. Then there exists a family $\mathfrak{N}(K) = \{(U_i, \varphi_i);\ i = 1, 2, \dots, v\}$ of distinguished charts (U_i, φ_i), $i = 1, 2, \dots, v$, with the following properties.*

(i) $K \subset \bigcup_{i=1}^{v} U_i$.

(ii) *For each U_i, $U_i \cap K$ is contained in a plaque Q_i of U_i:*

$$U_i \cap K \subset Q_i, \qquad i = 1, 2, \dots, v.$$

Here we may take $Q_i = \varphi_i^{-1}(\hat{\pi}^{-1}(0))$.

(iii) *If $U_i \cap U_j \neq \varnothing$, then $U_i \cap U_j \cap K \neq \varnothing$.*

(iv) *For U_i, U_j such that $U_i \cap U_j \neq \varnothing$, there exists a distinguished chart (U_{ij}, φ_{ij}) which satisfies:*

(a) $U_i \cup U_j \subset U_{ij}$.

(b) *Let Q_{ij} be a plaque of U_{ij}. Then $Q_{ij} \cap U_i$ is a plaque of U_i if $Q_{ij} \cap U_i \neq \varnothing$, and $Q_{ij} \cap U_j$ is a plaque of U_j if $Q_{ij} \cap U_j \neq \varnothing$.*

(v) *Let (U_{ij}, φ_{ij}), (U_{kl}, φ_{kl}) be distinguished charts specified in* (iv) *with $U_{ij} \cap U_{kl} \neq \varnothing$. Then we can select a distinguished chart $(U_{ijkl}, \varphi_{ijkl})$ satisfying the following.*

(a) $U_{ij} \cup U_{kl} \subset U_{ijkl}$.

(b) *Let Q_{ijkl} be a plaque of U_{ijkl}. Then, $Q_{ijkl} \cap U_{ij}$ is a plaque of U_{ij} if $Q_{ijkl} \cap U_{ij} \neq \varnothing$, and $Q_{ijkl} \cap U_{kl}$ is a plaque of U_{kl} if $Q_{ijkl} \cap U_{kl} \neq \varnothing$.*

The family $\mathfrak{N}(K)$ is called a *coherent system of charts* or *coherent chart-system* on K. The reader should keep in mind that by definition charts in a coherent system are distinguished.

PROOF. Since K is compact, we can choose distinguished charts (U_i, φ_i),

$i = 1, 2, \ldots, v$, such that $K \subset \bigcup_{i=1}^{v} Q_i$, where Q_i is a plaque of U_i, so (i) is satisfied.

Again, by the compactness of K, $U_i \cap K$ is contained in a finite union of plaques (cf. Theorem 4.12). Suppose that $U_i \cap K$ intersects plaques other than Q_i. Then choose a distinguished chart (U_i', φ_i') such that

$$Q_i \subset U_i' \subset U_i,$$

the plaque Q_i is also a plaque of U_i' and no plaques of U_i' but Q_i intersect K; further (U_i', φ_i') can be so chosen that $Q_i = \varphi_i'^{-1}(\hat{\pi}^{-1}(0))$. Renaming (U_i', φ_i') as (U_i, φ_i) we have verified (ii).

We may assume that the above U_i's have been chosen in such a way that the following holds:

$$\text{If } Q_i \cap Q_j = \varnothing, \text{ then } \overline{Q}_i \cap \overline{Q}_j = \varnothing, \quad i = 1, 2, \ldots, v.$$

Now suppose that $U_i \cap U_j \neq \varnothing$ and $U_i \cap U_j \cap K = \varnothing$ for some U_i, U_j. Then we can find distinguished charts (U_i', φ_i') and (U_j', φ_j') such that

$$Q_i \subset U_i' \subset U_i, \qquad Q_j \subset U_j' \subset U_j,$$

Q_i and Q_j are plaques of U_i' and U_j' respectively, and $U_i' \cap U_j' = \varnothing$. Rewriting (U_i', φ_i') as (U_i, φ_j) and (U_j', φ_j') as (U_j, φ_j) we have verified (iii).

Consider a new set of distinguished charts (U_i'', φ_i''), $i = 1, 2, \ldots, v'$, such that $K \subset \bigcup_{i=1}^{v'} U_i''$. We shall think of $\bigcup_{i=1}^{v'} U_i''$ as a metric space (cf. (7), §10), and apply Lemma 4.6 to U_i'': choose the U_i, $i = 1, 2, \ldots, v$, such that they satisfy Lemma 4.6(ii): if $U_i \cap U_j \neq \varnothing$, then

$$U_i \cup U_j \subset U_k'',$$

for some U_k''. Denote U_k'' by U_{ij}. Apply Lemma 4.4 to the open sets $U_i \subset U_{ij}$ and $U_j \subset U_{ij}$ to find distinguished charts $(U_{i'}, \varphi_{i'})$ and $(U_{j'}, \varphi_{j'})$ satisfying the condition (ii) of Lemma 4.4 and

$$Q_i \subset U_{i'} \subset U_i, \qquad Q_j \subset U_{j'} \subset U_j.$$

Relabel $(U_{i'}, \varphi_{i'})$ and $(U_{j'}, \varphi_{j'})$ by (U_i, φ_i) and (U_j, φ_j) respectively. This verifies (iv), and the proof of (v) is similar.

In what follows, we assume that K is a compact subset of a leaf $L_{\hat{\alpha}}$ and that $\mathfrak{N}(K)$ is a coherent system of charts on K. For $x \in K$, a chain $\mathscr{C} = \{U_{\lambda_1}, U_{\lambda_2}, \ldots, U_{\lambda_{m'}}\}$ at x is called a *coherent chain at* x if the distinguished charts $(U_{\lambda_i}, \varphi_{\lambda_i})$, $i = 1, 2, \ldots, m'$, belong to $\mathfrak{N}(K)$.

LEMMA 5.2. *Let* $\mathscr{C} = \{U_{\lambda_1}, U_{\lambda_2}, \ldots, U_{\lambda_{m'}}\}$ *be a coherent chain at* x. *Suppose that* $\mathscr{C}$ *also forms a chain at a point* $z \in U_{\lambda_1}$. *Then a plaque chain*

$$Q_i', \quad Q_i' \subset U_{\lambda_i}, \quad i = 1, 2, \ldots, m', \quad z \in Q_1',$$

associated with $\mathscr{C}$ *at* z *is uniquely determined by* z.

Proof. Evidently, Q'_1 is uniquely determined. By Theorem 5.1(iv), there exists a distinguished chart $(U_{\lambda_1\lambda_2}, \varphi_{\lambda_1\lambda_2})$ with $U_{\lambda_1} \cup U_{\lambda_2} \subset U_{\lambda_1\lambda_2}$ satisfying the condition in Theorem 5.1(iv)(b). Then the method used to prove Lemma 4.7(ii) produces a unique plaque Q'_2 of U_{λ_2} with respect to the property that $Q'_1 \cap Q'_2 \neq \varnothing$. Similarly the Q'_i, $i = 1, 2, \dots, m'$, are uniquely determined.

Let $x \in K$ and m be a positive integer. Suppose there exists a point z in M^n such that every coherent chain $\mathscr{C} = \{U_{\lambda_1}, U_{\lambda_2}, \dots, U_{\lambda_{m'}}\}$ at x with the length m', $m' \leq m$, also forms a chain at z. Then we say that z is an *admissible point* of the coherent chains at x of length at most m.

Lemma 5.3. *The set of all admissible points of the coherent chains at x of length at most m is an open set containing $\bigcap_{x\in U_i, U_i\in\mathfrak{N}(K)}(U_i \cap K)$ and contained in $\bigcap_{x\in U_i, U_i\in\mathfrak{N}(K)} U_i$.*

Proof. There are only finitely many coherent chains at x of length at most m. Consider the set O_1 of Lemma 4.5 for each such chain. Their intersection will be the desired open set.

Let $l\colon [0, 1] \to K$ be a C^0 curve in K. By a *coherent chain over l* we mean a coherent chain $\mathscr{C} = \{U_{\lambda_1}, U_{\lambda_2}, \dots, U_{\lambda_{m'}}\}$ at $l(0)$, for which there exist numbers t_i, $i = 0, 1, 2, \dots, m'$:

$$0 = t_0 < t_1 < \cdots < t_{m'} = 1$$

such that

$$l([t_{i-1}, t_i]) \subset U_{\lambda_i}, \qquad i = 1, 2, \dots, m'.$$

It is obvious that given a C^0 curve l in K, one can find a coherent chain $\mathscr{C}$ with $l(0) \in U_{\lambda_1}$ and $l(1) \in U_{\lambda_{m'}}$ as preassigned.

Lemma 5.4. *Let $\mathscr{C} = \{U_{i_1}, U_{i_2}, \dots, U_{i_{m'}}\}$, $\mathscr{C}' = \{U_{j_1}, U_{j_2}, \dots, U_{j_{m''}}\}$ be coherent chains over l such that $U_{i_1} = U_{j_1}$ and $U_{i_{m'}} = U_{j_{m''}}$. Suppose that m', $m'' \leq m$, and let $z \in U_{i_1}$ be an admissible point of the coherent chains at $l(0)$ of length at most m. Consider $\mathscr{C}$, $\mathscr{C}'$ as chains at z, and let their respective plaque chains be*

$$Q'_i,\ i = 1, 2, \dots, m', \qquad Q''_j,\ j = 1, 2, \dots, m'', \qquad Q'_1 = Q''_1.$$

Then we have that $Q'_{m'} = Q''_{m''}$.

Proof. Take subdivisions

$$0 = t_0 < t_1 < \cdots < t_{m'} = 1, \qquad 0 = t'_0 < t'_1 < \cdots < t'_{m''} = 1 \tag{1}$$

such that

$$l([t_{k-1}, t_k]) \subset U_{i_k}, \qquad k = 1, 2, \dots, m',$$
$$l([t'_{k-1}, t'_k]) \subset U_{j_k}, \qquad k = 1, 2, \dots, m''.$$

We combine the two subdivisions in (1) to get a new subdivision

$$0 = t''_0 < t''_1 < \cdots < t''_{m'''} = 1$$

in such a way that

$$[t''_{k''-1}, t''_{k''}] \subset [t_{k-1}, t_k], \qquad [t''_{k''-1}, t''_{k''}] \subset [t'_{k'-1}, t'_{k'}].$$

Then we have

$$l([t''_{k''-1}, t''_{k''}]) \subset l([t_{k-1}, t_k]) \subset U_{i_k},$$
$$l([t''_{k''-1}, t''_{k''}]) \subset l([t'_{k'-1}, t'_{k'}]) \subset U_{j_{k'}}.$$

Therefore, $U_{i_k} \cap U_{j_{k'}} \neq \varnothing$, and by Theorem 5.1(iv)(a), we can find a distinguished chart $(U_{i_k j_{k'}}, \varphi_{i_k j_{k'}})$ such that

$$U_{i_k} \cup U_{j_{k'}} \subset U_{i_k j_{k'}}.$$

Now notice that if $k'' = 1$ we may let $k = k' = 1$, and since we have $Q'_1 = Q''_1$ there exists a plaque $\widehat{Q}_{kk'}$ of $U_{i_k j_{k'}}$ such that $Q'_k \cup Q''_{k'} \subset \widehat{Q}_{kk'}$. We make an induction assumption that up to k, k', k'', there exist plaques $Q_{kk'}$ of $U_{i_k j_{k'}}$ with

$$Q'_k \cup Q''_{k'} \subset \widehat{Q}_{kk'}. \tag{2}$$

Letting $\delta, \delta' = 0$ or 1, we get

$$[t''_{k''}, t''_{k''+1}] \subset [t_{k-1+\delta}, t_{k+\delta}] \cap [t'_{k'-1+\delta'}, t'_{k'+\delta'}].$$

Thus $U_{i_{k+\delta}} \cap U_{j_{k'+\delta'}} \neq \varnothing$, so by Theorem 5.1(iv)(a), we can find a distinguished chart $(U_{i_{k+\delta}, j_{k'+\delta'}}, \varphi_{i_{k+\delta}, j_{k'+\delta'}})$ such that

$$U_{i_{k+\delta}} \cup U_{j_{k'+\delta'}} \subset U_{i_{k+\delta}, j_{k'+\delta'}}.$$

Since $Q'_{k+\delta}, Q''_{k'+\delta'} \subset U_{i_{k+\delta}, j_{k'+\delta'}}$, we have

$$\widehat{Q}_{kk'} \cap U_{i_{k+\delta}, j_{k'+\delta'}} \neq \varnothing.$$

Hence, by Theorem 5.1(v)(b), using the method of proof for Lemma 4.7(ii), we can find a unique plaque $\widehat{Q}_{k+\delta, k'+\delta'}$ of $U_{i_{k+\delta}, j_{k'+\delta'}}$ such that

$$\widehat{Q}_{kk'} \cap \widehat{Q}_{k+\delta, k'+\delta'} \neq \varnothing.$$

But $\widehat{Q}_{k+\delta, k'+\delta'} \cap U_{i_{k+\delta}}$ is a plaque of $U_{i_{k+\delta}}$ (cf. Theorem 5.1(iv)(b)); so it must be $Q'_{k+\delta}$. Similarly $\widehat{Q}_{k+\delta, k'+\delta'} \cap U_{j_{k'+\delta'}}$ is a plaque of $U'_{j_{k'+\delta'}}$, so it is the plaque $Q''_{k'+\delta'}$. Therefore we have

$$Q'_{k+\delta} \cup Q''_{k'+\delta'} \subset \widehat{Q}_{k+\delta, k'+\delta'}.$$

Hence by induction on k, k', k'', we have established (2) for all k, k', k''.

Let $k'' = m'''$, $k = m'$, $k' = m''$. Then (2) becomes

$$Q'_{m'} \cup Q''_{m''} \subset \widehat{Q}_{m'm''}.$$

But $U_{i_{m'}} = U_{j_{m''}}$. Hence by Theorem 5.1(iv)(b), $Q'_{m'} = Q''_{m''}$.

We now consider two C^0 curves in K:

$$l_0\colon [0, 1] \to K, \qquad l_1\colon [0, 1] \to K,$$

with $x = l_0(0) = l_1(0)$, $y = l_0(1) = l_1(1)$, which are homotopic, $l_0 \simeq l_1$, relative to x and y; in other words, there is a family of C^0 curves

$$l_s\colon [0, 1] \to K, \qquad 0 \leq s \leq 1,$$

such that $l_s(0) = x$, $l_s(1) = y$, and the change of l_s is continuous in parameter s, $0 \leq s \leq 1$.

Now Suppose $\mathscr{C} = \{U_{i_1}, U_{i_2}, \dots, U_{i_{m'}}\}$ is a coherent chain over l_0 and $\mathscr{C}' = \{U_{j_1}, U_{j_2}, \dots, U_{j_{m''}}\}$ a coherent chain over l_1, and assume that

$$U_{i_1} = U_{j_1}, \qquad U_{i_{m'}} = U_{j_{m''}}.$$

Clearly, if $\epsilon > 0$ is small enough, $\mathscr{C}$ is a coherent chain over l_s, $0 \leq s \leq \epsilon$. Hence, we can find a suitable subdivision $0 = s_0 < s_1 < \cdots < s_u = 1$ and coherent chains at x:

$$\mathscr{C}^{(k)} = \{U_1^{(k)}, U_2^{(k)}, \dots, U_{m_k}^{(k)}\}, \qquad k = 0, 1, 2, \dots, u-1,$$

such that

$$\mathscr{C}^{(0)} = \mathscr{C}, \qquad \mathscr{C}^{(u-1)} = \mathscr{C}', \qquad U_1^{(k)} = U_{i_1}, \qquad U_{m_k}^{(k)} = U_{i_{m'}},$$

and each $\mathscr{C}^{(k)}$ is a coherent chain over l_s for s, $s_k \leq s \leq s_{k+i}$. Such chains $\mathscr{C}^{(k)}$, $k = 0, 1, 2, \dots, u-1$ are said to be a *homotopy* between $\mathscr{C}$ and $\mathscr{C}'$ and $\max_k m_k$ the *length* of the homotopy.

THEOREM 5.5. *Let $\mathscr{C}^{(k)}$, $k = 0, 1, 2, \dots, u-1$, be a homotopy of length at most m between coherent chains $\mathscr{C}$ over l_0 and $\mathscr{C}'$ over l_1. Let z be an admissible point of the coherent chains at $l(0)$ of length at most m. Let*

$$Q'_i, \qquad i = 1, 2, \dots, m'$$

be the plaque chain associateRd with $\mathscr{C}$ at z and

$$Q''_j, \qquad j = 1, 2, \dots, m''$$

be the plaque chain associated with $\mathscr{C}'$ at z such that $Q'_1 = Q''_1$. Then $Q'_{m'} = Q''_{m''}$.

PROOF. Denote the plaque chain associated with $\mathscr{C}^{(1)}$ at z by

$$Q'''_k, \qquad k = 1, 2, \dots, m_1, \qquad Q'''_1 = Q'_1.$$

Since $\mathscr{C} = \mathscr{C}^{(0)}$ and $\mathscr{C}^{(1)}$ are two coherent chains over l_{s_1}, Lemma 5.4 implies that $Q'''_{m_1} = Q'_{m'}$. Apply this trick to the $\mathscr{C}^{(k)}$, $k = 0, 1, 2, \ldots$, $u-1$, step by step in this order to finish the proof.

§20. Local stability theory

We have already cited bundle foliations as a special example of foliations. When a foliation possesses a compact leaf which satisfies certain conditions, it becomes locally or globally a bundle foliation. This type of statement is referred to as a stability theorem. In this section and the next, we shall investigate local and global stability theorems, both of which are due to Reeb.

Let M^n be an n-dimensional C^s manifold and let $\mathscr{F} = \{L_\alpha ; \alpha \in A\}$ be a C^r codimension q foliation of M^n, $r \geq 1$. Let $L_{\hat{\alpha}}$ be a compact leaf whose fundamental group $\pi_1(L_{\hat{\alpha}})$ is finite of order ν (see Appendix V, Note 1).

We apply Theorem 5.1 to $K = L_{\hat{\alpha}}$. Let $\mathfrak{N}(L_{\hat{\alpha}})$ be a coherent system of charts on $L_{\hat{\alpha}}$:

$$\mathfrak{N}(L_{\hat{\alpha}}) = \{(U_i, \varphi_i); i = 1, 2, \ldots, v\}.$$

Then the statement of Theorem 5.1(ii) is: $U_i \cap L_{\hat{\alpha}} = Q_i$, $i = 1, 2, \ldots, v$.

For each Q_i, $i = 1, 2, \ldots, v$, we fix a point $x_i \in Q_i$ and choose a family of C^0 curves l_i on $L_{\hat{\alpha}}$ joining x_1 and x_i:

$$l_i: [0, 1] \to L_{\hat{\alpha}}, \qquad l_i(0) = x_1, \ l_i(1) = x_i, \ i = 1, 2, \ldots, v.$$

If $U_i \cap U_j \neq \varnothing$ for U_i, $U_j \in \mathfrak{N}(L_{\hat{\alpha}})$, then by Theorem 5.1(iii) we have

$$U_i \cap U_j \cap L_{\hat{\alpha}} = Q_i \cap Q_j \neq \varnothing.$$

For such U_i, U_j, consider a C^0 curve l_{ij} on $Q_i \cup Q_j$ joining x_i and x_j:

$$l_{ij}: [0, 1] \to Q_i \cup Q_j, \qquad l_{ij}(0) = x_i, \ l_{ij}(1) = x_j.$$

Here we choose l_{ij} so that

$$l_{ji}(t) = l_{ij}(1-t), \qquad 0 \leq t \leq 1.$$

Consider C^0 curves w_k on $L_{\hat{\alpha}}$ based at x:

$$w_k: [0, 1] \to L_{\hat{\alpha}}, \qquad w_k(0) = w_k(1) = x_1, \ k = 1, 2, \ldots, \nu,$$

such that the homotopy classes $\{w_k\}$, $k = 1, 2, \ldots, \nu$, run through all elements of $\pi_1(L_{\hat{\alpha}}, x_1)$. We now choose coherent chains over the families of the curves: l_i, $i = 1, 2, \ldots, v$, w_k, $k = 1, 2, \ldots, \nu$, and l_{ij} as follows:

(i) $\{U_1, U_2^{(i)}, \ldots, U_{m_i-1}^{(i)}, U_i\}$ over l_i, $i = 1, 2, \ldots, v$;
(ii) $\{U_i, U_j\}$ over l_{ij}, $i, j = 1, \ldots, v$;
(iii) $\{U_1, \widehat{U}_2^{(k)}, \ldots, \widehat{U}_{m'_k-1}^{(k)}, U_1\}$ over w_k, $k = 1, 2, \ldots, \nu$.

We shall denote by $w_k \cdot l_i$ the C^0 curves obtained by connecting the curves w_k and l_i:

$$w_k \cdot l_i \colon [0, 1] \to L_{\hat{\alpha}}, \qquad w_k \cdot l_i(t) = \begin{cases} w_k(2t), & 0 \le t \le 1/2, \\ l_i(2t-1), & 1/2 \le t \le 1. \end{cases}$$

Then $\mathscr{C}^{(k)(i)} = \{U_1, \widehat{U}_2^{(k)}, \dots, \widehat{U}_{m'_k-1}, U_1, U_2^{(i)}, \dots, U_{m_i-1}^{(i)}, U_i\}$ is a coherent chain over $w_k \cdot l_i$. Let $z \in U_1$. Suppose $\mathscr{C}^{(k)(i)}$ is a coherent chain at z; then z determines a unique plaque chain associated with $\mathscr{C}^{(k)(i)}$ at z (cf. Lemma 5.2). This plaque chain starts with Q'_1, $z \in Q'_1$, and ends with Q'_i which is a plaque of U_i. We denote Q'_i by $Q_{i,z}^{(\gamma)}$, where γ denotes the homotopy class $\{w_k\}$, an element of $\pi_1(L_{\hat{\alpha}}, x_1)$.

Suppose $U_i \cap U_j \neq \varnothing$. We consider a C^0 curve $w_k \cdot l_i \cdot l_{ij} \cdot l_j^{-1}$ which we obtain by connecting w_k, l_i, l_{ij}, l_j^{-1} in this order. Here $l_j^{-1} \colon [0, 1] \to L_{\hat{\alpha}}$ is defined by $l_j^{-1}(t) = l_j(1-t)$. This path starts at x_1 and ends at x_1. Choosing a suitable C^0 curve $w_{k'}$ and connecting w_k, l_i, l_{ij}, l_j^{-1}, $w_{k'}^{-1}$ in this order, we obtain a C^0 curve $w_k \cdot l_i \cdot l_{ij} \cdot l_j^{-1} \cdot w_{k'}^{-1}$ whose homotopy class is the unit element e of $\pi_1(L_{\hat{\alpha}}, x_1)$:

$$\{w_k \cdot l_i \cdot l_{ij} \cdot l_j^{-1} \cdot w_{k'}^{-1}\} = e.$$

Here $w_{k'}^{-1}(t) = w_{k'}(1-t)$. Therefore the C^0 curves $w_k \cdot l_i \cdot l_{ij}$ and $w_{k'} \cdot l_j$—both joining x_1 and x_j—are homotopic with x_1 and x_j kept fixed. There exists a homotopy between the coherent chain over $w_k \cdot l_i \cdot l_{ij}$:

$$\mathscr{C}^{(k)(i)(ij)} = \{U_1, \widehat{U}_2^{(k)}, \dots, \widehat{U}_{m'_k-1}^{(k')}, U_1, U_2^{(i)}, \dots, U_{m_i-1}^{(i)}, U_i, U_j\},$$

and the coherent chain over $w_{k'} \cdot l_j$:

$$\mathscr{C}^{(k')(j)} = \{U_1, \widehat{U}_2^{(k')}, \dots, \widehat{U}_{m'_{k'}-1}^{(k')}, U_1, U_2^{(j)}, \dots, U_{m_j-1}^{(j)}, U_j\}$$

(cf. §19). Denote by m_{ij} the length of this homotopy. Let $z \in U_1$ be an admissible point of the coherent chains at x_1 of length at most m_{ij}. Denote by

$$Q'_1, \dots, Q_{i,z}^{(\gamma)}, Q''_j, \qquad Q''_j \subset U_j, \quad \text{and} \quad Q'_1, \dots, Q_{j,z}^{(\gamma')}, \quad \gamma' = \{w_{k'}\},$$

the plaque chains associated with $\mathscr{C}^{(k)(i)(ij)}$ at z and $\mathscr{C}^{(k')(j)}$ at z respectively. Then by Theorem 5.5, we have $Q''_j = Q_{j,z}^{(\gamma')}$. Hence

$$Q_{i,z}^{(\gamma)} \cap Q_{j,z}^{(\gamma')} \neq \varnothing. \tag{1}$$

For i, j such that $U_i \cap U_j \neq \varnothing$, denote $\max_{i,j} m_{ij} = \hat{m}$. Then we have the following

Lemma 5.6. *Let $\widehat{V}$ be the set of all admissible points of the coherent chains at x_1 whose respective lengths are at most $\hat{m}$. Then there exists an open set $\widehat{V}'$, $x_1 \in \widehat{V}' \subset \widehat{V}$ such that for any z in $\widehat{V}'$ we have*

$$\overline{Q_{i,z}^{(\gamma)}} \subset \bigcup_{i=1}^{v} U_i, \qquad \gamma \in \pi_1(L_{\hat{\alpha}}, x_1), \; i = 1, 2, \dots, v.$$

Proof. For $m' \leq \hat{m}$, let $\mathscr{C} = \{U_{\lambda_1}, U_{\lambda_2}, \dots, U_{\lambda_{m'}}\}$ be a coherent chain of length at most m', $(U_{\lambda_i}, \varphi_{\lambda_i}) \in \mathfrak{N}(L_{\hat{\alpha}})$. Choose an open set $O_{m'}$ contained in $U_{\lambda_{m'}}$ such that $O_{m'}$ is a union of plaques of $U_{\lambda_{m'}}$ and such that

$$Q_{\lambda_{m'}} \subset O_{m'}, \qquad \overline{O}_{m'} \subset \bigcup_{i=1}^{v} U_i.$$

Define O_i, $i = m' - 1, m' - 2, \dots, 1$, inductively by

$$O_i = \varphi_{\lambda_i}^{-1}(\hat{\pi}^{-1} \circ \hat{\pi}(\varphi_{\lambda_i}(U_{\lambda_i} \cap O_{i+1}))),$$

(cf. Proof of Lemma 4.5). Denote O_1 by $O_{\mathscr{C}}$. Define $\widehat{V}'$ by $\widehat{V}' = \bigcap_{\mathscr{C}} O_{\mathscr{C}}$, where the intersection runs through all coherent neighborhood chains $\mathscr{C}$ of length at most $\hat{m}$, to finish the proof.

Let $\widehat{V}'$ be the open set of Lemma 5.6 and z a point of $\widehat{V}'$. Assume that z is on a leaf L_β. Denote by $L(z)$ the (finite) union of the plaques $Q_{i,z}^{(\gamma)}$, $\gamma \in \pi_1(L_{\hat{\alpha}}, x_1)$, $i = 1, 2, \dots, v$:

$$L(z) = \bigcup_{i,\gamma} Q_{i,z}^{(\gamma)}.$$

Evidently $L(z) \subset L_\beta$. Assume $y \in \overline{L(z)}$; a priori, $y \in \overline{Q_{i,z}^{(\gamma)}}$ for some i. Hence by Lemma 5.6 $y \in U_j$ for some j. If $i \neq j$, then by (1) a plaque of U_j containing y must be some plaque $Q_{j,z}^{(\gamma')}$ such that $Q_{i,z}^{(\gamma)} \cap Q_{j,z}^{(\gamma')} \neq \varnothing$. Therefore,

$$L_\beta = L(z) = \bigcup_{i,\gamma} \overline{Q_{i,z}^{(\gamma)}}, \quad \text{a finite union},$$

so we see that the leaf L_β is compact.

The union $\bigcup_{L_\beta \cap \widehat{V}' \neq \varnothing} L_\beta$ is an open set of M^n containing $L_{\hat{\alpha}}$ (cf. Theorem 4.10); it is a union of compact leaves. Given an open set U containing the leaf $L_{\hat{\alpha}}$, we may choose the U_i, $i = 1, 2, \dots, v$, with $\bigcup_{i=1}^{v} U_i \subset U$ so that $\bigcup_{L_\beta \cap \widehat{V}' \neq \varnothing} L_\beta \subset U$. Therefore, we have proved part (i) of the following

Theorem 5.7 (local stability theorem). *Let $L_{\hat{\alpha}}$ be a compact leaf of a C^r codimension q foliation of an n-dimensional C^s manifold M^n, $r \geq 1$. Suppose that the fundamental group $\pi_1(L_{\hat{\alpha}})$ is finite. Then, we have the following.*

(i) *Given an open set* U *containing* $L_{\hat{\alpha}}$ *in* M^n, *there exists an open set* U' *in* M^n, *which is a union of compact leaves, such that* $L_{\hat{\alpha}} \subset U' \subset U$.

(ii) *If* $r \geq 2$, *we may choose* U' *in* (i) *in such a way that any* L_β *in* U' *is a covering space of* $L_{\hat{\alpha}}$; *in particular* $\pi_1(L_\beta)$ *is a finite group.*

PROOF OF THEOREM 5.7 (ii). We shall first prove the case $q = 1$. Let U' be the open set defined in (i). By Theorem 4.2 there exists a C^{r-1} codimension $n-1$ foliation $\mathcal{F}'$ of U' which is transverse to the restriction of $\mathcal{F}$ to U' (see the note on Theorem 4.2 if $L_\beta \subset \partial M^n$). Let p be a point on $L_{\hat{\alpha}}$ and L'_p the leaf of $\mathcal{F}'$ through p. Recall that L'_p is locally a trajectory of a vector field; so if U' is chosen small enough, L'_p will intersect L_β, and for each $x \in L_\beta$ there exists a point $p \in L_{\hat{\alpha}}$ such that the point x is contained in L'_p. Since L_β is compact, if U' is sufficiently small L'_p and L_β meet in a finite number of points, say $x_1^{(p)}, x_2^{(p)}, \dots, x_{u(p)}^{(p)}$. Now define a map

$$\hat{\pi}\colon L_\beta \to L_{\hat{\alpha}}$$

by $\hat{\pi}(x_i^{(p)}) = p$, $i = 1, 2, \dots, u(p)$. Then by the construction of L'_p, L_β is a covering space of $L_{\hat{\alpha}}$. Here $u(p)$ is a constant independent of the choice of a point p. We notice that in general the fundamental group of a covering space is a subgroup of the fundamental group of its base space. Hence $\pi_1(L_\beta)$ is finite.

For the case $q > 1$, we need the exponential map Exp as in the case $q > 1$ of Theorem 4.3 to construct a leaf corresponding to L'_p of the case $q = 1$, and carry out the proof as above; however, we shall omit the details here.

EXAMPLE. Define a C^∞ diffeomorphism

$$h\colon S^{n-q} \times S^q \to S^{n-q} \times S^q$$

by

$$\begin{aligned} &h((x_1, x_2, \dots, x_{n-q+1}), (y_1, y_2, \dots, y_{q+1})) \\ &\quad = ((-x_1, -x_2, \dots, -x_{n-q+1}), (-y_1, -y_2, \dots, -y_q, y_{q+1})). \end{aligned}$$

With the identification $p \sim h(p)$, $p \in S^{n-q} \times S^q$, we obtain an n-dimensional C^∞ manifold $M^n = S^{n-q} \times S^q / \sim$. Denote by L_y a subset of M^n corresponding to $S^{n-q} \times \{y\}$, $y \in S^q$. Then $\{L_y ; y \in S^q\}$ is a C^∞ codimension q foliation of M^n. When $y = (0, 0, \dots, 0, \pm 1)$, L_y is the $(n-q)$-dimensional projective space; otherwise, L_y is S^{n-q}.

If $q \geq 2$, Theorem 5.7 (i) is valid only in some open set containing $L_{\hat{\alpha}}$ even when M^n is compact; in general, the global result that every leaf of a foliation is compact is not attainable (see Appendix V, Note 2). On the other hand, when $q = 1$ we have a global result which will be discussed in §21.

§21. Global stability theorem

In considering codimension one foliations we come up with global results which are more intricate than the local stability theorem (Theorem 5.7). In the following, let M^n be a C^s n-manifold and let $\mathscr{F} = \{L_\alpha ; \alpha \in A\}$ be a C^r codimension one foliation of M^n, $r \geq 1$. Consider a coherent system of charts $\mathfrak{N}(L_{\hat{\alpha}}) = \{(U_i, \varphi_i); i = 1, 2, \dots, v\}$ on a compact leaf $L_{\hat{\alpha}}$.

Since $\mathscr{F}$ is of codimension one, we have $\hat{\pi} \circ \varphi_i : U_i \to (-1, 1)$, and a plaque of U_i takes the form $(\hat{\pi} \circ \varphi_i)^{-1}(a)$, for some a, $-1 < a < 1$ (see §18 on $\hat{\pi}$).

A coherent chart-system $\mathfrak{N}(L_{\hat{\alpha}})$ is said to be *two-sided* if given distinguished charts U_i and $U_j \in \mathfrak{N}(L_{\hat{\alpha}})$ with $U_i \cap U_j \neq \varnothing$, $\hat{\pi} \circ \varphi_i(z)$ and $\hat{\pi} \circ \varphi_j(z)$ have the same sign for each $z \in U_i \cap U_j$. We say that the leaf $L_{\hat{\alpha}}$ is *two-sided* if it has a two-sided coherent system of charts. When no such $\mathfrak{N}(L_{\hat{\alpha}})$ exists we say that $L_{\hat{\alpha}}$ is *one-sided.*

Following the proof of Theorem 4.1 we can routinely show that a necessary and sufficient condition for $L_{\hat{\alpha}}$ to be two-sided is that there be a C^{r-1} vector field of M^n on $L_{\hat{\alpha}}$ which is transverse to $L_{\hat{\alpha}}$.

The leaf constructed from $I \times \{1/2\}$ (the bold line in Figure 4.7) on the Möbius strip (cf. Example 4, §17) is one-sided. In the example in §20, L_y is one-sided for $y = (0, \pm 1)$ and two-sided for $y \neq (0, \pm 1)$.

Lemma 5.8. *For $L_{\hat{\alpha}}$, $\mathfrak{N}(L_{\hat{\alpha}})$ as above, there exists an open set V in the manifold M^n containing $L_{\hat{\alpha}}$ such that each leaf L_β with $V \cap L_\beta \neq \varnothing$ must be one of the following.*

(i) *L_β is compact, and $L_\beta \subset \bigcup_{i=1}^{v} U_i$. Further, $L_\beta \cap U_i$ consists of one plaque if $L_{\hat{\alpha}}$ is two-sided, and $L_\beta \cap U_i$ consists of two plaques if $L_{\hat{\alpha}}$ is one-sided and $L_\beta \neq L_{\hat{\alpha}}$. In the case $r \geq 2$, if $L_{\hat{\alpha}}$ is two-sided then L_β is C^r diffeomorphic to $L_{\hat{\alpha}}$, and if $L_{\hat{\alpha}}$ is one-sided and $L_\beta \neq L_{\hat{\alpha}}$ then L_β is a double covering space of $L_{\hat{\alpha}}$, that is L_β is the total space, which is arcwise connected, of a C^r bundle with the fiber consisting of two points.*

(ii) *L_β is noncompact, and $\overline{L}_\beta \cap \bigcup_{i=1}^{v} U_i$ contains a compact leaf.*

Proof. Let $\hat{x} \in L_{\hat{\alpha}}$ and let $\widehat{O}$ be the set of all admissible points of the coherent chains at $\hat{x}$ of lengths at most v (Lemma 5.3). Following the construction of $\widehat{V}'$ from $\widehat{V}$ in the proof of Lemma 5.6, we construct $\widehat{O}'$ from $\widehat{O}$; here $\widehat{O} \cap L_{\hat{\alpha}} \neq \varnothing$. Then for a leaf L_β with $L_\beta \cap \widehat{O}' \neq \varnothing$, we have

$$L_\beta \cap U_i \neq \varnothing, \qquad i = 1, 2, \dots, v,$$

and for each i, there exists a plaque Q_i' of U_i such that

$$Q_i' \subset L_\beta, \qquad \overline{Q_i'} \subset \bigcup_{i=1}^{v} U_i. \tag{1}$$

Let $\{Q_i^{(\lambda)}; \lambda \in \Lambda_i\}$ be the set of all plaques of U_i satisfying

$$Q_i^{(\lambda)} \subset L_\beta, \qquad \overline{Q_i^{(\lambda)}} \subset \bigcup_{i=1}^{v} U_i, \qquad i = 1, 2, \ldots, v.$$

By (1) we have $\Lambda_i \neq \varnothing$.

First we assume that $L_{\hat{\alpha}}$ is two-sided. We may also assume that $\mathfrak{N}(L_{\hat{\alpha}})$ is two-sided. Define a subset Λ_i' of Λ_i by $\Lambda_i' = \{\lambda \in \Lambda_i; \hat{\pi} \circ \varphi_i(Q_i^{(\lambda)}) \geq 0\}$. By changing the sign of the image of φ_i if necessary, we may assume that $\Lambda_i' \neq \varnothing$, $i = 1, 2, \ldots, v$. For each i, $i = 1, 2, \ldots, v$, define a plaque Q_i'' of U_i by

$$\hat{\pi} \circ \varphi_i(Q_i'') = \inf_{\lambda \in \Lambda_i'} \hat{\pi} \circ \varphi_i(Q_i^{(\lambda)}), \qquad i = 1, 2, \ldots, v.$$

Evidently, $\overline{Q_i''} \subset \bigcup_{k=1}^{v} U_k$. Now if $y \in \overline{Q_i''}$ then $y \in U_j$ for some j, so y is in one of the plaques of U_j. But then by the definition of Q_i'' we must have that $y \in Q_j''$. Therefore,

$$L' = \bigcup_{i=1}^{v} Q_i'' = \bigcup_{i=1}^{v} \overline{Q_i''}$$

is a compact leaf.

In the case that L_β is compact Λ_i is a finite set (cf. Theorem 4.12), and thus $L' \subset L_\beta$, that is $L' = L_\beta$. Further, by the condition (ii) of Theorem 5.1 for coherent neighborhoods, we must have $L_\beta \cap U_i = Q_i''$.

In the case that L_β is not compact, $L' \subset \overline{L}_\beta$. Therefore,

$$V = \bigcup_{\widehat{O'} \cap L_\beta \neq \varnothing} L_\beta,$$

is the desired open set.

We next suppose that $L_{\hat{\alpha}}$ is one-sided. Decompose Λ_i into

$$\Lambda_i = \Lambda_i' \cup \Lambda_i'', \quad \hat{\pi} \circ \varphi_i(Q_i^{(\lambda)}) \geq 0, \quad \lambda \in \Lambda_i', \quad \hat{\pi} \circ \varphi_i(Q_i^{(\lambda)}) \leq 0, \quad \lambda \in \Lambda_i'',$$

where $i = 1, 2, \ldots, v$. Define plaques Q_i'', $Q_i''' \subset L_\beta$ of U_i by

$$\hat{\pi} \circ \varphi_i(Q_i'') = \inf_{\lambda \in \Lambda_i'} \hat{\pi} \circ \varphi_i(Q_i^{(\lambda)}), \qquad \hat{\pi} \circ \varphi_i(Q_i''') = \sup_{\lambda \in \Lambda_i''} \hat{\pi} \circ \varphi_i(Q_i^{(\lambda)}),$$
$$i = 1, 2, \ldots, v.$$

Then by the identical argument for the two-sided case, we see that

$$L' = \bigcup_{i=1}^{v} (Q_i'' \cup Q_i''')$$

is a compact leaf. If L_β is compact, then $L_\beta = L'$; if L_β is not compact, we must have $L' \subset \overline{L}_\beta$. So once again $V = \bigcup_{\widehat{O'} \cap L_\beta \neq \varnothing} L_\beta$ is the desired set.

Finally we follow the proof of Theorem 5.7(ii) to verify the statement for the case $r \geq 2$ in (i).

As an example of a compact leaf $L_{\hat{\alpha}}$ for which a leaf L_β in Theorem 5.8(ii) exists, we mention the leaf $S^1 \times S^1$ of the Reeb foliation (Example B, §16). From Lemma 5.8 we get the following

THEOREM 5.9. *Let M^n be a compact, connected n-dimensional C^s manifold and $\mathscr{F} = \{L_\alpha; \alpha \in A\}$ a C^r codimension one foliation of M^n, $r \geq 1$. If every leaf of $\mathscr{F}$ is compact we have one of the following (cf. the example in* §20):

(i) $\mathscr{F} = \{\pi^{-1}(p); p \in S^1\}$ *for some C^r map $\pi: M^n \to S^1$. In this case, every leaf of $\mathscr{F}$ is two-sided. If $r \geq 2$, π is a C^r bundle and $\mathscr{F}$ is a bundle foliation*;

(ii) *there exists a C^r map $\pi: M^n \to [0, 1]$ such that*

$$\mathscr{F} = \{\pi^{-1}(p); p \in [0, 1]\}.$$

In this case, all leaves of $\mathscr{F}$ with the exception of $\pi^{-1}(0)$, $\pi^{-1}(1)$ are two-sided. If $r \geq 2$, all two-sided leaves are C^r diffeomorphic and they form double covering spaces of each of $\pi^{-1}(0)$ and $\pi^{-1}(1)$.

PROOF. Define $\pi: M^n \to A$ by $\pi(L_\alpha) = \alpha$. For a leaf $L_{\hat{\alpha}}$ let V be the set constructed in Lemma 5.8. Then since every leaf of $\mathscr{F}$ is compact we get only case (i) in Lemma 5.8. Write $V = \bigcup_{\beta \in A_{\hat{\alpha}}} L_\beta$, $A_{\hat{\alpha}} \subset A$. Define

$$\eta_{\hat{\alpha}}: A_{\hat{\alpha}} \to \hat{\pi} \circ \varphi_1(V \cap U_1)$$

by $\eta_{\hat{\alpha}}(\beta) = \hat{\pi} \circ \varphi_1(L_\beta \cap U_1)$, if $L_{\hat{\alpha}}$ is two-sided, and

$$\eta_{\hat{\alpha}}(\beta) = \hat{\pi}(\varphi_1(L_\beta \cap U_1) \cap \hat{\pi}^{-1}([0, 1))),$$

if $L_{\hat{\alpha}}$ is one-sided. Then $\eta_{\hat{\alpha}}$ is a one-to-one map. Give A the topology in which the $(A_{\hat{\alpha}}, \eta_{\hat{\alpha}})$, $\hat{\alpha} \in A$, form a C^r coordinate neighborhood system on A, and regard A as a one-dimensional C^r manifold. Since M^n is compact and connected, A is also compact and connected. Hence A must be either the unit circle S^1 or the interval $[0, 1]$. The proof of the rest of the theorem is similar to the proof of Theorem 5.7(ii).

Finally we shall discuss a global stability theorem which was proved by Reeb.

THEOREM 5.10 (global stability theorem) (Reeb). *Let M^n be a compact, connected n-dimensional C^s manifold and let $\mathscr{F} = \{L_\alpha; \alpha \in A\}$ be a C^r codimension one foliation of M^n, $r \geq 2$. Suppose $\mathscr{F}$ has a compact leaf whose fundamental group is finite. Then every leaf of $\mathscr{F}$ is compact and its fundamental group is finite.*

PROOF. Let $\{L_\alpha; \alpha \in A'\}$ be the set of all compact leaves with finite fundamental groups. Set $\bigcup_{\alpha \in A'} L_\alpha = W$. By Theorem 5.7 if $L_{\hat{\alpha}} \in W$, there exists an open set U' in M^n such that $L_{\hat{\alpha}} \subset U'$, and U' is a union of compact leaves each having a finite fundamental group. Thus W is open in M^n. Suppose that $M^n - W \neq \varnothing$. Then $\overline{W} - W \neq \varnothing$ because M^n is connected. Let W' be a connected component of W and y a point of $\overline{W}' - W'$. If $y \in L_\gamma$ for some γ, then $L_\gamma \subset \overline{W'}$ by Theorem 4.9. Let $(U_\lambda, \varphi_\lambda)$ be a distinguished chart such that connected components of $W' \cap U_\lambda$ are

$$\varphi_\lambda^{-1} \circ \hat{\pi}^{-1}((a_\sigma, b_\sigma)), \qquad a_\sigma < b_\sigma, \ \sigma \in \Sigma.$$

For an arbitrary $\sigma \in \Sigma$, let W_σ be the set of all leaves which intersect $\varphi_\lambda^{-1} \circ \hat{\pi}^{-1}((a_\sigma, b_\sigma))$. Then by Theorem 4.10, W_σ is an open set of W'. On the other hand, a subset F of U_λ:

$$F = \{\varphi_\lambda^{-1}(0, 0, \dots, 0, x_n); a_\sigma < x_n < b_\sigma\}$$

is closed in $W' \cap U_\lambda$, so W_σ, the set of all leaves which intersect F, is also closed in W'. Hence $W_\sigma = W'$. Therefore, for a leaf $L_\beta \subset W'$ the "cardinality" of the number of the connected components of $L_\beta \cap U_\lambda$ is at least the "cardinality" of Σ. Since L_β is compact, by Theorem 4.12 the number of the connected components is finite, so Σ is a finite set. Hence it follows that the set of connected components of $L_\gamma \cap U_\lambda$ is finite. Since M^n is compact there exists a finite set of distinguished charts $(U_{\lambda_i}, \varphi_{\lambda_i})$, $i = 1, 2, \dots, m$, such that $\bigcup_{i=1}^m U_{\lambda_i} = M^n$. But the connected components of $L_\gamma \cap U_{\lambda_i}$ are finite in number, $i = 1, 2, \dots, m$; hence L_γ is compact. Applying Lemma 5.8(i) to L_γ, we conclude that the fundamental group of L_γ is finite. Hence we have

$$y \in L_\gamma \subset W' \subset W,$$

which is a contradiction. Therefore we conclude that $W = M^n$.

From Theorem 5.9 we obtain the following

THEOREM 5.11. *Under the hypothesis of Theorem* 5.10, (i) *or* (ii) *in Theorem* 5.9 *occurs.*

§22. Holonomies

We shall denote by O the origin of Euclidean q-space $\mathbf{R}^q$. Let f be a C^r map of an open subset U of $\mathbf{R}^q$, containing O, to $\mathbf{R}^q$:

$$f\colon U \to \mathbf{R}^q,$$

such that $f(O) = O$. Suppose that $f\colon U \to f(U)$ is a C^r diffeomorphism. Then f is called a *local C^r diffeomorphism* of $\mathbf{R}^q$ at the origin. Here $r = 0, 1, 2, \dots, \infty$.

Suppose that

$$f_1\colon U_1 \to \mathbf{R}^q, \qquad f_2\colon U_2 \to \mathbf{R}^q$$

are local diffeomorphisms at the origin. We write $f_1 \sim f_2$ if there exists an open set U' in $\mathbf{R}^q$ with $O \in U' \subset U_1 \cap U_2$ such that

$$f_1|U' = f_2|U'.$$

The relation $\sim$ is obviously an equivalence relation. An equivalence class in the relation $\sim$ is called a *germ* of local diffeomorphisms of $\mathbf{R}^q$ at the origin. We denote by G_q^r the set of all germs of C^r diffeomorphisms of $\mathbf{R}^q$ at the origin and by $[f]$ the equivalence class of f.

Let $[f]$, $[g] \in G_q^r$ be represented by $f: U \to \mathbf{R}^q$, $g: U' \to \mathbf{R}^q$ respectively. Define a C^r map

$$g \circ f: f^{-1}(f(U) \cap U') \to \mathbf{R}^q$$

by $g \circ f(x) = g(f(x))$. Then $g \circ f$ is a local diffeomorphism of $\mathbf{R}^q$ at the origin. Define the product $[f]\cdot[g]$ of $[f]$ and $[g]$ by

$$[f]\cdot[g] = [g \circ f].$$

Then $[f]\cdot[g]$ is determined by $[f]$ and $[g]$ independent of the choice of their representatives. With this product, G_q^r becomes a group: the unit element of G_q^r is $[\mathrm{id}]$, where $\mathrm{id}: \mathbf{R}^q \to \mathbf{R}^q$ is the identity map. The inverse of $[f]$ is represented by $f^{-1}: f(U) \to U$, where $f: U \to \mathbf{R}^q$ is a representative of $[f]$; $[f]^{-1} = [f^{-1}]$.

Let M^n be an n-dimensional C^s manifold and let $\mathscr{F} = \{L_\alpha; \alpha \in A\}$ be a C^r codimension q foliation of M^n. Fix a point $\hat{x}$ on a leaf L_α and consider the fundamental group $\pi_1(L_\alpha, \hat{x})$ of L_α based at $\hat{x}$.

Let γ be an element of $\pi_1(L_\alpha, \hat{x})$ represented by a C^0 curve

$$w: [0, 1] \to L_\alpha, \qquad w(0) = w(1) = \hat{x}$$

so that $\gamma = \{w\}$. Take a compact subset K of L_α with $w([0, 1]) \subset K$, and consider a coherent system $\mathfrak{M}(K) = \{(U_i, \varphi_i); i = 1, 2, \dots, v\}$ of charts on K.

Choose a coherent chain $\mathscr{C}$ over w:

$$\mathscr{C} = \{U_{\lambda_1}, U_{\lambda_2}, \dots, U_{\lambda_{m-1}}, U_{\lambda_1}\}, \qquad \hat{x} \in U_{\lambda_1}.$$

The set $\widehat{O}$ of all $z \in U_{\lambda_1}$ such that $\mathscr{C}$ is a chain at z is, by Lemma 4.5, open in U_{λ_1} containing Q_1, $U_{\lambda_1} \cap K \subset Q_1$, and it is a union of plaques of the chart U_{λ_1}. By Lemma 5.2, the plaque chain associated with $\mathscr{C}$ at z:

$$Q_i^{(z)}, \qquad i = 1, 2, \dots, m, \qquad z \in Q_1^{(z)}, \qquad Q_1^{(z)}, Q_m^{(z)} \subset U_{\lambda_1},$$

is uniquely determined by z. Define a map $f: \hat{\pi} \circ \varphi_{\lambda_1}(\widehat{O}) \to \mathbf{R}^q$ by

$$f(\hat{\pi} \circ \varphi_{\lambda_1}(Q_1^{(z)})) = \hat{\pi} \circ \varphi_{\lambda_1}(Q_m^{(z)}), \qquad z \in \widehat{O}, \tag{1}$$

where $\hat{\pi}: (-1, 1)^n \to (-1, 1)^q$ is the projection map. Then the map f is a local C^r diffeomorphism of $\mathbf{R}^q$ at the origin.

Let $\mathscr{C}' = \{U_{\mu_1}, U_{\mu_2}, \dots, U_{\mu_{m'-1}}, U_{\mu_1}\}$ be a coherent chain over w different from $\mathscr{C}$ with $\hat{x} \in U_{\mu_1}$, and define the local diffeomorphism f' of $\mathbf{R}^q$ as we defined f in (1). Now define $h: \hat{\pi} \circ \varphi_{\lambda_1}(U_{\lambda_1} \cap U_{\mu_1}) \to \mathbf{R}^q$ by

$$h(\hat{\pi} \circ \varphi_{\lambda_1}(Q_1' \cap U_{\lambda_1})) = \hat{\pi} \circ \varphi_{\mu_1}(Q_1' \cap U_{\mu_1}),$$

where Q_1' is a plaque. Then h is a local diffeomorphism of $\mathbf{R}^n$ at the origin satisfying the identity

$$[f] = [h][f'][h]^{-1}.$$

Hence it follows that w determines an element of G_q^r uniquely up to the inner automorphisms of G_q^r.

Next we pick another C^0 curve

$$\overline{w}: [0, 1] \to L_\alpha, \qquad \overline{w}(0) = \overline{w}(1) = \hat{x},$$

such that $\gamma = \{w\} = \{\overline{w}\}$. Choose K so that K contains a homotopy between the curves w and $\overline{w}$. Consider a coherent chain over $\overline{w}$:

$$\overline{\mathscr{C}} = \{U_{\lambda_1}, U_{\lambda_2'}, \dots, U_{\lambda'_{m''-1}}, U_{\lambda_1}\},$$

and let $\overline{f}$ be the local diffeomorphism of $\mathbf{R}^n$ defined in the same manner as the map f in (1). By Theorem 5.5, we have $[f] = [\overline{f}]$; so for each element γ of $\pi_1(L_\alpha, \hat{x})$ there is an element of G_q^r uniquely defined up to the inner automorphisms. If we write $\Psi(\{w\}) = [f]$ for this correspondence, we have a map

$$\Psi: \pi_1(L_\alpha, \hat{x}) \to G_q^r.$$

It is immediate from the definition of f that Ψ is a homomorphism. To see this, fix a U_{λ_1} with $\hat{x} \in U_{\lambda_1}$ and always take a coherent chain over the path w starting from and ending in U_{λ_1}. For a change of the base point $\hat{x}$ the change in Ψ is by inner automorphisms of G_q^r: we have the well-defined homomorphism $\Psi: \pi_1(L_\alpha) \to G_q^r$ up to the inner automorphisms of the group G_q^r. The homomorphism Ψ is called the *holonomy* of the leaf L_α, and the subgroup $\Psi(\pi_1(L_\alpha, \hat{x}))$ of G_q^r is called the *holonomy group* of L_α.

It goes without saying that the holonomy group of a leaf L_α whose fundamental group is finite is finite. Also trivial is the fact that the holonomy group of any leaf in a bundle foliation $\mathscr{F}$ consists of the unit element alone. The holonomy group of the leaf $S^1 \times S^1$ in the Reeb foliation (Example B, §16) is the direct sum of two infinite cyclic groups.

For the local stability theorem, Theorem 5.12, we assumed that $\pi_1(L_{\hat{\alpha}})$ was finite; however, the proof depended only on the fact that the family of plaques $Q_{i,z}^{(\gamma)}$, $\gamma \in \pi_1(L_{\hat{\alpha}})$, was finite for a fixed z. Now observe that even if $\pi_1(L_{\hat{\alpha}})$ is not finite, the family $Q_{i,z}^{(\gamma)}$, $\gamma \in \pi_1(L_{\hat{\alpha}})$, will be finite so long as the holonomy group of $L_{\hat{\alpha}}$ is finite. Hence we may generalize Theorem 5.7 in the following

THEOREM 5.12 (local stability theorem). *In Theorem 5.7, replace the assumption that $\pi_1(L_{\hat{\alpha}})$ be finite by the assumption that the holonomy group of $L_{\hat{\alpha}}$ be finite. Then we have the conclusion of Theorem 5.7.*

CHAPTER VI

The Existence of Compact Leaves

§23. Foliations without compact leaves

The C^∞ codimension two foliation of S^3 whose leaves are the trajectories of the C^∞ vector field X_H constructed via the Hopf map of (1) in §14 (cf. Example A, §16) has the property that all of its leaves are compact and homeomorphic to S^1. Notice that this foliation is a bundle foliation (cf. Example 3, §17). In a sharp contrast with this is the codimension two foliation of S^3 in Example A, §16, whose leaves are the trajectories of the Schweitzer dynamical system; this foliation has no compact leaves. In the language of foliations, the construction of the Schweitzer dynamical system goes as follows: start with the C^∞ foliation defined from X_H, modify this foliation, using the vector field X_1 defined in §14 by (2), to a foliation with a single compact leaf, then remove the compact leaf using the C^1 foliation defined by the Denjoy vector field X_D on the torus. The result is a C^1 codimension two foliation of S^3 without compact leaves. Whether there exists a C^r codimension two foliation on S^3 without compact leaves for $r \geq 2$ is an open question at present([1]).

In §15, we saw how to eliminate periodic trajectories from a closed manifold using a special C^∞ vector field on $D^q \times [-3, 3]$ to obtain the Wilson vector field X_W. In the following we shall discuss an application of X_W to eliminate compact leaves of a foliation.

Let M^n be a C^s manifold and $\mathscr{F} = \{L_\alpha ; \alpha \in A\}$ a C^r codimension q foliation of M^n, $r \geq 1$. Let $(U_\lambda, \varphi_\lambda)$ be a distinguished chart, and position $D^{n-q} \times D^q$ in $\varphi_\lambda(U_\lambda) = (-1, 1)^n$ so that

$$D^{n-q} \subset (-1, 1)^{n-q}, \qquad D^q \subset (-1, 1)^q.$$

Then for a leaf L_α with $L_\alpha \cap U_\lambda \neq \varnothing$, the intersection of a connected component of $\varphi_\lambda(L_\alpha \cap U_\lambda)$ and $D^{n-q} \times D^q$ is $D^{n-q} \times \{y\}$, $y \in D^q$, that is, the restriction of $\mathscr{F}$ to $\varphi_\lambda^{-1}(D^{n-q} \times D^q)$ is locally $\{D^{n-q} \times \{y\}; y \in D^q\}$.

For Wilson's dynamical system we defined the C^∞ vector field Y_{k+1} on

([1])Translator's note. Recently J. Harrison solved the problem for the case $r = 2$. See Topology 27 (1988), 249-278.

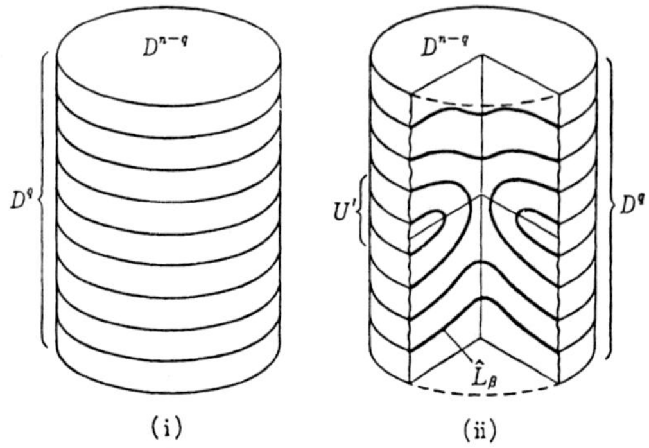

FIGURE 6.1.

$D^{m+k} \times [-3, 3]$. If we let $m = 1$ and $k = q - 1$, $n - 1 \geq q \geq 2$, we get the C^∞ vector field Y_q on $D^q \times [-3, 3]$, and the trajectories of Y_q decompose $D^q \times [-3, 3]$ into the family $\{l_\beta ; \beta \in B\}$ of one-dimensional manifolds. Let U' be a sufficiently small neighborhood of the origin in D^q. For $x \in U'$, the trajectory l_β containing $(x, \pm 3)$ is noncompact. In the case $k = 1$, that is $q = 2$, only four trajectories in $\{l_\beta ; \beta \in B\}$ are contained in $\operatorname{Int} D^q \times (-3, 3)$; these four are homeomorphic to S^1 and hence are compact. However, for $k \geq 2$, that is $q \geq 3$, there is no compact trajectory l_β contained in $\operatorname{Int} D^q \times (-3, 3)$.

We shall express an element $x = (x_1, x_2, \dots, x_{n-q})$ of D^{n-q} by

$$x_i = tx'_i, \qquad i = 1, 2, \dots, n - q,$$

where $x' = (x'_1, x'_2, \dots, x'_{n-q}) \in S^{n-q-1}$, and $0 \leq t \leq 1$; set $x = (x', t)$ with $x' \in S^{n-q-1}$ and $t \in [0, 1]$. Define a subset $\hat{L}_\beta$ of $D^{n-q} \times D^q$ by

$$\hat{L}_\beta = \{((x', t), y) \in D^{n-q} \times D^q ; (y, 3t) \in l_\beta , x' \in S^{n-q-1}\}.$$

Then $\hat{L}_\beta$ is a C^∞ $(n - q)$-manifold and $D^{n-q} \times D^q$ decomposes into the family $\{\hat{L}_\beta ; \beta \in B\}$ (Figure 6.1 (ii)). We can see in Figure 6.1, (i) and (ii), that the intersection of $\hat{L}_\beta$ and the boundary $(\partial D^{n-q} \times D^q) \cup (D^{n-q} \times \partial D^q)$ of $D^{n-q} \times D^q$ is either $\partial D^{n-q} \times \{y\}$, $y \in D^q$, or $D^{n-q} \times \{y\}$, $y \in \partial D^q$.

Thus, one obtains a new C^r foliation of M^n upon replacing a subset of $\varphi_\lambda^{-1}(D^{n-q} \times D^q)$, which is locally the family $\{D^{n-q} \times \{y\} ; y \in D^q\}$, by the family $\{\hat{L}_\beta ; \beta \in B\}$ of subsets $\hat{L}_\beta$.

If in addition M^n is closed, by performing the above replacement finitely many times one can construct a C^r foliation $\mathcal{F}' = \{L'_{\alpha'} ; \alpha' \in A'\}$, with each leaf $L'_{\alpha'}$ containing at least one $\hat{L}_\beta$. Hence by the properties of the $\hat{L}_\beta$ one establishes the following

THEOREM 6.1. *Let M^n be a closed n-dimensional C^s manifold and $\mathcal{F}$ a C^r codimension q foliation of M^n, $r \geq 1$, $n - 1 \geq q \geq 2$. Then we* *may modify* $\mathscr{F}$ *to a* C^r *foliation* $\mathscr{F}'$ *in such a way that if* $q \geq 3$, $\mathscr{F}'$ *has no compact leaves, and if* $q = 2$, $\mathscr{F}'$ *has only finitely many compact leaves which are* C^r *diffeomorphic to* $S^1 \times S^{n-3}$.

In the case of $q = 2$, one can apply the Schweitzer method developed in §14 to a portion of the C^∞ vector field Y_2 on $D^q \times [-3, 3]$ to remove the four circles from the family $\{l_\beta; \beta \in B\}$ thereby constructing a C^1 vector field on $D^q \times [-3, 3]$ whose trajectories are all noncompact (cf. Theorem 3.7). Starting with this vector field, one constructs the family $\{\widehat{L}_\beta; \beta \in B\}$ in a like manner as above, and performing a finite number of replacements one gets the following

THEOREM 6.2. *Let* M^n *be a closed n-dimensional* C^s *manifold and* $\mathscr{F}$ *a* C^r *codimension two foliation of* M^n, $r \geq 1$. *Then we can modify* $\mathscr{F}$ *to a* C^1 *foliation without compact leaves.*

We shall show in §25 that a result of this nature is in general not valid for codimension one foliations.

§24. The holonomy lemma

The following holonomy lemma will be used repeatedly throughout the rest of the present chapter.

THEOREM 6.3 (the holonomy lemma). *Let* M^n *be an n-dimensional* C^s *manifold and* $\mathscr{F} = \{L_\alpha; \alpha \in A\}$ *a* C^r *codimension one foliation of* M^n, *where* $r \geq 2$. *Let* E *be a compact connected* $C^{s'}$ *manifold,* $s' \geq r$, *and let*

$$f : E \to L_{\hat{\alpha}}$$

be a $C^{r'}$ *map from* E *to a leaf* $L_{\hat{\alpha}}$, $0 \leq r' \leq \min(r, s)$ (2). *Suppose that* f *is homotopic to zero, that is,* $f \simeq h$ *in* $L_{\hat{\alpha}}$, *where* $h: E \to L_{\hat{\alpha}}$ *is a constant map. Then there exists a family of* $C^{r'}$ *maps, continuous in* t:

$$f_t : E \to M^n, \qquad -\epsilon < t < \epsilon,$$

satisfying the following:

(i) $f_0 = f$.

(ii) *For a fixed* t, $f_t(E)$ *is contained in a leaf* $L_{\alpha(t)}$ *of* $\mathscr{F}$:

$$f_t(E) \subset L_{\alpha(t)}.$$

(iii) *For a fixed* p *in* E, *define a map*

$$l_p : (-\epsilon, \epsilon) \to M^n$$

by $l_p(t) = f_t(p)$. *Then* l_p *is a* C^r *curve transverse to the leaves of* $\mathscr{F}$: *the tangent vector to* l_p *at* $l_p(t)$ *is not contained in* $T_{l_p(t)}(L_{\alpha(t)})$.

(2)Translator's note. The notation here is adopted from the Russian translation of this book.

(iv) *If* f *is a* $C^{r'}$ *immersion,* f_t *is also a* $C^{r'}$ *immersion for each* t, *and a map*

$$\hat{f}: E \times (-\epsilon, \epsilon) \to M^n$$

defined by $\hat{f}(y, t) = f_t(y)$ *is a* C^0 *immersion.*

PROOF. We consider a continuous homotopy between f and h:

$$F: E \times I \to L_{\hat{\alpha}},$$

such that $F|E \times \{1\} = f$, $F|E \times \{0\} = h$. The set $F(E \times I)$ is compact in $L_{\hat{\alpha}}$. Let $\mathfrak{N}(F(E \times I)) = \{(U_i, \varphi_i); i = 1, 2, \dots, v\}$ be a coherent chart system on $F(E \times I)$. Denote $h(E) = \{\hat{x}\}$ and assume that $\hat{x} \in U_v$. Choose an open cover of E:

$$\{V_k; k = 1, 2, \dots, u\}, \qquad E = \bigcup_{k=1}^{u} V_k,$$

in such a way that each V_k is arcwise connected and that $f(\overline{V}_k) \subset U_{i_k}$ for some $U_{i_k} \in \mathfrak{N}(F(E \times I))$.

Select a point $y_k \in V_k$ for each k, $k = 1, 2, \dots, u$, and define a family of C^0 curves in $F(E \times I)$:

$$l_k: [0, 1] \to F(E \times I), \qquad k = 1, 2, \dots, u$$

by $l_k(\tau) = F(y_k, \tau)$, $0 \le \tau \le 1$. Each path l_k joins $\hat{x}$ and $f(y_k)$. Choose a coherent chain $\mathscr{C}^{(k)} = \{U_v, U_2^{(k)}, U_3^{(k)}, \dots, U_{m_k-1}^{(k)}, U_{i_k}\}$ over l_k.

If $V_k \cap V_{k'} \neq \varnothing$, we select a C^0 curve $\hat{l}_{kk'}$ in $V_k \cup V_{k'}$ joining y_k and $y_{k'}$:

$$\hat{l}_{kk'}: [0, 1] \to V_k \cup V_{k'}, \qquad \hat{l}_{kk'}(0) = y_k, \ \hat{l}_{kk'}(1) = y_{k'}.$$

The C^0 curve $l_k \cdot (f \circ \hat{l}_{kk'})$ obtained by connecting l_k and $f \circ \hat{l}_{kk'}$ is clearly homotopic to $l_{k'}$ with $\hat{x}$ and $f(y_{k'})$ kept fixed:

$$l_k \cdot (f \circ \hat{l}_{kk'}) \simeq l_{k'}.$$

Set $\mathscr{C}_{k'}^{(k)} = \{U_v, U_2^{(k)}, U_3^{(k)}, \dots, U_{m_k-1}^{(k)}, U_{i_k}, U_{i_{k'}}\}$. Then $\mathscr{C}_{k'}^{(k)}$ is a coherent chain over $l_k \cdot (f \circ \hat{l}_{kk'})$ and there is a homotopy between $\mathscr{C}_{k'}^{(k)}$ and $\mathscr{C}^{(k')}$, whose length we shall denote by $m_{kk'}$ (cf. §19). Let m be an integer satisfying $m > m_{kk'}$ for all pairs k, k' with $V_k \cap V_{k'} \neq \varnothing$.

Denote by $Q_{v,t}$ the plaque $\varphi_v^{-1} \circ (\hat{\pi}^{-1}(t))$ of U_v, $-1 < t < 1$. If we take $\epsilon > 0$ small enough, then for each t, $-\epsilon < t < \epsilon$, the plaque $Q_{v,t}$ will contain an admissible point of the coherent chains at $\hat{x}$ whose respective lengths are at most m. Let $z \in Q_{v,t}$ be such an admissible point, and denote by

$$Q_{v,t}, Q_{2,t}^{(k)}, \dots, Q_{m_k-1,t}^{(k)}, Q_{i_k,t}$$

the plaque chain associated with the chain $\mathscr{C}^{(k)}$ at z, $k = 1, 2, \ldots, u$. If we have $V_k \cap V_{k'} \neq \varnothing$, denote by

$$Q_{v,t}, Q^{(k)}_{2,t}, \ldots, Q^{(k)}_{m_k-1,t}, Q_{i_k,t}, Q^{(k,k')}_t$$

the plaque chain associated with the coherent chain $\mathscr{C}^{(k)}_{k'}$ at z. Then by Theorem 5.5, $Q^{(k,k')}_t = Q_{i_{k'},t}$. Thus the $Q_{i_k,t}$, $k = 1, 2, \ldots, u$, are on a single leaf $L_{\alpha(t)}$:

$$\bigcup_{i=1}^{u} Q_{i_k,t} \subset L_{\alpha(t)}.$$

Consider a C^{r-1} codimension $n-1$ foliation $\mathscr{F}'$ transverse to $\mathscr{F}$; such a foliation $\mathscr{F}'$ exists by virtue of Theorem 4.2. In the case $\partial M^n \neq \varnothing$, $\mathscr{F}'$ should be a *foliation transverse to the boundary* to be defined in §29 (see the remark after Theorem 4.2).

Take ϵ sufficiently small, and for t with $-\epsilon < t < \epsilon$, and $p \in E$ define $f_t(p)$ by

$$f_t(p) = (\text{leaf of } \mathscr{F}' \text{ through } f(p)) \cap Q_{i_k,t},$$

where $p \in V_k$. Since ϵ was chosen small enough, $f_t(p)$ is well defined. We have thus constructed a family of maps $f_t\colon E \to M^n$ which clearly satisfy (i), (ii), (iv). Also immediate is (iii) by the definition of $\mathscr{F}'$.

In Theorem 6.3, E was a compact $C^{s'}$ manifold, $s' \geq 0$; however, the proof works just as well if E is a finite complex.

As a corollary to the holonomy lemma, we shall prove the following theorem which will give us a criterion for the compactness of a leaf.

Theorem 6.4. *Let M^n be a C^s manifold and let $\mathscr{F} = \{L_\alpha; \alpha \in A\}$ be a C^r codimension one foliation of M^n, $r \geq 2$. If $\mathscr{F}$ has a leaf L which is not closed in M^n, for instance if L is a noncompact leaf in a compact manifold M^n, then there exists a simple closed C^1 curve l in M^n:*

$$l\colon [0, 1] \to M^n, \qquad l(0) = l(1),$$

which is transverse to the leaves of $\mathscr{F}$ such that $l([0, 1]) \cap L \neq \varnothing$.

Proof. Assume first that $\mathscr{F}$ is transversely orientable. Then by Theorem 4.1, there exists a C^{r-1} vector field $Y = \{Y(p); p \in M^n\}$ transverse to $\mathscr{F}$. The integral curves of Y intersect the leaves of $\mathscr{F}$ transversely. Since we have $\bar{L} - L \neq \varnothing$, there is a point p in $\bar{L} - L$. The integral curve $\varphi_{\{p\}}$ of Y through the point p is transverse to the leaves and intersects L at infinitely many points. Select a C^1 curve $g\colon [0, 1] \to M^n$ as a suitable restriction of $\varphi_{\{p\}}$ such that $g(0), g(1) \in L$, and $g(0) \neq g(1)$ (Figure 6.2).

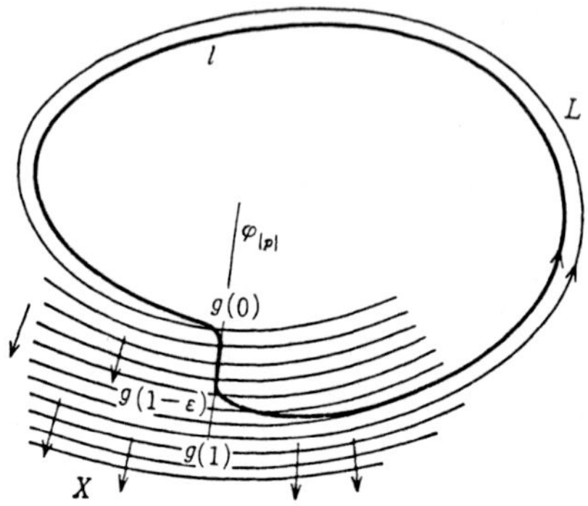

FIGURE 6.2.

Consider a C^1 curve f on L:

$$f\colon [0,1] \to L,$$

such that $f(0) = g(1)$, $f(1) = g(0)$ and if $t \neq t'$, $f(t) \neq f(t')$. Then f is homotopic to zero, so by the holonomy lemma, we can construct a family of C^1 curves satisfying the conditions (i)–(iv) of Theorem 6.3:

$$f_t\colon [0,1] \to L_{\alpha(t)}, \qquad f_0 = f,$$

where $0 \leq t < \epsilon$ and ϵ is sufficiently small. We may assume $f_t(0) = g(1-t)$; this can be done by choosing a suitable $\mathcal{F}'$ in the proof of Theorem 6.3; by choosing the vector field Z suitably in the proof of Theorem 4.2. Let l be a C^0 curve which is the sum of $g([0, 1-(\epsilon/2)])$ and $\bigcup_{0\leq\tau\leq 1} f_{\epsilon(1-\tau)/2}(\tau)$, with the corners smoothed out at $g(0)$ and $g(1-\epsilon)$ (Figure 6.2). Then l is the desired C^1 curve.

When $\mathcal{F}$ is not transversely orientable, we pass to a double covering space $\pi\colon \hat{M}^n \to M^n$ in which the C^r foliation:

$$\hat{\mathcal{F}} = \{\text{the arcwise connected components of } \pi^{-1}(L_\alpha);\ L_\alpha \in \mathcal{F}\}$$

is transversely orientable. We apply the above result to $\hat{M}^n$ and $\hat{\mathcal{F}}$, and obtain a C^1 curve l on $\hat{M}^n$. Then $\pi \circ l$ is the desired simple closed curve in M^n – if $\pi \circ l$ is not simple, we shake it out in a small neighborhood of each intersection point to general position.

§25. Compact leaves of codimension one foliations of S^3 (theorem of Novikov)

The Reeb foliation of S^3 has a compact leaf $S^1 \times S^1$, and when we actually try to construct a C^r codimension one foliation of S^3 we invariably end up with a compact leaf. Based on this observation there had been a conjecture, called the *Ehresmann conjecture*, that every C^r codimension one foliation

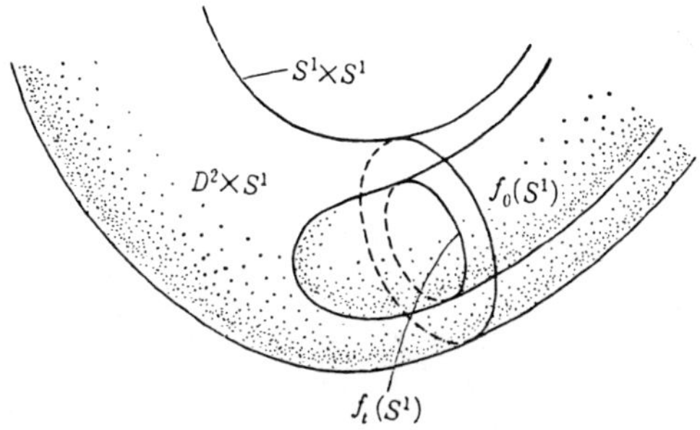

FIGURE 6.3

of S^3 has a compact leaf. In 1964, Novikov proved the conjecture in the following

THEOREM 6.5. *A C^r codimension one foliation $\mathscr{F}$ of the three-sphere S^3 has a compact leaf, $r \geq 2$.*

The proof requires several lemmas. In the following we assume that $\mathscr{F}$ is a C^r codimension one foliation of S^3, $r \geq 2$.

Fix a leaf L_0 of $\mathscr{F}$. Consider a closed $C^{r'}$ curve on L_0:

$$f_0 : S^1 \to L_0,$$

which is not null-homotopic, $0 \leq r' \leq r$. We say that f_0 is a *vanishing cycle* of $\mathscr{F}$ if there exists a family of closed $C^{r'}$ curves varying continuously in t:

$$f_t : S^1 \to S^3, \qquad 0 \leq t \leq \epsilon,$$

satisfying the following:

(i) $f_t(S^1)$ is contained in a leaf $L_{\alpha(t)}$ of $\mathscr{F}$.

(ii) For a fixed x, $f_t(x)$ is a C^r curve in parameter t, $0 \leq t \leq \epsilon$, transverse to the leaves of $\mathscr{F}$.

(iii) f_0 is not null-homotopic on L_0, but f_t is null-homotopic on $L_{\alpha(t)}$ for $0 < t \leq \epsilon$.

Let f_0 be a $C^{r'}$ curve on the compact leaf $S^1 \times S^1$ of the Reeb foliation defined by $f_0(S^1) = S^1 \times \{p\}$, $p \in S^1$. Then f_0 is a vanishing cycle (Figure 6.3).

LEMMA 6.6. *Let $\mathscr{F}$ be a C^r codimension one foliation of S^3, $r \geq 2$. Then there exists at least one vanishing cycle of $\mathscr{F}$.*

PROOF. Since S^3 is simply connected, $\mathscr{F}$ is transversely orientable. Hence by Theorem 4.1, S^3 has a C^{r-1} vector field Y which intersects every leaf of $\mathscr{F}$ transversely. Suppose that the integral curve $\varphi_{\{p\}}$ of Y through a point p on S^3 is periodic. Then $\varphi_{\{p\}}$ is a simple closed C^r curve intersecting the leaves of $\mathscr{F}$ transversely. If $\varphi_{\{p\}}$ is not periodic, $\{\varphi_{\{p\}}(i);\, i \in \mathbf{Z}\}$

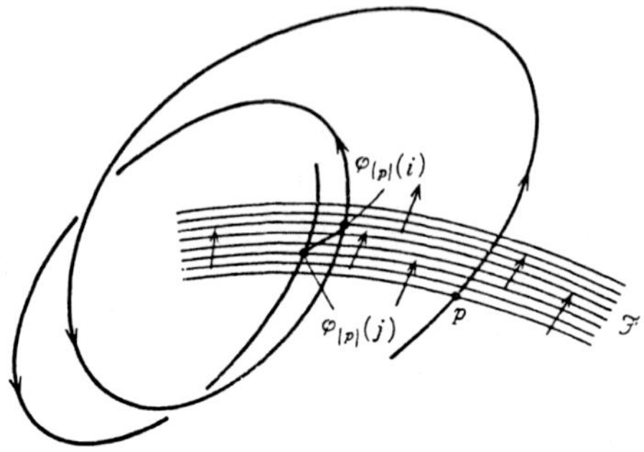

FIGURE 6.4.

is an infinite set, so it has an accumulation point, say a. Now follow the proof of Lemma 1.13; take suitable i and j, $i < j$, and join the end points $\varphi_{\{p\}}(i)$ and $\varphi_{\{p\}}(j)$ of $\bigcup_{i \le t \le j} \varphi_{\{p\}}(t)$ as shown in Figure 6.4. The resulting simple closed C^r curve intersects the leaves of $\mathcal{F}$ transversely.

Let $l \colon S^1 \to S^3$ be a simple closed C^r curve which intersects the leaves of $\mathcal{F}$ transversely. The construction in the preceding paragraph guarantees the existence of such an l. Since S^3 is simply connected, there is a continuous map $g \colon D^2 \to S^3$ such that $g|\partial D^2 = l$. We may assume that g is an immersion of class C^r as follows. Consider $\mathbf{R}^3 = S^3 - \{\text{a point}\}$; then l is a simple closed C^r curve in $\mathbf{R}^3$. The simplest situation for l in $\mathbf{R}^3$ is shown in Figure 6.5 (i) together with a choice of g. If l is a trefoil knot, separate it into two segments by points a, b on l (Figure 6.5). Choose

$$g_1 \colon D^2 \to \mathbf{R}^3, \qquad g_2 \colon D^2 \to \mathbf{R}^3,$$

as shown in Figure 6.5 (iii), (iv), and define $g \colon D^2 \to \mathbf{R}^3$ in such a way that

$$g(D^2) = g_1(D^2) \cup g_2(D^2).$$

The same argument applies to more complicated knots l.

Let $(U_\lambda, \varphi_\lambda)$ be a distinguished chart; then $\varphi_\lambda(g(D^2) \cap U_\lambda)$ is a surface in the open cube $(-1, 1)^3$ (Figure 6.6 (i)). We may assume that $g(D^2)$ is in a general position with respect to $\mathcal{F}$, perturbing $g(D^2)$ ($g|\partial D^2$ stays fixed) slightly if necessary so that $\varphi_\lambda(g(D^2) \cap U_\lambda) \cap \hat{\pi}^{-1}(t)$, $-1 < t < 1$, is locally one of the following, that is, given $x \in U_\lambda$ there is a small open set $U_{\lambda, x}$, $x \in U_{\lambda, x} \subset U_\lambda$, and given $p \in D^2$ there is a small neighborhood U of p such that $\varphi_\lambda(g(U) \cap U_{\lambda, x}) \cap \hat{\pi}^{-1}(t)$ is one of the following.

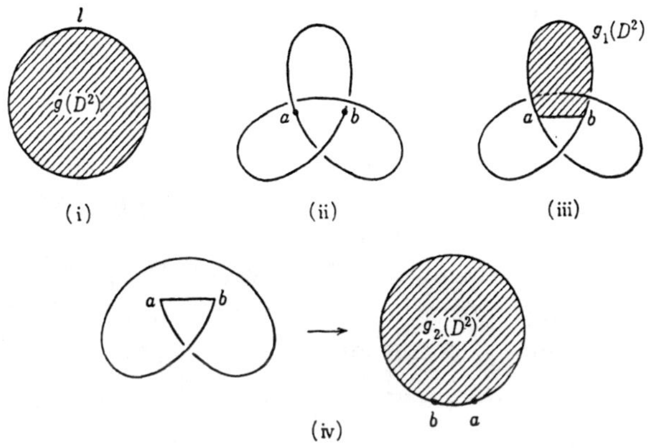

FIGURE 6.5

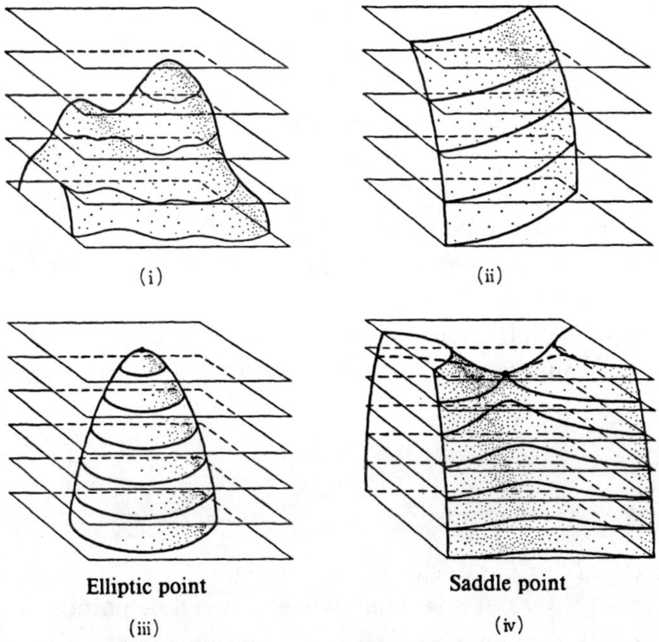

FIGURE 6.6

(i) The empty set.

(ii) A C^r curve (Figure 6.6 (ii)).

(iii) A single point p_0 in a neighborhood of which $\varphi_\lambda(g(U) \cap U_{\lambda,x})$ is ellipsoidal with the elliptic point p_0 as shown in Figure 6.6 (iii).

(iv) Two C^r curves intersecting in a point q_0 in a neighborhood of which the surface $\varphi_\lambda(g(U) \cap U_{\lambda,x})$ is hyperboloidal with the saddle point q_0 as shown in Figure 6.6 (iv).

Suppose now that for a given distinguished chart $(U_\lambda, \varphi_\lambda)$, we have positioned $g(D^2)$ so that $\varphi_\lambda(g(D^2)\cap U_\lambda)\cap\hat{\pi}^{-1}(t)$, $-1<t<1$, is locally one of (i), (ii), (iii), or (iv), and suppose that (U_μ, φ_μ) is a another distinguished chart with $U_\lambda \cap U_\mu \neq \varnothing$; then the above situations hold for $g(D^2)\cap U_\lambda \cap U_\mu$. Hence if we position $g(D^2)$ in such a way that (i), (ii), (iii), or (iv) hold for a finite number of distinguished charts covering $g(D^2)$, these conditions hold for any distinguished chart $(U_\lambda, \varphi_\lambda)$.

The family $\{g^{-1}(L_\alpha); L_\alpha \in \mathcal{F}\}$ of subsets in D^2 gives us a 'pattern' as shown in Figure 6.7. The pattern in Figure 6.7 consists of C^r curves, the points, c_1, c_2, c_3, and c_4, corresponding to the elliptic points in $g(D^2)$, and the points, c_1', c_2', and c_3', corresponding to the saddle points in $g(D^2)$. Through a slight perturbation of $g(D^2)$ if necessary, we may assume each leaf of $\mathcal{F}$ has no more than one saddle point, and there is no C^r curve joining two points in D^2 corresponding to saddle points.

Now choose a C^{r-1} vector field X on D^2 which satisfies the following.

(i) The singular points of X correspond to the elliptic points or the saddle points in $g(D^2)$.

(ii) The trajectories of X without singular points are the connected components of $g^{-1}(L_\alpha)$, $L_\alpha \in \mathcal{F}$—the connected components of $g^{-1}(L_\alpha)$ minus {the points in (i)}, to be a bit more precise.

(iii) X points inward along ∂D^2 (Figure 6.7).

Notice that a pattern as above can be generated by a vector field because D^2 is simply connected.

Let p be a point of D^2, and consider the ω-limit set $L^+(p)$ of the integral curve $\varphi_{\{p\}}$ of X through p. Since there is no way that $\varphi_{\{p\}}(t)$ goes outside D^2, $t>0$, we have

$$L^+(p)\subset D^2.$$

If $p'\in\partial D^2$, $L^+(p')$ is one of the following.

(a) $L^+(p')$ consists of one point, which is a saddle point.

(b) $L^+(p')$ is a subset of D^2 with at least two points.

In case (a), $L^+(p')$ is like c_2' in Figure 6.7, and one of the two C^r curves issuing from c_2' might come back to c_2'—but even then the other curve will never go to a saddle point. The ω-limit set of this curve is thus a subset of D^2 with at least two points. Hence there is a point $\hat{p}$ in D^2 such that $L^+(\hat{p})$ consists of at least two points.

By the Poincaré-Bendixson theorem (Theorem 3.3), if $L^+(\hat{p})$ has no singular points then $L^+(\hat{p})$ is a periodic trajectory (the swirl on the right in Figure 6.7). If $L^+(\hat{p})$ has a singular point, by an argument similar to that used in the proof of Theorem 3.3, $L^+(\hat{p})$ consists of a point—corresponding to a saddle point in $g(D^2)$—and one or both of the two curves issues from

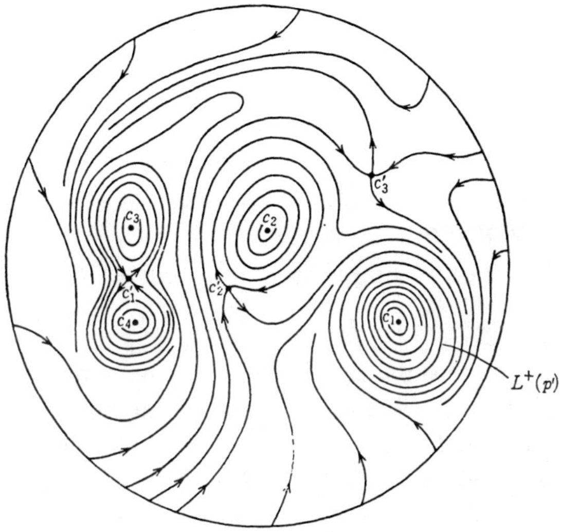

FIGURE 6.7

and returns to it (the point c_1' in Figure 6.7). In any case, we have

$$L^+(\hat{p}) \subset g^{-1}(L_\alpha),$$

for some leaf L_α and $\varphi_{\{\hat{p}\}}$ spirals toward $L^+(\hat{p})$ from the outside.

Fix an orientation on S^1, and choose a C^0 map

$$h: S^1 \to D^2 \tag{1}$$

such that $h(S^1) = L^+(\hat{p})$ and h maps S^1 in the direction of the arrow on the trajectory in $L^+(\hat{p})$. Then the map

$$g \circ h: S^1 \to L_\alpha$$

is not homotopic to zero on L_α. To see this, suppose $g \circ h$ were null-homotopic. Then by the holonomy lemma (Theorem 6.3) there exists a family of continuous maps

$$h_t : S^1 \to S^3, \qquad -\epsilon < t < \epsilon,$$

such that h_t varies continuously with respect to t, $h_t(x)$ is transverse to the leaves for a fixed $x \in S^1$, $h_t(S^1)$ is on a single leaf $L_{\alpha(t)}$ of $\mathcal{F}$ and $h_0 = g \circ h$. Choose $\mathcal{F}'$ in the proof of Theorem 6.3 suitably—special care is necessary when $L^+(\hat{p})$ contains a saddle point—so that

$$h_t(S^1) \subset g(D^2) \cap L_{\alpha(t)}, \qquad -\epsilon < t \leq 0,$$

and $g^{-1}(h_t(S^1))$ is outside $L^+(\hat{p})$, $-\epsilon < t < 0$. This shows that just outside of $L^+(\hat{p})$ the 'pattern' of D^2 is concentric. But this contradicts the fact that $\varphi_{\{\hat{p}\}}$ spirals toward $L^+(\hat{p})$ from the outside. Hence $g \circ h$ is not null-homotopic on the leaf L_α. Hence if $L^+(\hat{p})$ is a periodic trajectory or if $L^+(\hat{p})$ consists of a saddle point and a curve issuing from and returning to this point, $L^+(\hat{p})$ ($g|L^+(\hat{p})$ to be more precise) is not null-homotopic on the leaf L_α, and if $L^+(\hat{p})$ consists of a saddle point and two curves issuing from and returning to this point, the image of one of the two loops under g is not null-homotopic on L_α.

Let H be the set of all simple closed C^0 curves $f: S^1 \to D^2$ such that

$$g \circ f: S^1 \to g(D^2) \cap L_\alpha$$

is not homotopic to zero on L_α. Then from the above discussion, we have that $H \neq \varnothing$. In the following we will identify all curves in H whose images are identical. By the Jordan curve theorem $f(S^1)$ bounds a ball in D^2, which will be denoted by $D(f)$, $f \in H$. Notice that both of the sets $D(f)$ and $D^2 - D(f)$ are unions of trajectories. We order H by $\prec$ as follows: for the curves f_1, $f_2 \in H$, $f_1 \prec f_2$ if and only if $D(f_1) \subset D(f_2)$.

We shall first show that $(H, \prec)$ is an inductive ordered set. Suppose that a subset $\{f_\sigma; \sigma \in \Sigma\}$ of H satisfies the following.

$$\text{If } \sigma, \sigma' \in \Sigma, \text{ then } f_\sigma \prec f_{\sigma'} \text{ or } f_{\sigma'} \prec f_\sigma.$$

Then the set

$$D_\Sigma = \bigcap_{\sigma \in \Sigma} D(f_\sigma)$$

is the intersection of compact sets with the finite intersection property, and so it is a nonempty closed set.

The set D_Σ contains more than one point. For, if D_Σ were a singleton set, then this point would have to correspond to an elliptic point—but in a neighborhood of an elliptic point, $g(D^2)$ looks like (iii) in Figure 6.6, in which every $g(D^2) \cap L_\alpha$ is homotopic to zero. This contradicts the definition of D_Σ.

Evidently D_Σ is a union of trajectories, and it is easy to prove that $\operatorname{Int} D_\Sigma$ is also a union of trajectories (cf. Theorem 3.1). Thus $D_\Sigma - \operatorname{Int} D_\Sigma$ is a union of trajectories; we want to show that indeed it is one of the following.

(a) A simple closed curve.

(b) The union of a saddle point and a trajectory issuing from and returning to it.

(c) The union of a saddle point and two trajectories issuing from and returning to it.

Suppose that $D_\Sigma - \operatorname{Int} D_\Sigma$ contains a trajectory $C(p)$, $p \in D_\Sigma - \operatorname{Int} D_\Sigma$, other than (a), (b), or (c). By the argument employed above, either $L^+(p)$ or $L^-(p)$ is of the form (a), (b), or (c). Assume that $L^+(p)$ is such; then as $C(p) \subset D_\Sigma - \operatorname{Int} D_\Sigma$, we also have

$$L^+(p) \subset D_\Sigma - \operatorname{Int} D_\Sigma.$$

By assumption

$$D_\Sigma - \operatorname{Int} D_\Sigma - L^+(p) \neq \varnothing, \tag{2}$$

but by the method of the proof used for Lemma 1.16 we see that (2) contradicts the definitions of D_Σ and $L^+(p)$. Hence $D_\Sigma - \operatorname{Int} D_\Sigma$ must be one of (a), (b), or (c).

Thus we can take a closed C^0 curve $\hat{f}\colon S^1 \to D^2$ such that $\hat{f}(S^1) = D_\Sigma - \operatorname{Int} D_\Sigma$, in exactly the same way as we defined the map $h\colon S^1 \to D^2$ in (1). Consider $g \circ \hat{f}\colon S^1 \to S^3$; $g \circ \hat{f}(S^1)$ is on L_β, but $g \circ \hat{f}$ is not null-homotopic on L_β. For suppose $g \circ \hat{f}$ were homotopic to zero; then by the holonomy lemma we can show that for every $\sigma \in \Sigma$ with $D_\Sigma \subset D(f_\sigma)$, $D(f_\sigma) - D_\Sigma$ contains a common nonempty set, which would contradict the definition of D_Σ. We take a simple closed C^0 curve

$$\hat{f}'\colon S^1 \to D^2$$

such that $\hat{f}'(S^1) \subset D_\Sigma - \operatorname{Int} D_\Sigma$ and $g \circ \hat{f}'$ is not null-homotopic on the leaf containing $g \circ \hat{f}'(S^1)$: for instance we let $\hat{f}' = \hat{f}$ if $D_\Sigma - \operatorname{Int} D_\Sigma$ is (a) or (b), and let $\hat{f}'$ be one of the two trajectories whose image under g is not null-homotopic in the case of (c). Clearly $\hat{f}' \in H$ and $\hat{f}' \prec f_\sigma$ for every f_σ, $\sigma \in \Sigma$. Hence we have shown that $(H, \prec)$ is an inductive set.

By Zorn's lemma, $(H, \prec)$ has a minimal element, say f_0. We claim that the map $g \circ f_0$ is a vanishing cycle. A trajectory contained in $\operatorname{Int} D(f_0)$ is either a simple closed curve, a trajectory issuing from and returning to a saddle point, a vertex, or a saddle point. For suppose there is a trajectory in $\operatorname{Int} D(f_0)$ which is different from any of the above; then we can construct $f_0' \in H$ with $f_0' \prec f_0$ from a limit set of this trajectory just as we did earlier. Since the number of saddle points is finite, we can choose a neighborhood U_0 of $\partial D(f_0)$ in $D(f_0)$ so small that the trajectories which intersect $U_0 \cap \operatorname{Int} D(f_0)$ are all periodic. If we denote these periodic trajectories by f_t, $0 \le t \le \epsilon$, we see by the minimality of f_0 that each $g \circ f_t$ is null-homotopic on the leaf containing $g \circ f_t(S^1)$. Hence $g \circ f_0$ is a vanishing cycle.

LEMMA 6.7. *Let $\mathcal{F}$ be a C^r codimension one foliation of S^3, $r \ge 2$, and let L_0 be a leaf of $\mathcal{F}$. Suppose that a map $f_0\colon S^1 \to L_0$ is a vanishing cycle on L_0. Then there exist a leaf L_0' arbitrarily close to L_0 and a family of closed C^0 curves*

$$g_t\colon S^1 \to S^3, \qquad 0 \le t \le \epsilon,$$

which vary continuously in t and satisfy the following.

(i) *$g_t(S^1)$ is on a leaf $L'_{\alpha(t)}$. Here $L'_{\alpha(0)} = L_0'$.*

(ii) *For a fixed x, $g_t(x)$ is a C^r curve with parameter t, $0 \le t \le \epsilon$, which is transverse to the leaves of $\mathcal{F}$. If for $x, y \in S^1$, $g_0(x) \neq g_0(y)$, then $g_t(x) \neq g_t(y)$, $0 \le t \le \epsilon$.*

(iii) g_0 *is a vanishing cycle:* g_0 *is not null-homotopic on* L'_0 *but* g_t *is null-homotopic on* $L'_{\alpha(t)}$ *for* $0 < t \le \epsilon$.

(iv) *Let* $\tilde{L}'_{\alpha(t)}$ *be the universal cover of* $L'_{\alpha(t)}$ *and* $\tilde{g}_t : S^1 \to \tilde{L}'_{\alpha(t)}$ *the lift of* g_t. *Then* $\tilde{g}_t$ *is a simple closed curve* (cf. Appendix VI, Note 1).

PROOF. Varying f_0 slightly if necessary, we may assume that $f_0 : S^1 \to L_0$ is a C^r immersion and that if $f_0(x') = f_0(x'')$ for some x', $x'' \in S^1$, the paths

$$\{f_0(x); x' - \delta < x < x' + \delta\} \quad \text{and} \quad \{f_0(x); x'' - \delta < x < x'' + \delta\}$$

intersect transversely at $f_0(x') = f_0(x'')$. Since $f_0 : S^1 \to L_0$ is a vanishing cycle, there exists a family of closed C^0 curves $f_t : S^1 \to S^3$ varying continuously in t, $0 \le t \le \epsilon'$, which satisfy the conditions (i), (ii), and (iii) in the definition of vanishing cycles. Here we take ϵ' sufficiently small and readjust the parameter t if necessary so that in (ii) we have: if for x, $y \in S^1$ $f_0(x) \neq f_0(y)$ then $f_t(x) \neq f_t(y)$, $0 \le t \le \epsilon'$.

Since f_t is homotopic to zero on $L_{\alpha(t)}$, the lift $\tilde{f}_t$ of $f_t : S^1 \to L_{\alpha(t)}$ to $\tilde{L}_{\alpha(t)}$ is a map

$$\tilde{f}_t : S^1 \to \tilde{L}_{\alpha(t)}, \qquad 0 < t \le \epsilon'.$$

There is a finite number w of pairs (x, y), x, $y \in S^1$ with $x \neq y$, such that $\tilde{f}_t(x) = \tilde{f}_t(y)$, for some t, $0 < t \le \epsilon'$. Now let (x, y) be one of these pairs, and define U and K by

$$U = \{t \in (0, \epsilon']; \tilde{f}_t(x) = \tilde{f}_t(y)\}, \qquad K = \{t \in [0, \epsilon']; f_t(x) = f_t(y)\}.$$

Then trivially $\varnothing \neq U \subset K$. Notice that $0 \in K$ but $0 \notin U$. Further by the continuity of f_t with respect to t, K is a closed set in $[0, \epsilon']$.

We show next that U is open in the closed interval $[0, \epsilon']$. Were U not open, there would exist a $\hat{t}$ in U such that for $t \in (0, \epsilon']$ arbitrarily close to $\hat{t}$ we have $\tilde{f}_t(x) \neq \tilde{f}_t(y)$. Choose a base point x_0 in S^1 and let xy be the arc containing x_0. Then $f_{\hat{t}}|xy$ is a closed curve of $L_{\alpha(\hat{t})}$, and since we have $\tilde{f}_{\hat{t}}(x) = \tilde{f}_{\hat{t}}(y)$, it is null-homotopic on $L_{\alpha(\hat{t})}$. Now let

$$F : xy \times I \to L_{\alpha(\hat{t})}, \qquad F|xy \times \{0\} = f_{\hat{t}}|xy$$

be a homotopy between $f_{\hat{t}}|xy$ and 0. Then since F is null-homotopic on the leaf $L_{\alpha(\hat{t})}$, by the holonomy lemma (Theorem 6.3) there exists a family of maps

$$F_\tau : xy \times I \to S^3, \qquad \hat{t} - \epsilon'' < \tau < \hat{t} + \epsilon'',$$

varying continuously in τ such that $F_\tau(xy \times I)$ is on a leaf $L_{\beta(\tau)}$ and $F_{\hat{t}} = F$. This implies that $f_t : xy \to L_{\alpha(t)}$ is null-homotopic on $L_{\alpha(t)}$ and that $\tilde{f}_t(x) = \tilde{f}_t(y)$ for t sufficiently close to $\hat{t}$. But this is a contradiction. Hence U is an open set in K.

We now take $(t', t'') \subset U$, $t' \notin U$. Since K is closed, $t' \in K$. One of the closed curves

$$f_{t'}|xy, \qquad f_{t'}|yx$$

is not null-homotopic on $L_{\alpha(t')}$. For if $t' = 0$ we know that f_0 is not null-homotopic on L_0, and if $t' > 0$, $\tilde{f}_{t'}(x) \neq \tilde{f}_{t'}(y)$ implies that $f_{t'}|xy$ is not null-homotopic on $L_{\alpha(t')}$.

Define a family of maps

$$f'_t \colon S^1 \to L_{\alpha(t'+t)}, \qquad 0 \le t \le t'' - t'$$

by

$$f'_t = \begin{cases} f_{t'+t}|xy, & \text{if } f_{t'}|xy \text{ is not null-homotopic on } L_{\alpha(t')}, \\ f_{t'+t}|yx, & \text{if } f_{t'}|yx \text{ is not null-homotopic on } L_{\alpha(t')}. \end{cases}$$

Then f'_0 is a vanishing cycle, and the number of pairs (x, y), $x, y \in S^1$ with $x \neq y$, such that $\tilde{f}'_t(x) = \tilde{f}'_t(y)$ for some $0 < t \le t'' - t'$, is less than w. Since ϵ' can be as small as we want, we can continue lowering w until we get $g_t \colon S^1 \to S^3$, $0 \le t \le \epsilon$ which satisfy (iv).

LEMMA 6.8. *For the family of curves* $g_t \colon S^1 \to S^3$, $0 \le t \le \epsilon$, *of Lemma* 6.7, there exists a C^0 immersion

$$G \colon (0, \hat{\epsilon}] \times D^2 \to S^3, \qquad 0 < \hat{\epsilon} < \epsilon$$

which satisfies the following.

(i) $G(t, x) = g_t(x)$ if $x \in S^1$.

(ii) $G(\{t\} \times D^2) \subset L'_{\alpha(t)}$.

(iii) With $x \in D^2$ fixed, $G(t, x)$ is a C^r curve transverse to the leaves, $0 < t \le \hat{\epsilon}$.

(iv) Let V be a subset of D^2 consisting of all x for which $\lim_{t \to 0} G(t, x)$ exists, then V is an open set in D^2 which contains S^1 and $V \neq D^2$ (see Figure 6.8).

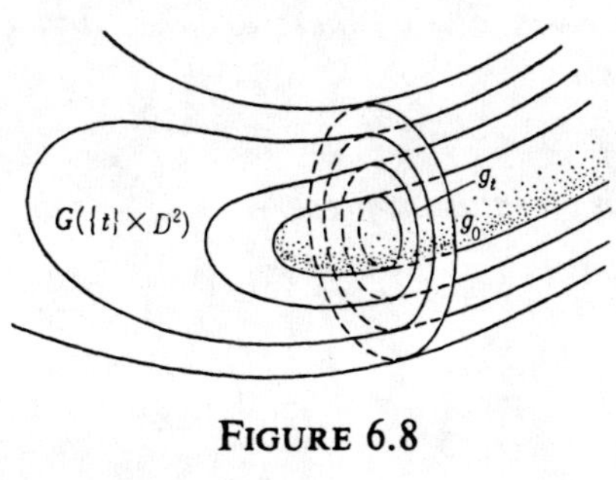

FIGURE 6.8

PROOF. The universal covering space $\tilde{L}_\alpha$ of a leaf L_α is a C^r two-manifold; so $\tilde{L}_\alpha$ is S^2 or $\mathbf{R}^2$. If $\tilde{L}_\alpha$ is S^2, L_α is embedded in S^3 and orientable, so that $L_\alpha = S^2$; however, if a leaf L_α is S^2, by Theorem 5.11 $\mathcal{F}$ is a bundle foliation given by a C^r bundle whose base space is S^1 with the fiber S^2. But then the total space cannot be S^3. Therefore $\tilde{L}_\alpha = \mathbf{R}^2$.

Choose $\hat{\epsilon}$ with $0 < \hat{\epsilon} < \epsilon$. By the Jordan curve theorem, the image of the map $\tilde{g}_{\hat{\epsilon}}\colon S^1 \to \tilde{L}'_{\alpha(\hat{\epsilon})} = \mathbf{R}^2$ bounds a ball D^2; therefore there exists a C^0 embedding $\tilde{G}_{\hat{\epsilon}}\colon D^2 \to \tilde{L}'_{\alpha(\hat{\epsilon})}$ such that $\tilde{G}_{\hat{\epsilon}}|S^1 = \tilde{g}_{\hat{\epsilon}}$. Let $\pi\colon \tilde{L}'_{\alpha(\hat{\epsilon})} \to L'_{\alpha(\hat{\epsilon})}$ be the projection and define

$$G_{\hat{\epsilon}}\colon D^2 \to L'_{\alpha(\hat{\epsilon})}$$

by $G_{\hat{\epsilon}}(x) = \pi \circ \tilde{G}_{\hat{\epsilon}}(x)$. The map $G_{\hat{\epsilon}}$ is a C^0 immersion. By the nolonomy lemma (Theorem 6.3) applied to $G_{\hat{\epsilon}}\colon D^2 \to L'_{\alpha(\hat{\epsilon})}$, there exists a family of C^0 immersions

$$G_t\colon D^2 \to S^3, \qquad \epsilon_1 < t \le \hat{\epsilon},$$

such that $G_t(D^2)$ is contained in a single leaf. Here $0 \le \epsilon_1 < \hat{\epsilon}$.

Now define

$$G\colon (\epsilon_1, \hat{\epsilon}] \times D^2 \to S^3$$

by $G(t, x) = G_t(x)$. Then G is a C^0 immersion—Theorem 6.3 (iv)—and $G(t, x) = g_t(x)$ for $x \in S^1$. If $\epsilon_1 \neq 0$, we apply the above procedure to the map $\tilde{g}_{\epsilon_1}\colon S^1 \to L'_{\alpha(\epsilon_1)}$, and extend the immersion $G\colon (\epsilon_1, \hat{\epsilon}] \times D^2 \to S^3$ to an immersion $G\colon (\epsilon_2, \hat{\epsilon}] \times D^2 \to S^3$. Here $\epsilon_2 < \epsilon_1$. By repeating this procedure we can extend G to $G\colon (0, \hat{\epsilon}] \times D^2 \to S^3$. The reader will have little trouble in seeing that G is a C^0 immersion and satisfies the properties (i), (ii), and (iii).

By the continuity of the map G it is evident that V is open in D^2, and it is equally obvious that V contains S^1. Further suppose that $V = D^2$ and define a map $G_0\colon D^2 \to L'_{\alpha(0)}$ by $G_0(x) = \lim_{t\to 0} G(t, x)$; then through the map G_0, the map g_0 becomes null-homotopic on $L'_{\alpha(0)}$, which contradicts our assumption. Hence we have established (iv).

LEMMA 6.9. *Let $G\colon (0, \hat{\epsilon}] \times D^2 \to S^3$ be the C^0 immersion constructed in Lemma* 6.8. *Then given $\delta > 0$, there exist ϵ' and ϵ'', $0 < \epsilon' < \epsilon'' < \delta$, which satisfy the following.*

(i) $L'_{\alpha(\epsilon')} = L'_{\alpha(\epsilon'')}$.

(ii) *There exists an orientation-preserving C^0 embedding*

$$\hat{h}\colon D^2 \to \operatorname{Int} D^2$$

such that

$$G(\epsilon'', x) = G(\epsilon', \hat{h}x).$$

Proof. Let V be the set in Lemma 6.8 (iv), and fix $x \in D^2 - V$. Then we have a C^r curve $G(t, x)$, $0 < t \leq \hat{\epsilon}$, of infinite length which intersects the leaves transversely. Select a sequence of points in $(0, \epsilon']$

$$t_1' > t_2' > \cdots > t_i' > \cdots, \qquad \lim_{i\to\infty} t_i' = 0$$

such that $G(t_1', x)$, $G(t_2', x)$, ... converges, and set

$$\lim_{i\to\infty} G(t_i', x) = z, \qquad z \in S^3.$$

Let L' be the leaf containing z. Since the curve $G(t, x)$, $0 < t \leq \hat{\epsilon}$, is transverse to the leaves, for a sufficiently large i we can take t_i near t_i' such that $G(t_i, x) \in L'$ and $\lim_{i\to\infty} t_i = 0$.

Consider the Riemannian metric on L' induced by the standard Riemannian metric on S^3, so that one has a distance on L'. If $y \in V$ we have $\lim_{i\to\infty} G(t_i, y) \in L'_{\alpha(0)}$; so the C^0 curves on L':

$$\{ G(t_i, y); y \in S^1 \}, \qquad i = 1, 2, \ldots,$$

are apart from each other with respect to this metric. Hence for a small enough t_i we have that $z \in G(\{t_i\} \times D^2)$. Let $\tilde{L}'$ be the universal cover of the leaf L', $\hat{z} \in \tilde{L}'$ one of the lifts of z and denote by $\tilde{G}_{t_i}: D^2 \to \tilde{L}'$ the lift of $G|\{t_i\} \times D^2$ containing $\hat{z}$ in its image. Then we saw in the proof of Lemma 6.8 that $\tilde{G}_{t_i}$ is a C^0 embedding, and $\tilde{G}_{t_i}(D^2)$ can contain only finitely many $\tilde{G}_{t_j}(D^2)$'s. Hence there is a number t_k, $t_k < t_i$, such that the set inclusion $\tilde{G}_{t_k}(D^2) \subset \tilde{G}_{t_i}(D^2)$ holds. Define $\hat{h}$ by $\tilde{G}_{t_k}(\hat{h}(x)) = \tilde{G}_{t_i}(x)$ to complete the proof.

Lemma 6.10. *There exists no closed C^1 curve on S^3 transverse to the leaves of $\mathcal{F}$ which intersects L_0'.*

Proof. Let us assume the existence of a closed C^1 curve $l: S^1 \to S^3$ transverse to the leaves of $\mathcal{F}$ such that $l(S^1) \cap L_0' \neq \varnothing$. Since $\mathcal{F}$ is transversely orientable and L_0' is also orientable, we may assume that l passes through L_0' always from the back to the front of the leaf.

Let $g_t: S^1 \to S^3$, $0 \leq t \leq \epsilon$, be the family of closed curves in Lemma 6.7. Then $l(S^1) \cap g_0(S^1)$ is finite (possibly empty).

Suppose first that l intersects $g_0(S^1)$ at $l(s_1) = p_1$, $l(s_2) = p_2$, where $s_1 < s_2$ (Figure 6.9). Let $\hat{l}: [0, 1] \to L_0'$ be a C^1 curve such that $\hat{l}(0) = p_1$, $\hat{l}(1) = p_2$ with neither self-intersections nor any other intersections with l. Using the holonomy lemma (Theorem 6.3) construct a family of C^1 curves

$$\hat{l}_t: [0, 1] \to S^3, \qquad -\epsilon < t < \epsilon$$

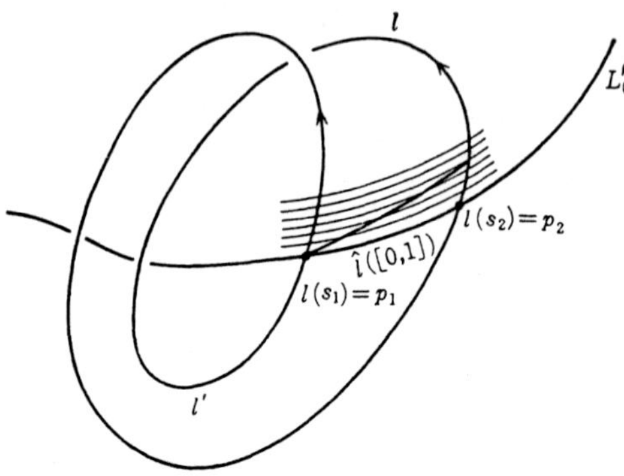

FIGURE 6.9

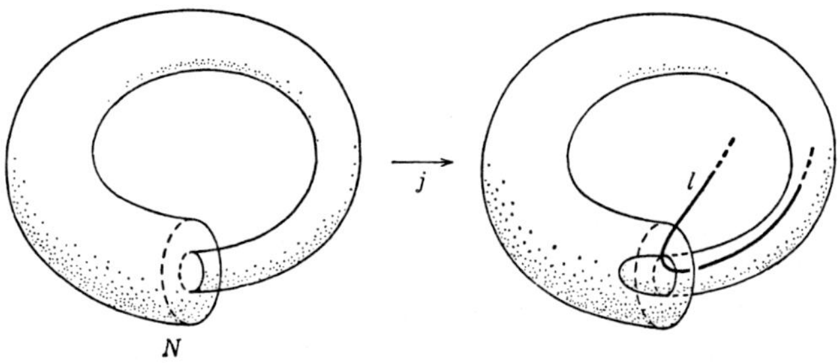

FIGURE 6.10

varying continuously in t, such that $\hat{l}_0 = \hat{l}$, $\hat{l}_t([0, 1])$ sits on a leaf, and with the parameter of l readjusted $\hat{l}_t(0) = l(s_1 + t)$, $\hat{l}_t(1) = l(s_2 + t)$ (Figure 6.9). By smoothing out the corners of the broken curve

$$l(s_2 + \epsilon/2, s_1) \cup \{\, \hat{l}_t(2t/\epsilon) ; 0 \le t \le \epsilon/2 \}$$

at p_1 and $l(s_2 + (\epsilon/2))$ we obtain a C^1 curve l' (Figure 6.9). The number of intersections of l' and $g_0(S^1)$ is one less than the number of the intersections of l and $g_0(S^1)$, so repeating this process if necessary, we may assume that the curves l and $g_0(S^1)$ meet exactly in one point. A similar argument also enables us to choose l so that $l(s_0)$ is a preassigned point on $g_0(S^1)$. Consequently we may assume that the intersection $l(S^1) \cap g_0(S^1)$ is a singleton set $\{l(s_0) = g_0(S^1)\}$. Modifying l a bit more using the holonomy lemma once again, we may further assume that

$$g_t(x_0) = l(s_0 + t).$$

Then we can choose $\hat{\epsilon}$ sufficiently small so that l and $G((0, \hat{\epsilon}\,] \times S^1)$ meet only in $G((0, \hat{\epsilon}\,] \times \{x_0\})$, where G is the immersion of Lemma 6.8.

Choose ϵ', ϵ'' as in Lemma 6.9, and denote by N the C^0 manifold obtained from $[\epsilon', \epsilon''] \times D^2$ by identifying (ϵ'', x) with $(\epsilon', \hat{h}(x))$, $x \in D^2$,

(Figures 6.8, 6.10). Define $j : N \to S^3$ by $j(t, x) = G(t, x), x \in D^2$, $\epsilon' \leq t \leq \epsilon''$. Then j is a C^0 immersion. We notice that l enters $j(N)$ through $G((0, \hat{\epsilon}] \times \{x_0\})$. But $\partial N = ([\epsilon', \epsilon''] \times S^1) \cup (\{\epsilon'\} \times (D^2 - \hat{h}(D^2)))$, and since $j(\{\epsilon'\} \times (D^2 - \hat{h}(D^2))) \subset L'_{\alpha(\epsilon')}$, the curve l would not leave $j(N)$ through this portion as it passes through the leaf from the back to the front. Also $j([\epsilon', \epsilon''] \times S^1)$ intersects l only in $j([\epsilon', \epsilon''] \times \{x_0\})$. Hence l cannot leave $j(N)$. But this contradicts the assumption that l is a closed curve.

PROOF OF THEOREM 6.5. By Theorem 6.4, the leaf L'_0 in Lemma 6.10 is compact.

We can extend Theorem 6.5 to the following

THEOREM 6.11. *Let M^3 be a closed three-dimensional C^∞ manifold with a finite fundamental group and $\mathcal{F}$ a C^r codimension one foliation of M^n, $r \geq 2$. Then $\mathcal{F}$ must contain a compact leaf.*

For the proof, it suffices to apply the method of proof used in Theorem 6.5 to the universal cover of M^3. Recall the C^∞ foliation $\mathcal{F}_{a,b} = \{L_\alpha ; \alpha \in A\}$ of $S^1 \times S^1$ in Example A, §16. Let $\mathcal{F}' = \{L_\alpha \times S^1 ; \alpha \in A\}$. Then $\mathcal{F}'$ is a C^∞ codimension one foliation of $S^1 \times S^1 \times S^1$. If b/a is irrational $\mathcal{F}'$ has no compact leaves. Hence the existence of a compact leaf requires the finiteness assumption on $\pi_1(M^3)$.

Let L be a compact leaf in Theorem 6.5. Then by Theorem 4.1 there exists a C^{r-1} vector field on S^3 transverse to $\mathcal{F}$; hence the Euler number $\chi(L)$ of L must be zero. Thus such an L is always homeomorphic to $S^1 \times S^1$. A similar statement applies to compact leaves in Theorem 6.11.

CHAPTER VII

Foliations and Differential Forms

§26. Differential forms

Let V be an n-dimensional vector space over $\mathbf{R}$. A map $\varphi: V \to \mathbf{R}$ is called a *one-form* or a *linear form* on V if φ is linear, that is

$$\varphi(\lambda v + \mu w) = \lambda\varphi(v) + \mu\varphi(w), \qquad v, w \in V, \quad \lambda, \mu \in \mathbf{R}.$$

We denote by V^* the set of all one-forms on V. For $\varphi, \varphi' \in V^*$, $\lambda, \mu \in \mathbf{R}$ we define a linear $\lambda\varphi + \mu\varphi' \in V^*$ by

$$(\lambda\varphi + \mu\varphi')(v) = \lambda\varphi(v) + \mu\varphi'(v), \qquad v \in V.$$

Then the set V^* becomes a vector space called the *dual space* of V.

For a basis $e_1, e_2, \ldots, e_n$ in V, we define one-forms $\beta_1, \beta_2, \ldots, \beta_n$ in V^* by

$$\beta_j(e_i) = \delta_{ij}, \qquad i, j = 1, 2, \ldots, n.$$

Then the $\beta_1, \beta_2, \ldots, \beta_n$ form a basis in the dual space V^*. Thus V^* is an n-dimensional vector space. We call the $\beta_1, \beta_2, \ldots, \beta_n$ the *dual basis* of the basis $e_1, e_2, \ldots, e_n$.

Let f be a map defined on the direct product of q copies of V, $V \times V \times \cdots \times V = \{(v_1, v_2, \ldots, v_q); v_i \in V\}$, to $\mathbf{R}$:

$$f: V \times V \times \cdots \times V \to \mathbf{R}.$$

If f is linear in each component:

$$\begin{aligned} f(v_1, v_2, \ldots, v_{i-1}, \lambda v_i + \mu w_i, v_{i+1}, \ldots, v_q) \\ = \lambda f(v_1, v_2, \ldots, v_{i-1}, v_i, v_{i+1}, \ldots, v_q) \\ + \mu f(v_1, v_2, \ldots, v_{i-1}, w_i, v_{i+1} \ldots, v_q), \end{aligned}$$

we say that f is a *q-form* on V. Denote by $T^{(q)}(V^*)$ the set of all q-forms on V. For $f, f' \in T^{(q)}(V^*)$ and $\lambda, \mu \in \mathbf{R}$, define $\lambda f + \mu f' \in T^{(q)}(V^*)$ by

$$(\lambda f + \mu f')(v_1, v_2, \ldots, v_q) = \lambda f(v_1, v_2, \ldots, v_q) + \mu f'(v_1, v_2, \ldots, v_q).$$

With these operations, the set $T^{(q)}(V^*)$ becomes a vector space over $\mathbf{R}$. We observe that $T^{(1)}(V^*) = V^*$. We set $T^{(0)}(V^*) = \mathbf{R}$.

Using the dual basis β_j, $j = 1, 2, \ldots, n$, we define

$$\beta_{j_1 j_2 \cdots j_q} \in T^{(q)}(V^*), \qquad 1 \le j_k \le n,\ k = 1, 2, \ldots, q$$

by

$$\beta_{j_1 j_2 \cdots j_q}(v_1, v_2, \ldots, v_q) = \beta_{j_1}(v_1) \cdot \beta_{j_2}(v_2) \ldots \beta_{j_q}(v_q).$$

Then the $\beta_{j_1 j_2 \cdots j_q}$, $1 \le j_k \le n$, $k = 1, 2, \ldots, q$, form a basis in $T^{(q)}(V^*)$. Thus the vector space $T^{(q)}(V^*)$ is n^q-dimensional.

For $f \in T^{(q)}(V^*)$ and $g \in T^{(s)}(V^*)$, define an $(q+s)$-form $f \otimes g$ in the vector space $T^{(q+s)}(V^*)$ by

$$(f \otimes g)(v_1, v_2, \ldots, v_{q+s}) = f(v_1, v_2, \ldots, v_q) \cdot g(v_{q+1}, v_{q+2}, \ldots, v_{q+s}).$$

The $q+s$-form $f \otimes g$, the *tensor product* of f and g, satisfies the following

$$(\lambda f + \mu f') \otimes g = \lambda(f \otimes g) + \mu(f' \otimes g),$$
$$f \otimes (\lambda g + \mu g') = \lambda(f \otimes g) + \mu(f \otimes g'),$$

and for $h \in T^{(r)}(V^*)$,

$$(f \otimes g) \otimes h = f \otimes (g \otimes h).$$

Set $T^* = \sum_{q=0}^{\infty} T^{(q)}(V^*)$. A typical element of T^* has the form $(f_0, f_1, \ldots, f_q, \ldots)$, where all but finitely many $f_q \in T^{(q)}(V^*)$ are zero; we shall denote this element by $\sum_q f_q$. We define the sum and the scalar product in T^* as follows: for $\sum_q f_q$, $\sum_q g_q \in T^*$, and λ, $\mu \in \mathbf{R}$,

$$\lambda \sum_q f_q + \mu \sum_q g_q = \sum_q (\lambda f_q + \mu g_q).$$

The set T^* becomes a vector space over $\mathbf{R}$ with these operations. Further with the product $\otimes$ defined by

$$(\sum_q f_q) \otimes (\sum_q g_q) = \sum_r (\sum_{q+s=r} f_q \otimes g_s),$$

the vector space T^* becomes an algebra over $\mathbf{R}$.

For the dual basis $\beta_1, \beta_2, \ldots, \beta_n$, the tensor products $\beta_{j_1} \otimes \beta_{j_2} \otimes \cdots \otimes \beta_{j_q}$, $1 \le j_k \le n$, form a basis in $T^{(q)}(V^*)$. Notice that $\beta_{j_1} \otimes \beta_{j_2} \otimes \cdots \otimes \beta_{j_q}$ is nothing but the q-form $\beta_{j_1 j_2 \cdots j_q}$ defined above.

Let S_q be the group of all permutations on q letters, $1, 2, \ldots, q$. For a permutation $\sigma \in S_q$, denote by $\sigma(k)$ the image of k under σ, $1 \le k \le q$. Let $\epsilon_\sigma = \pm 1$ be the sign of permutation σ.

For $f \in T^{(q)}(V^*)$ and $\sigma \in S_q$, define a q-form $\sigma(f)$ in $T^{(q)}(V^*)$ by

$$\sigma(f)(v_1, v_2, \ldots, v_q) = f(v_{\sigma(1)}, v_{\sigma(2)}, \ldots, v_{\sigma(q)}).$$

We say that a q-form $f \in T^{(q)}(V^*)$ is *alternating* if f satisfies

$$\sigma(f) = \epsilon_\sigma f, \qquad \sigma \in S_q.$$

The set of all alternating q-forms is a subspace of $T^{(q)}(V^*)$, which is denoted by $\bigwedge^q V^*$. We set $\wedge^0 V^* = \mathbf{R}$. Since $\wedge^1 V^* = T^{(1)}(V^*) = V^*$, we shall identify $\wedge^1 V^*$ with V^*. When $q > n$, $\bigwedge^q V^* = \{0\}$ by definition.

Define a linear map $A_q : T^{(q)}(V^*) \to T^{(q)}(V^*)$ by

$$A_q(f) = \frac{1}{q!} \sum_{\sigma \in S_q} \epsilon_\sigma \sigma(f).$$

Then for every q-form f in $T^{(q)}(V^*)$, the q-form $A_q(f)$ is alternating.

For $f \in \bigwedge^q V^*$, $g \in \bigwedge^s V^*$, we define the *exterior product* or the *wedge product* $f \wedge g \in \bigwedge^{q+s} V^*$ of f and g as follows:

$$f \wedge g = \frac{(q+s)!}{q!s!} A_{q+s}(f \otimes g).$$

The wedge product satisfies

$$(\lambda f + \mu f') \wedge g = \lambda(f \wedge g) + \mu(f' \wedge g),$$
$$f \wedge (\lambda g + \mu g') = \lambda(f \wedge g) + \mu(f \wedge g'),$$

for $f \in \bigwedge^q V^*$, $g \in \bigwedge^s V^*$, $\lambda \in \mathbf{R}$, and

$$(f \wedge g) \wedge h = f \wedge (g \wedge h),$$

for $h \in \bigwedge^r V^*$. We omit the proof.

Set $\bigwedge V^* = \sum_{q=0}^n \bigwedge^q V^*$, and define the wedge product of $\sum_q f_q$ and $\sum_q g_q$ in $\bigwedge V^*$, where f_q, $g_q \in \bigwedge^q V^*$, by

$$\Big(\sum_q f_q\Big) \wedge \Big(\sum_q g_q\Big) = \sum_r \Big(\sum_{q+s=r} f_q \wedge g_s\Big).$$

Then $\bigwedge V^*$ becomes an algebra over $\mathbf{R}$, the *exterior algebra* of V^*.

The proof of the next lemma routinely follows by induction on q, and we omit it here.

LEMMA 7.1. *For $\varphi_1, \varphi_2, \dots, \varphi_q \in V^*$, we have the identity*

$$(\varphi_1 \wedge \varphi_2 \wedge \cdots \wedge \varphi_q)(v_1, v_2, \dots, v_q) = \begin{vmatrix} \varphi_1(v_1) & \varphi_1(v_2) & \dots & \varphi_1(v_q) \\ \varphi_2(v_1) & \varphi_2(v_2) & \dots & \varphi_2(v_q) \\ \vdots & \vdots & & \vdots \\ \varphi_q(v_1) & \varphi_q(v_2) & \dots & \varphi_q(v_q) \end{vmatrix}, \quad (1)$$

where $v_1, v_2, \dots, v_q \in V$.

Let $\varphi_1, \varphi_2, \dots, \varphi_q$ be one-forms in V^* and $\sigma \in S_q$. From the definition of the tensor product, we get

$$(\varphi_{\sigma(1)} \otimes \varphi_{\sigma(2)} \otimes \cdots \otimes \varphi_{\sigma(q)})(v_1, v_2, \dots, v_q)$$
$$= \varphi_{\sigma(1)}(v_1) \cdot \varphi_{\sigma(2)}(V_2) \cdots \varphi_{\sigma(q)}(v_q),$$

where $v_1, v_2, \dots, v_q \in V$, so

$$\left(\sum_{\sigma\in S_q} \epsilon_\sigma \varphi_{\sigma(1)} \otimes \varphi_{\sigma(2)} \otimes \cdots \otimes \varphi_{\sigma(q)}\right)(v_1, v_2, \dots, v_q) \\ = \sum_{\sigma\in S_q} \epsilon_\sigma \varphi_{\sigma(1)}(v_1)\cdot \varphi_{\sigma(2)}(v_2)\cdots \varphi_{\sigma(q)}(v_q). \tag{2}$$

The right-hand side of (2) is equal to the right-hand side of (1). Hence we obtain the identity

$$\varphi_1 \wedge \varphi_2 \wedge \cdots \wedge \varphi_q = \sum_{\sigma\in S_q} \epsilon_\sigma \varphi_{\sigma(1)} \otimes \varphi_{\sigma(2)} \otimes \cdots \otimes \varphi_{\sigma(q)}. \tag{3}$$

Let $e_1, e_2, \dots, e_n$ be a basis in V and $\beta_1, \beta_2, \dots, \beta_n$ be its dual basis. Let f be an alternating q-form, $f \in \bigwedge^q(V^*)$. Since $f \in \bigwedge^q V^* \subset T^{(q)}(V^*)$, setting $f_{j_1 j_2 \dots j_q} = f(e_{j_1}, e_{j_2}, \dots, e_{j_q})$ we can express f as

$$f = \sum_{j_1, j_2, \dots, j_q} f_{j_1 j_2 \cdots j_q} \beta_{j_1} \otimes \beta_{j_2} \otimes \cdots \otimes \beta_{j_q},$$

which we rewrite as

$$f = \sum_{j_1<j_2<\cdots<j_q} \Big(\sum_{\sigma\in S_q} f_{j_{\sigma(1)} j_{\sigma(2)} \cdots j_{\sigma(q)}} \beta_{j_{\sigma(1)}} \otimes \beta_{j_{\sigma(2)}} \otimes \cdots \otimes \beta_{j_{\sigma(q)}}\Big). \tag{4}$$

Now f is an alternating q-form so that

$$f_{j_{\sigma(1)} j_{\sigma(2)} \cdots j_{\sigma(q)}} = \epsilon_\sigma f_{j_1 j_2 \cdots j_q}.$$

Hence by (3) the term inside the parentheses of (4) is $f_{j_1 j_2 \dots j_q} \beta_{j_1} \wedge \cdots \wedge \beta_{j_q}$. Therefore we get the identity

$$f = \sum_{j_1<j_2<\cdots<j_q} f_{j_1 j_2 \cdots j_q} \beta_{j_i} \wedge \beta_{j_2} \wedge \cdots \wedge \beta_{j_q}.$$

On the other hand, $\binom{n}{q}$ many elements of $\bigwedge^q V^*$, $\beta_{j_1} \wedge \beta_{j_2} \wedge \cdots \wedge \beta_{j_q}$ with $j_1 < j_2 < \cdots < j_q$, are linearly independent, because if

$$\sum_{j_1<j_2<\cdots j_q} \lambda_{j_1 j_2 \cdots j_q} \beta_{j_1} \wedge \beta_{j_2} \wedge \cdots \wedge \beta_{j_q} = 0,$$

then from (3):

$$\sum_{j_1<j_2<\cdots<j_q} \lambda_{j_1 j_2 \cdots j_q} \sum_{\sigma\in S_q} \epsilon_\sigma \beta_{j_{\sigma(1)}} \otimes \beta_{j_{\sigma(2)}} \otimes \cdots \otimes \beta_{j_{\sigma(q)}} = 0,$$

and so $\lambda_{j_1 j_2 \cdots j_q} = 0$.

Thus we have shown that $\binom{n}{q}$ many

$$\beta_{j_1} \wedge \beta_{j_2} \wedge \cdots \wedge \beta_{j_q}, \qquad j_1 < j_2 < \cdots < j_q$$

form a basis in $\bigwedge^q V^*$. Hence the vector space $\bigwedge^q V^*$ is $\binom{n}{q}$-dimensional and consequently the exterior algebra $\bigwedge V^*$ is 2^n-dimensional as a vector space.

By Lemma 7.1, we have the identity

$$\varphi_1 \wedge \varphi_2 \wedge \cdots \wedge \varphi_q \wedge \psi_i = (-1)^q \psi_i \wedge \varphi_1 \wedge \varphi_2 \wedge \cdots \wedge \varphi_q, \tag{5}$$

for $\varphi_1, \varphi_2, \ldots, \varphi_q, \psi_1, \psi_2, \ldots, \psi_s \in V^*$, which implies that

$$\begin{aligned}&(\varphi_1 \wedge \varphi_2 \wedge \cdots \wedge \varphi_q) \wedge (\psi_1 \wedge \psi_2 \wedge \cdots \wedge \psi_s)\\ &\quad = (-1)^{qs} (\psi_1 \wedge \psi_2 \wedge \cdots \wedge \psi_s) \wedge (\varphi_1 \wedge \varphi_2 \wedge \cdots \wedge \varphi_q).\end{aligned}$$

Thus for $f \in \bigwedge^q V^*$, $g \in \bigwedge^s V^*$, we have established the identity

$$f \wedge g = (-1)^{qs} g \wedge f. \tag{6}$$

In particular we have

$$\varphi \wedge \psi = -\psi\varphi, \qquad \varphi \wedge \varphi = 0, \qquad \varphi, \psi \in V^*. \tag{7}$$

The following is an easy consequence of Lemma 7.1; we omit the proof.

LEMMA 7.2. *Suppose elements* $\varphi_1, \varphi_2, \ldots, \varphi_n, \psi_1, \psi_2, \ldots, \psi_n$ *in* V^* *are related by*

$$\psi_i = \sum_{j=1}^n a_{ij}\varphi_j, \qquad i = 1, 2, \ldots, n.$$

Then

$$\psi_1 \wedge \psi_2 \wedge \cdots \wedge \psi_n = \det(a_{ij})\varphi_1 \wedge \varphi_2 \wedge \cdots \wedge \varphi_n.$$

Let V, V' be vector spaces over $\mathbf{R}$ and $F: V \to V'$ a linear map:

$$F(\lambda v + \mu w) = \lambda F(v) + \mu F(w), \qquad v, w \in V, \ \lambda, \mu \in \mathbf{R}.$$

Define a map

$$F^*: T^{(q)}(V'^*) \to T^{(q)}(V^*), \qquad q = 0, 1, 2, \ldots$$

by

$$(F^*(f'))(v_1, v_2, \ldots, v_q) = f'(F(v_1), F(v_2), \ldots, F(v_q)), \qquad v_i \in V,$$

where $f' \in T^{(q)}(V'^*)$. Then F^* is a linear map of the vector space $T^{(q)}(V'^*)$ to the vector space $T^{(q)}(V^*)$. The restriction of F^* to $\wedge^q V'^*$ is a linear map

$$F^*: \bigwedge^q V'^* \to \bigwedge^q V^*, \qquad q = 0, 1, 2, \ldots, \dim V'.$$

Summing over q we get a linear map denoted again by F^*:

$$F^*: \bigwedge V'^* \to \bigwedge V^*.$$

Let M^n be a C^s n-manifold, $s \geq 1$. For the tangent space $T_p(M^n)$ of the manifold M^n at a point $p \in M^n$, consider the vector space $\bigwedge^q(T_q(M))^*$ consisting of all q-forms on $T_q(M)$. A map

$$\omega\colon M^n \to \bigcup_{p \in M^n} (\bigwedge^q(T_p(M))^*), \tag{8}$$

assigning to each $p \in M^n$ an alternating q-form $\omega(p) \in \bigwedge_q(T_p(M))^*$, is called a *differential q-form* or simply a *q-form* on M^n.

The unit vectors, $\partial/\partial x_1, \partial/\partial x_2, \dots, \partial/\partial x_n$, form a basis in Euclidean space $\mathbf{R}^n$. Let $(\mathbf{R}^n)^*$ be the dual space of $\mathbf{R}^n$ and let $dx_1, dx_2, \dots, dx_n$ be the dual basis of $\partial/\partial x_1, \partial/\partial x_2, \dots, \partial/\partial x_n$:

$$dx_i\left(\frac{\partial}{\partial x_j}\right) = \delta_{ij}.$$

Then $\bigwedge^q(\mathbf{R}^n)^*$ is a vector space over $\mathbf{R}$ endowed with a basis $dx_{j_i} \wedge dx_{j_2} \wedge \cdots \wedge dx_{j_q}$, $j_1 < j_2 < \cdots < j_q$.

Let $(U_\lambda, \varphi_\lambda)$ be a chart in M^n. For each point $p \in U_\lambda$ the isomorphism $\Phi_\lambda\colon T_p(M^n) \to \mathbf{R}^n$ was defined in §10 by (3). We now define a map $\Phi_\lambda^*: (\mathbf{R}^n)^* \to (T_p(M^n))^*$ by

$$\Phi_\lambda^*(dx_i)(v) = dx_i(\Phi_\lambda(v)), \qquad v \in T_p(M^n).$$

Then Φ_λ^* is an isomorphism. The inverse $(\Phi_\lambda^*)^{-1}\colon (T_p(M^n))^* \to (\mathbf{R}^n)^*$ induces a natural isomorphism also denoted by $(\Phi_\lambda^*)^{-1}$:

$$(\Phi_\lambda^*)^{-1}\colon \bigwedge^q(T_p(M^n))^* \to \bigwedge^q(\mathbf{R}^n)^*. \tag{9}$$

Using the isomorphism (9), a q-form ω in (8) can be locally defined by

$$(\Phi_\lambda^*)^{-1}\omega(p) = \sum_{j_1<j_2<\cdots<j_q} a_{j_1j_2\cdots j_q}(\varphi_\lambda(p))\, dx_{j_1} \wedge dx_{j_2} \wedge \cdots \wedge dx_{j_q},$$

where $p \in U_\lambda$, and each $a_{j_1j_2\cdots j_q}$ is a real-valued function defined on $\varphi_\lambda(U_\lambda)$. We say that ω is *differentiable of class* C^r at p if the $a_{j_1j_2\cdots j_q}$ are differentiable of class C^r at $\varphi_\lambda(p)$, $r \leq s-1$. Clearly this definition is independent of the choice of $(U_\lambda, \varphi_\lambda)$. If the ω is C^r at each point of M^n, we say that it is a C^r *q-form* on M^n. A C^r zero-form is a C^r function on M^n. We shall denote by $A^q_{(r)}(M^n)$ or simply by $A^q(M^n)$ the set of all C^r q-forms on M^n, $q = 0, 1, 2, \dots, n$.

For q-forms $\omega_1, \omega_2 \in A^q(M^n)$, we define a q-form $\omega_1 + \omega_2 \in A^q(M^n)$ by

$$(\omega_1 + \omega_2)(p) = \omega_1(p) + \omega_2(p).$$

For a C^r function $g: M^n \to \mathbf{R}$ on M^n we define a q-form $g\omega_1 \in A^q(M^n)$ by

$$(g\omega_1)(p) = g(p)\omega_1(p).$$

We define the *exterior product* or *wedge product* $\omega \wedge \theta \in A^{q+s}$ of $\omega \in A^q(M^n)$ and $\theta \in A^s(M^n)$ by

$$(\omega \wedge \theta)(p) = \omega(p) \wedge \theta(p).$$

Here we set $A^q(M^n) = 0$ if $q > n$. The wedge product satisfies the identity

$$\omega \wedge \theta = (-1)^{qs}\theta \wedge \omega.$$

Denote by $A(M^n)$ the direct sum of $A^q(M^n)$, $q = 0, 1, 2, \dots, n$:

$$A(M^n) = \sum_{q=0}^{n} A^q(M^n),$$

and define the wedge product of elements $\sum_q \omega^q$ and $\sum_q \theta^q$ by

$$(\sum_q \omega^q) \wedge (\sum_q \theta^q) = \sum_l (\sum_{q+k=l} \omega^q \wedge \theta^k),$$

where $\omega^q, \theta^q \in A^q(M^n)$. Then $A(M^n)$ is an algebra over $\mathbf{R}$.

THEOREM 7.3. *There exists a unique linear map*

$$d: A^q_{(r)}(M^n) \to A^{q+1}_{(r-1)}(M^n), \qquad q = 0, 1, 2, \dots, n,$$

which satisfies the following.

(i) $d(\omega_1 + \omega_2) = d\omega_1 + d\omega_2$, $\omega_1, \omega_2 \in A^q_{(r)}(M^n)$.

(ii) $d(\omega \wedge \theta) = d\omega \wedge \theta + (-1)^q \omega \wedge d\theta$, $\omega \in A^q_{(r)}(M^n)$, $\theta \in A^k_{(r)}(M^n)$.

(iii) *For* $f \in A^0_{(r)}(M^n)$ *and a chart* $(U_\lambda, \varphi_\lambda)$ *of* M^n:

$$(\Phi_\lambda^*)^{-1} df(p) = \sum_{i=1}^{n} \frac{\partial(f \circ \varphi_\lambda^{-1})}{\partial x_i}(\varphi_\lambda(p))\, dx_i, \qquad p \in U_\lambda.$$

(iv) *If* $r \geq 2$, *then for* $f \in A^0_{(r)}(M^n)$, $d(df) = 0$.

The map d is called an *exterior derivative* and $d\omega$ the *exterior derivative* of ω.

PROOF. In a chart $(U_\lambda, \varphi_\lambda)$, $\omega \in A^q_{(r)}(M^n)$ takes on the form

$$(\Phi_\lambda^*)^{-1} \omega(p) = \sum_{j_1 < j_2 < \cdots < j_q} a_{j_1 j_2 \cdots j_q}(\varphi_\lambda(p))\, dx_{j_1} \wedge dx_{j_2} \wedge \cdots \wedge dx_{j_q}.$$

If a map d satisfies (i), (ii), (iii), and (iv), $d\omega$ has the expression

$$(\Phi_\lambda^*)^{-1} d\omega(p) = \sum_{j_1 < j_2 < \cdots < j_q} \left(\sum_{i=1}^{n} \frac{\partial}{\partial x_i} a_{j_1 j_2 \cdots j_q}(\varphi_\lambda(p))\, dx_i \right) \wedge dx_{j_1} \wedge dx_{j_2} \wedge \cdots \wedge dx_{j_q}. \tag{10}$$

Hence if d exists, it is unique.

Now for a C^r q-form ω define $d\omega(p)$, $p \in U_\lambda$, by (10). We then have a linear map $d: A^q_{(r)}(U_\lambda) \to A^{q+1}_{(r-1)}(U_\lambda)$, which obviously satisfies (i) and (iii). Further, since

$$(\Phi_\lambda^*)^{-1} d(df) = \sum_{i<j} \left(\frac{\partial^2 (f \circ \varphi_\lambda^{-1})}{\partial x_j \partial x_i} - \frac{\partial^2 (f \circ \varphi_\lambda^{-1})}{\partial x_i \partial x_j} \right) dx_i \wedge dx_j = 0,$$

the map d also satisfies (iv). In order to show that d satisfies (ii), it is enough to consider the case

$$(\Phi_\lambda^*)^{-1}\omega = f\, dx_{i_1} \wedge dx_{i_2} \wedge \cdots \wedge dx_{i_q}, \qquad (\Phi_\lambda^*)^{-1}\theta = g\, dx_{j_1} \wedge dx_{j_2} \wedge \cdots \wedge dx_{j_k},$$

since both the map d and the wedge product are linear. Computing $d(\omega \wedge \theta)$, we get

$$\begin{aligned}
&(\Phi_\lambda^*)^{-1} d(\omega \wedge \theta) \\
&\quad = \left(\sum_{i=1}^n \frac{\partial}{\partial x_i}(fg)\, dx_i \right) dx_{i_1} \wedge \cdots \wedge dx_{i_q} \wedge dx_{j_1} \wedge \cdots \wedge dx_{j_k} \\
&\quad = \sum_{i=1}^n \left(\frac{\partial f}{\partial x_i} g + f \frac{\partial g}{\partial x_i} \right) dx_i \wedge dx_{i_1} \wedge \cdots \wedge dx_{i_q} \wedge dx_{j_1} \wedge \cdots \wedge dx_{j_k} \\
&\quad = \left(\sum_{i=1}^n \frac{\partial f}{\partial x_i} dx_i \wedge dx_{i_1} \wedge \cdots \wedge dx_{i_q} \right) \wedge (g\, dx_{j_1} \wedge \cdots \wedge dx_{j_k}) \\
&\qquad + (-1)^q (f\, dx_{i_1} \wedge \cdots \wedge dx_{i_q}) \wedge \left(\sum_{i=1}^n \frac{\partial g}{\partial x_i} dx_i \wedge dx_{j_1} \wedge \cdots \wedge dx_{j_k} \right) \\
&\quad = (\Phi_\lambda^*)^{-1} (d\omega \wedge \theta + (-1)^q \omega \wedge d\theta).
\end{aligned}$$

Hence d satisfies (ii).

Now for each chart $(U_\lambda, \varphi_\lambda)$ we define a map $d: A^q_{(r)}(U_\lambda) \to A^{q+1}_{(r-1)}(U_\lambda)$ by (10). By the uniqueness of such d, if $U_\lambda \cap U_\mu \neq \varnothing$, the d defined on the chart $(U_\lambda, \varphi_\lambda)$ and the d defined on (U_μ, φ_μ) agree on $U_\lambda \cap U_\mu$. Therefore we have shown the existence of the desired map d.

Theorem 7.4. *The operator* d *of Theorem* 7.3 *satisfies the identity*

$$d(d\omega) = 0.$$

Here we assume that $r \geq 2$.

Proof. It is enough to compute $d(d\omega)$ for ω with $(\Phi_\lambda^*)^{-1}\omega = f\, dx_{i_1} \wedge dx_{i_2} \wedge \cdots \wedge dx_{i_q}$:

$$\begin{aligned}
(\Phi_\lambda^*)^{-1} d(d\omega) = \sum_{i<j} &\left(\frac{\partial^2 (f \circ \varphi_\lambda^{-1})}{\partial x_j \partial x_i} - \frac{\partial^2 (f \circ \varphi_\lambda^{-1})}{\partial x_i \partial x_j} \right) dx_i \\
&\wedge dx_j \wedge dx_{i_1} \wedge dx_{i_2} \wedge \cdots \wedge dx_{i_q} = 0;
\end{aligned}$$

therefore $d(d\omega) = 0$.

Let ω be a C^r q-form on M^n: $\omega \in A^q_{(r)}(M^n)$. We say that ω is *closed* if $d\omega = 0$. We say that a q-form $\theta \in A^q_{(r)}(M^n)$ is *exact* if there exists a differentiable C^r $(q-1)$-form $\hat{\theta}$ in $A^{q-1}_{(r+1)}(M^n)$ such that $d\hat{\theta} = \theta$. By Theorem 7.4, an exact differential form is closed.

Given a C^r q-form ω on M^n, the closure of $\{p \in M^n;\ \omega(p) \neq 0\}$ is called the *support* of ω.

Let M, M' be C^s manifolds and $h: M \to M'$ a C^r map, $1 \leq r < s$. We define a map

$$h^*: A^q_{(r)}(M') \to A^q_{(r)}(M), \qquad q = 0, 1, 2, \ldots$$

by

$$((h^*(\omega'))(p))(v_1, v_2, \ldots, v_q) = (\omega'(h(p)))(h_*(v_1), h_*(v_2), \ldots, h_*(v_q)),$$
$$\omega' \in A^q_{(r)}(M'), \qquad p \in M^n, \qquad v_1, v_2, \ldots, v_q \in T_p(M).$$

The map h^* is a homomorphism of exterior algebras.

§27. Integrals of differential forms

Let M^n be a compact oriented n-dimensional C^s manifold with or without boundary, $s \geq 1$. Since M^n is orientable, a C^r atlas $\mathscr{S}$ of M^n has a subfamily $\mathscr{S}'$ which satisfies the orientability conditions (i) and (ii) of §10. We take $\mathscr{S}'$ to be the maximal subfamily with respect to the inclusion relation. Thus for each point p in M^n the tangent space $T_p(M^n)$ at p is oriented. Let $(U_\lambda, \varphi_\lambda)$ be a chart of $\mathscr{S}'$ about p, and $\pi: T(M) \to M$ the standard projection of the tangent space. Via the map $\tilde{\Phi}_\lambda: \pi^{-1}(U_\lambda) \to \varphi_\lambda(U_\lambda) \times \mathbf{R}^n$ defined in §10, the orientation of $T_p(M^n)$ induces an orientation of $\mathbf{R}^n$. We choose $\mathscr{S}'$ in such a way that the induced orientation of $\mathbf{R}^n$ agrees with the basis $\partial/\partial x_1, \partial/\partial x_2, \ldots, \partial/\partial x_n$.

Let $(U_\lambda, \varphi_\lambda), (U_\mu, \varphi_\mu) \in \mathscr{S}'$ be charts about p, $p \in U_\lambda \cap U_\mu$, and for $v \in T_p(M^n)$ set

$$\tilde{\Phi}_\lambda(v) = (\varphi_\lambda(p), (v_1, v_2, \ldots, v_n)), \quad \tilde{\Phi}_\mu(v) = (\varphi_\mu(p), (v_1', v_2', \ldots, v_n')).$$

Setting $\varphi_\lambda(p) = (x_1, x_2, \ldots, x_n)$, $\varphi_\mu(p) = (z_1, z_2, \ldots, z_n)$, we have by (4) in §10:

$$v_i' = \sum_{j=1}^n \frac{\partial z_i}{\partial x_j}(\varphi_\lambda(p))v_j, \qquad i = 1, 2, \ldots, n. \tag{1}$$

Let $\hat{\omega}$ be a C^r n-form on M^n such that the support of $\hat{\omega}$ is contained in U_λ for some chart $(U_\lambda, \varphi_\lambda)$ of $\mathscr{S}'$, $0 \leq r < s$.

Then for $p \in U_\lambda$, $\hat{\omega}$ can be written:

$$(\Phi_\lambda^*)^{-1}\hat{\omega}(p) = \hat{f}(\varphi_\lambda(p))\, dx_1 \wedge dx_2 \wedge \cdots \wedge dx_n.$$

We define the *integral* $\int_{U_\lambda}\hat{\omega}$ of $\hat{\omega}$ on U_λ by

$$\int_{U_\lambda}\hat{\omega} = \int_{\varphi_\lambda(U_\lambda)} \hat{f}(\varphi_\lambda(p))\, dx_1 dx_2 \cdots dx_n. \tag{2}$$

LEMMA 7.5. *Let $(U_\lambda, \varphi_\lambda)$ and (U_μ, φ_μ) be charts in $\mathscr{S}'$ with $U_\lambda = U_\mu$. Then*

$$\int_{U_\lambda}\hat{\omega} = \int_{U_\mu}\hat{\omega}.$$

PROOF. Write $\varphi_\mu \circ \varphi_\lambda^{-1} : \varphi_\lambda(U_\lambda) \to \varphi_\mu(U_\mu)$ in terms of coordinate functions as $(\varphi_\mu \circ \varphi_\lambda^{-1})(x_1, x_2, \dots, x_n) = (z_1, z_2, \dots, z_n)$, and express $\hat{\omega}$ by

$$(\Phi_\mu^*)^{-1}\hat{\omega}(p) = \hat{g}(\varphi_\mu(p))\, dz_1 \wedge dz_2 \wedge \cdots \wedge dz_n.$$

Then by definition,

$$\int_{U_\mu}\hat{\omega} = \int_{\varphi_\mu(U_\mu)} \hat{g}(\varphi_\mu(p))\, dz_1 dz_2 \cdots dz_n.$$

Through the transformation formula for multiple integrals, the right-hand side of this equation becomes

$$\int_{\varphi_\lambda(U_\lambda)} \hat{g}(\varphi_\mu(p)) \left| \frac{\partial(z_1, z_2, \dots, z_n)}{\partial(x_1, x_2, \dots, x_n)} \right| dx_1 dx_2 \cdots dx_n.$$

But from (1) we get

$$(\Phi_\lambda^*)^{-1}\Phi_\mu^*(dz_i) = \sum_{j=1}^{n} \frac{\partial z_i}{\partial x_j} dx_j, \qquad i = 1, 2, \dots, n.$$

Hence by Lemma 7.2 we have

$$\hat{g}(\varphi_\mu(p)) \left| \frac{\partial(z_1, z_2, \dots, z_n)}{\partial(x_1, x_2, \dots, x_n)} \right| = \hat{f}(\varphi_\lambda(p)).$$

Thus we have established $\int_{U_\mu}\hat{\omega} = \int_{U_\lambda}\hat{\omega}$.

Let ω be a C^r n-form on M^n, $0 \le r < s$. Choose $(U_i, \varphi_i) \in \mathscr{S}'$, $i = 1, 2, \dots, m$, such that $\{U_i\}$ is an open cover of M^n: $M^n = \bigcup_{i=1}^m U_i$. Let μ_i, $i = 1, 2, \dots, m$, be a partition of unity subordinate to the open cover $\{U_i\}$ so that $\mu_i: M^n \to \mathbf{R}$ is a C^r function, $\sum_{i=1}^m \mu_i(p) = 1$, and the support of μ_i is contained in U_i. We define the *integral* $\int_{M^n}\omega$ of ω on M^n by

$$\int_{M^n}\omega = \sum_{i=1}^{n} \int_{U_i} \mu_i \omega. \tag{3}$$

Each summand in the right-hand side of this equation is defined by (2).

We shall show that the definition (3) depends neither on the choice of an open cover of M^n nor on a particular partition of unity. First we take a second partition of unity ν_i, $i = 1, 2, \cdots, m$, subordinate to $\{U_i\}$. By

Lemma 7.5, the integral $\int_{U_i \cap U_j} \mu_i \nu_j \omega$ is independent of the choice of a chart, so we have

$$\sum_{i=1}^{m} \int_{U_i} \mu_i \omega = \sum_{i=1}^{m} \int_{U_i} \left(\sum_{j=1}^{m} \mu_i \nu_j \omega \right) = \sum_{i,j=1}^{m} \int_{U_i \cap U_j} \mu_i \nu_j \omega$$
$$= \sum_{j=1}^{m} \int_{U_j} \left(\sum_{i=1}^{m} \mu_i \nu_j \omega \right) = \sum_{j=1}^{m} \int_{U_j} \nu_j \omega.$$

Hence the integral ω on M^n does not depend on the choice of a partition of unity.

We next consider $(U'_j, \varphi'_j) \in \mathscr{S}'$, $j = 1, 2, \dots, m'$, such that $\{U'_j\}$ is an open cover of M^n, and a partition of unity μ'_j, $j = 1, 2, \dots m'$, subordinate to $\{U'_j\}$. Combine $\{U_i\}$ and $\{U'_j\}$ to give M^n an open cover $\{U''_k\}$. Then we may regard $\{\mu_i\}$ and $\{\nu_j\}$ each as a partition of unity subordinate to $\{U''_k\}$ in a natural way, and so from the independence of the choice of a partition of unity we have

$$\sum_{i=1}^{m} \int_{U_i} \mu_i \omega = \sum_{k} \int_{U''_k} \mu_k \omega = \sum_{k} \int_{U''_k} \mu'_k \omega = \sum_{j=1}^{m'} \int_{U'_j} \mu'_j \omega.$$

Hence we see that the integral does not depend on the choice of an open cover of M^n.

Let M^n be a compact oriented C^s manifold with $\partial M^n \neq \varnothing$, $s \geq 1$. Then ∂M^n is a compact C^s $(n-1)$-manifold without boundary and ∂M^n has the induced orientation (cf. §11).

Let η be a C^r $(n-1)$-form on M^n, $1 \leq r < s$. Then for the inclusion map $\iota\colon \partial M^n \to M^n$, $\iota^* \eta$ is a C^r $(n-1)$-form on ∂M^n and we have the following

THEOREM 7.6 (theorem of Stokes).

$$\int_{\partial M^n} \iota^* \eta = \int_{M^n} d\eta.$$

PROOF. Choose a cover $\{U_i\}$ of M^n with $(U_i, \varphi_i) \in \mathscr{S}'$, $i = 1, 2, \dots, m$, and a partition of unity μ_i, $i = 1, 2, \dots, m$, subordinate to $\{U_i\}$.

We shall first show that if $U_i \cap \partial M^n = \varnothing$ then,

$$\int_{U_i} d(\mu_i \eta) = 0.$$

Setting

$$(\Phi_i^*)^{-1} \mu_i \eta = \sum_{j=1}^{n} g_j \, dx_1 \wedge dx_2 \wedge \cdots \wedge dx_{j-1} \wedge dx_{j+1} \wedge \cdots \wedge dx_n, \tag{4}$$

we get

$$(\Phi_i^*)^{-1} d(\mu_i \eta) = \sum_{j=1}^{n} (-1)^{j-1} \frac{\partial g_j}{\partial x_j} dx_1 \wedge dx_2 \wedge \cdots \wedge dx_n, \tag{5}$$

and since g_j takes on the value zero in a neighborhood of $\overline{\varphi_i(U_i)} - \varphi_i(U_i)$, $j = 1, 2, \ldots, n$, we have

$$\int_{U_i} d(\mu_i \eta) = \sum_{j=1}^{n} (-1)^{j-1} \int_{\varphi_i(U_i)} \frac{\partial g_j}{\partial x_j} dx_1 dx_2 \cdots dx_n = 0. \tag{6}$$

Hence we have obtained the desired result.

We now assume that $U_i \cap \partial M^n \neq \varnothing$ and wish to show that

$$\int_{U_i \cap \partial M^n} \mu_i (\iota^* \eta) = \int_{U_i} d(\mu_i \eta).$$

In this case we have $\varphi_i(U_i) \cap \mathbf{R}^{n-1} \neq \varnothing$, so using (4) for $(\Phi_i^*)^{-1} \mu_i \eta$, we get

$$(\Phi_i^*)^{-1} \mu_i (\iota^* \eta) = g_n(x_1, x_2, \ldots, x_{n-1}, 0)\, dx_1 \wedge dx_2 \wedge \cdots \wedge dx_{n-1}.$$

Now $(\Phi_i^*)^{-1} d(\mu_i \eta)$ satisfies (5), but if $j \neq n$, g_j takes on the value zero in a neighborhood of $\overline{\varphi_i(U_i)} - \varphi_i(U_i)$; hence we have on the right-hand side of equation (6)

$$\int_{\varphi_i(U_i)} \frac{\partial g_j}{\partial x_j} dx_1 dx_2 \cdots dx_n = 0.$$

Furthermore, since

$$\int_{\varphi_i(U_i)} \frac{\partial g_n}{\partial x_n} dx_1 dx_2 \cdots dx_n$$
$$= -\int_{\varphi_i(U_i) \cap \mathbf{R}^{n-1}} g_n(x_1, x_2, \ldots, x_{n-1}, 0)\, dx_1 dx_2 \cdots dx_{n-1},$$

we have

$$\int_{U_i} d(\mu_i \eta) = (-1)^n \int_{\varphi_i(U_i) \cap \mathbf{R}^{n-1}} g_n(x_1, x_2, \ldots, x_{n-1}, 0)\, dx_1 dx_2 \cdots dx_{n-1}.$$

On the other hand, by the choice of the orientation of ∂M^n we have

$$\int_{U_i \cap \partial M^n} \mu_i (\iota^* \eta)$$
$$= (-1)^n \int_{\varphi_i(U_i) \cap \mathbf{R}^{n-1}} g_n(x_1, x_2, \ldots, x_{n-1}, 0)\, dx_1 dx_2 \cdots dx_{n-1}.$$

This takes care of the case that $U_i \cap \partial M^n \neq \varnothing$. Hence we have shown

$$\int_{\partial M^n} \iota^* \eta = \sum_{i=1}^{m} \int_{U_i \cap \partial M^n} \mu_i \iota^* \eta = \sum_{i=1}^{m} \int_{U_i} d(\mu_i \eta) = \int_{M^n} d\eta.$$

§28. Foliations and tangent plane fields

Let M^n be an n-dimensional C^s manifold, $s \geq 1$. Given $p \in M^n$, denote by G_p^m the set of all m-dimensional subspaces—considered as vector spaces—of the tangent space $T_p(M)$. A map

$$\mathscr{D}^m : M^n \to \bigcup_{p \in M^n} G_p^m, \qquad \mathscr{D}^m(p) \in G_p^m,$$

which assigns to each point p of M^n an m-dimensional subspace of $T_p(M)$ is called an *m-dimensional tangent plane field*, a *tangent m-plane field* or simply an *m-plane field* on M^n, $m \leq n$.

If $\partial M^n \neq \emptyset$ we shall assume that for $p \in \partial M^n$, $\mathscr{D}^m(p)$ is an m-dimensional subspace of $T_p(\partial M^n)$.

We say that $\mathscr{D}^m$ is a *differentiable m-plane field of class C^r* or a *C^r m-plane field*, if for each $p \in M^n$, there exists a neighborhood U of p and C^r vector fields on U. $X_1, X_2, \ldots, X_m$, such that for any $p' \in U$, $X_1(p')$, $X_2(p')$, $\ldots$, $X_m(p')$ form a basis in the subspace $\mathscr{D}^m(p)$. The vector fields X_1, X_2, $\ldots$, X_m *generate* $\mathscr{D}^m$ in U.

Let $\mathscr{D}^m$ be an m-dimensional C^r plane field. We give M^n a Riemannian metric $\langle\ ,\ \rangle$. For $p \in M^n$ denote by $\bar{\mathscr{D}}^{n-m}(p)$ an $(n-m)$-dimensional subspace of $T_p(M^n)$ consisting of all vectors which are orthogonal to every vector in the subspace $\mathscr{D}^m(p)$. Then $\bar{\mathscr{D}}^{n-m} = \{\bar{\mathscr{D}}^{n-m}(p); p \in M^n\}$ is a C^r $(n-m)$-plane field, which is called the *dual tangent plane field* or *dual plane field* of $\mathscr{D}^m$.

Suppose $\omega_1, \omega_2, \ldots, \omega_q$ to be C^r one-forms on M^n such that for each point $p \in M^n$, the $\omega_1(p)$, $\omega_2(p)$, $\ldots$, $\omega_q(p)$ are linearly independent. Define a subspace of $T_p(M^n)$ by

$$\mathscr{D}^{n-q}(p) = \{v \in T_p(M^n); (\omega_i(p))(v) = 0, \quad i = 1, 2, \ldots, q\}.$$

Then $\mathscr{D}^{n-q} = \{\mathscr{D}^{n-q}(p); p \in M^n\}$ is an $(n-q)$-dimensional C^r plane field, the *tangent plane field defined by one-forms* ω_1, ω_2, $\ldots$, ω_q.

It is not always the case that a given C^r $(n-q)$-plane field is defined by one-forms ω_1, ω_2, $\ldots$, ω_q; however if $q = 1$ any $(n-1)$-plane field $\mathscr{D}^{n-1}$ can be defined by a certain C^r one-form over an open cover of M^n (Lemma 7.7).

Let $\{V_\sigma; \sigma \in \Sigma\}$ be an open cover of M^n and let ω_σ be a C^r q-form defined on V_σ, $\sigma \in \Sigma$, such that if $V_\sigma \cap V_\nu \neq \emptyset$,

$$\omega_\sigma = \omega_\nu \quad \text{or} \quad \omega_\sigma = -\omega_\nu$$

on $V_\sigma \cap V_\nu$. Then the set $\{\omega_\sigma; \sigma \in \Sigma\}$ is called a *C^r q-form over the open cover* $\{V_\sigma; \sigma \in \Sigma\}$ of M^n. If $\omega_\sigma = \omega_\nu$ on $V_\sigma \cap V_\nu$ for any pair σ, $\nu \in \Sigma$, $\{\omega_\sigma; \sigma \in \Sigma\}$ defines a C^r q-form on M^n.

Let $\{\omega_\sigma; \sigma \in \Sigma\}$ be a C^r one-form over $\{V_\sigma; \sigma \in \Sigma\}$ with $\omega_\sigma(p) \neq 0$ for every $p \in V_\sigma$, $\sigma \in \Sigma$. For each $p \in M^n$, choose an open set containing the point p, and define a subspace $D^{n-1}(p)$ by

$$\mathscr{D}^{n-1}(p) = \{ v \in T_p(M^n); (\omega_\sigma(p))(v) = 0 \}.$$

Then $\mathscr{D}^{n-1}(p)$ is uniquely determined independent of the choice of an open cover and $\mathscr{D}^{n-1} = \{\mathscr{D}^{n-1}(p); p \in M^n\}$ is a C^r $(n-1)$-plane field, which is called the *tangent plane field defined by* $\{\omega_\sigma; \sigma \in \Sigma\}$. The following lemma shows that the converse is also true.

LEMMA 7.7. *Given a* C^r $(n-1)$*-plane field* $\mathscr{D}^{n-1}$ *on* M^n, *there exists a* C^r *one-form over some open cover of* M^n *which defines* $\mathscr{D}^{n-1}$.

PROOF. Consider the dual C^r one-plane field $\bar{\mathscr{D}}^1$ of $\mathscr{D}^{n-1}$. For $p \in M^n$ there exist a neighborhood V_p of p and a C^r vector field $X^{(p)}$ on V_p such that $\| X^{(p)}(p') \| = 1$ for each $p' \in V_p$ and $\bar{\mathscr{D}}^1(p') = \{aX^{(p)}(p'); a \in \mathbf{R}\}$. Now define a C^r one-form ω_p on V_p by

$$(\omega_p(p'))(v) = \langle v, X^{(p)}(p')\rangle_{p'}, \qquad p' \in V_p, \ v \in T_{p'}(M^n).$$

Then the C^r differential form $\{\omega_p; p \in M^n\}$ over the open cover $\{V_p; p \in M^n\}$ gives the desired result.

Let $\mathscr{F} = \{L_\alpha; \alpha \in A\}$ be a C^r codimension q foliation of M^n, $r \geq 1$. For $p \in M^n$, let $p \in L_\alpha$ and consider the $(n-q)$-subspace $T_p(L_\alpha)$ of the tangent space $T_p(M^n)$ which consists of all vectors tangent to L_α. Thus we have obtained a C^{r-1} $(n-q)$-plane field on M^n which assigns to each point p in M^n the subspace $T_p(L_\alpha)$. This plane field will be denoted by $\mathscr{D}(\mathscr{F})$. If $\partial M^n \neq \varnothing$, then we have $\mathscr{D}(\mathscr{F})(p) \subset T_p(\partial M^n)$ for each $p \in \partial M^n$.

In general $\mathscr{D}(\mathscr{F})$ is of class C^{r-1}; however, in Example A in §16, we saw that $\mathscr{D}(\mathscr{F})$ is of class C^r, where $\mathscr{F}$ is a foliation defined by the trajectories of a nonsingular C^r vector field.

Let $\mathscr{D}^m$ be a C^r m-plane field on M^n. If ∂M^n is nonempty, we assume that $\mathscr{D}^m(p) \subset T_p(\partial M^n)$ for $p \in \partial M^n$. We say that $\mathscr{D}^m$ is *completely integrable* if there exists a $C^{r'}$ codimension $n-m$ foliation $\mathscr{F}$ of M^n such that $\mathscr{D}(\mathscr{F}) = \mathscr{D}^m$. Here r' is $r+1$ or r.

We now prove that if $\mathscr{D}^m$ is completely integrable there is a unique foliation $\mathscr{F}$ such that $\mathscr{D}(\mathscr{F}) = \mathscr{D}^m$. Suppose we have foliations $\mathscr{F}_1$, $\mathscr{F}_2$ such that

$$\mathscr{D}(\mathscr{F}_1) = \mathscr{D}(\mathscr{F}_2) = \mathscr{D}^m.$$

Let $(U_\lambda, \varphi_\lambda)$ be a foliated chart of $\mathscr{F}_1$, and for a point x in $\varphi_\lambda(U_\lambda)$ define $\mathscr{D}'(x)$ by

$$\mathscr{D}'(x) = \{ (\varphi_\lambda)_*(v); v \in \mathscr{D}^m(\varphi_\lambda^{-1}(x)) \}.$$

Then $\mathscr{D}' = \{\mathscr{D}'(x); x \in \varphi_\lambda(U_\lambda)\}$ is a C^r m-plane field on $\varphi_\lambda(U_\lambda)$. Denote

$$\mathscr{F}_1' = \{\text{arcwise connected components of } \varphi_\lambda(U_\lambda \cap L_\alpha);\, L_\alpha \in \mathscr{F}_1\};$$

(the family) $\mathscr{F}_1'$ is a $C^{r'}$ codimension $(n-m)$ foliation of $\varphi_\lambda(U_\lambda)$, and evidently we have $\mathscr{D}(\mathscr{F}_1') = \mathscr{D}'$. On the other hand if we let

$$\mathscr{F}_2' = \{\text{arcwise connected components of } \varphi_\lambda(U_\lambda \cap L'_{\alpha'});\, L'_{\alpha'} \in \mathscr{F}_2\},$$

again we have that $\mathscr{D}(\mathscr{F}_2') = \mathscr{D}'$. Since $(U_\lambda, \varphi_\lambda)$ is a foliated chart of $\mathscr{F}_1$, setting $x = (x_1, x_2, \dots, x_n)$ we get

$$\mathscr{D}'(x) = T_x(\varphi_\lambda(U_\lambda) \cap (\mathbf{R}^m \times \{(x_{m+1}, x_{m+2}, \dots, x_n)\})).$$

Hence, $\mathscr{F}_1' = \mathscr{F}_2'$, and so $\mathscr{F}_1$ and $\mathscr{F}_2$ agree.

A tangent plane field is not always completely integrable as we see in the following

EXAMPLE. Let $M^3 = \mathbf{R}^3 - \{0\}$, where 0 is the origin of $\mathbf{R}^3$. Using the Cartesian coordinates of $\mathbf{R}^3$ we define a C^∞ one-form θ by $\theta = x\,dy + y\,dz + z\,dx$. Since θ is never zero on M^3, we obtain the C^∞ two-plane field $\mathscr{D}^2$ on M^3 defined by θ as indicated above.

We shall show that $\mathscr{D}^2$ is not completely integrable. Suppose that it is completely integrable; then there is a C^∞ codimension one foliation of M^3 such that $\mathscr{D}(\mathscr{F}) = \mathscr{D}^2$. Let $(U_\lambda, \varphi_\lambda)$ be a distinguished chart of $\mathscr{F}$, with $\varphi_\lambda : U_\lambda \to (-1, 1)^3$, and define projection $\hat{\pi}: (-1, 1)^3 \to (-1, 1)$ as in §18. For simplicity we set $g = \hat{\pi} \circ \varphi_\lambda$. Then for each t, $-1 < t < 1$, $g^{-1}(t)$ is on a leaf of $\mathscr{F}$; so by Theorem 7.3 (iii), $dg(v) = 0$ for $v \in \mathscr{D}^2(p)$, $p \in U_\lambda$. Hence for some suitable nonzero C^∞ function $h: U_\lambda \to \mathbf{R}$, we have

$$\theta = h\,dg$$

on U_λ. Thus,

$$d\theta = dh \wedge dg = \frac{dh}{h} \wedge h\,dg = \frac{dh}{h} \wedge \theta.$$

Hence,

$$\theta \wedge d\theta = \theta \wedge \frac{dh}{h} \wedge \theta = 0.$$

But by a direct computation of $\theta \wedge d\theta$ we get

$$\theta \wedge d\theta = (x + y + z)\,dx \wedge dy \wedge dz,$$

so $\theta \wedge d\theta \neq 0$, which is a contradiction. Hence $\mathscr{D}^2$ is not completely integrable.

The following Frobenius Theorem gives us a necessary and sufficient condition for a C^r m-plane field to be completely integrable.

Let X and Y be C^r vector fields on M^n, $r \geq 1$. Let $(U_\lambda, \varphi_\lambda)$ be a chart of M^n. On U_λ, X and Y can be expressed by

$$\Phi_\lambda(X(p)) = \sum_{i=1}^{n} a_i(\varphi_\lambda(p))\frac{\partial}{\partial x_i},$$

$$\Phi_\lambda(Y(p)) = \sum_{i=1}^{n} b_i(\varphi_\lambda(p))\frac{\partial}{\partial x_i}, \qquad p \in U_\lambda,$$

where a_i, b_i are C^r functions defined on $\varphi_\lambda(U_\lambda)$. We define a map $[X, Y]$ on U_λ by

$$[X, Y](p) = \Phi_\lambda^{-1}\left(\sum_i\left(\sum_j\left(a_j\frac{\partial b_i}{\partial x_j} - b_j\frac{\partial a_i}{\partial x_j}\right)(\varphi_\lambda(p))\right)\frac{\partial}{\partial x_i}\right), \qquad p \in U_\lambda.$$

Then $[X, Y]$ is a C^{r-1} vector field on U_λ. We can easily see from definition that if we use a different chart $(U'_{\lambda'}, \varphi'_{\lambda'})$ to define $[X, Y]$ we will have the same result. Hence we obtain a C^{r-1} vector field $[X, Y]$ on M^n by considering $[X, Y]$ on each coordinate neighborhood in M^n. We call $[X, Y]$ the *bracket product* of X and Y. It is immediate from the definition that the bracket product satisfies the following:

(i) $[X, X] = 0$, $[X, Y] = -[Y, X]$,

(ii) For a C^r vector field Z on M^n, $r \geq 2$,

$[X, [Y, Z]] + [Y, [Z, X]] + [Z, [X, Y]] = 0$, the Jacobi identity. Let $\mathscr{D}^m$ be a C^r m-plane field on M^n. We say that $\mathscr{D}^m$ is *involutive* if for every $p \in M^n$, there exist a neighborhood U of p and m C^r vector fields $X_1, X_2, \ldots, X_m$ which generate $\mathscr{D}^m$ such that

$$[X_i, X_j](p) \in \mathscr{D}^m(p), \qquad p \in U,\ i, j = 1, 2, \ldots, m.$$

THEOREM 7.8 (Frobenius theorem; first form). *A C^r m-plane field $\mathscr{D}^m$ on a C^s manifold M^n is completely integrable if and only if $\mathscr{D}^m$ is involutive. Here $r \geq 2$.*

PROOF. First let us assume that $\mathscr{D}^m$ is completely integrable. Let $\mathscr{F}$ be a $C^{r'}$ codimension $n-m$ foliation of M^n such that $\mathscr{D}^m = \mathscr{D}(\mathscr{F})$, $r' = r+1$ or r. Let $(U_\lambda, \varphi_\lambda)$ be a distinguished chart of $\mathscr{F}$, and let $(x_1, x_2, \ldots, x_n) \in \varphi_\lambda(U_\lambda)$. The C^r vector fields $X_1, X_2, \ldots, X_m$ on U_λ defined by

$$\Phi_\lambda(X_i) = \frac{\partial}{\partial x_i}, \qquad i = 1, 2, \ldots, m$$

generate $\mathscr{D}^m$ on U_λ and $[X_i, X_j](p) = 0$. Therefore $\mathscr{D}^m$ is involutive.

Conversely assume that $\mathscr{D}^m$ is involutive. For an arbitrary point $p \in M^n$ there exist a neighborhood U of p and C^r vector fields $X_1, X_2, \ldots, X_m$ generating $\mathscr{D}^m$ in U such that $[X_i, X_j](p) \in \mathscr{D}^m$ for $p \in U$. Since complete integrability is a local property, it is enough to take a chart $(U_\lambda, \varphi_\lambda)$

with $p \in U_\lambda \subset U$, and show that $\mathscr{D}^m$ is completely integrable on U_λ. For suppose $\mathscr{F}_\lambda$ and $\mathscr{F}_\mu$ are C^r codimension $n-m$ foliations of U_λ and U_μ respectively such that

$$\mathscr{D}(\mathscr{F}_\lambda) = \mathscr{D}^m | U_\lambda, \qquad \mathscr{D}(\mathscr{F}_\mu) = \mathscr{D}^m | U_\mu.$$

If $U_\lambda \cap U_\mu \neq \varnothing$, then by the uniqueness of $\mathscr{F}_\lambda$ and $\mathscr{F}_\mu$ they agree on $U_\lambda \cap U_\mu$, so we can define a C^r codimension $n-m$ foliation on M^n.

For $m = 1$, we use the vector field X_1 of Example A in §16 to construct a foliation $\mathscr{F}_\lambda$ on U_λ. We then have $\mathscr{D}^1 = \mathscr{D}(\mathscr{F}_\lambda)$ on U_λ; hence $\mathscr{D}^1$ is completely integrable.

We shall finish the proof by induction on m. Suppose that we have shown the complete integrability of $\mathscr{D}^i$, $i = 1, 2, \dots, m-1$. We now prove the theorem for $\mathscr{D}^m$.

Using the image of the trajectories of X_1 under φ_λ we replace φ_λ by $\varphi_{\bar\lambda}$ and replace U_λ by $U_{\bar\lambda}$ accordingly, and obtain a new chart $(U_{\bar\lambda}, \varphi_{\bar\lambda})$ in place of $(U_\lambda, \varphi_\lambda)$ such that

$$\varphi_{\bar\lambda}(p) = (0, x_2^{(0)}, x_3^{(0)}, \dots, x_n^{(0)}) \in \varphi_{\bar\lambda}(U_{\bar\lambda}) \subset \mathbf{R}_+^n,$$

$$\Phi_{\bar\lambda}(X_1(\bar p)) = \frac{\partial}{\partial x_1}, \qquad \bar p \in U_{\bar\lambda},$$

where $(x_1, x_2, \dots, x_n) \in \varphi_{\bar\lambda}(U_{\bar\lambda})$. Further, we replace $X_2, X_3, \dots, X_m$ by $X_2', X_3', \dots, X_m'$ such that the C^{r-1} vector fields $X_1, X_2', X_3', \dots, X_m'$ generate $\mathscr{D}^m$ on $U_{\bar\lambda}$, and such that

$$\Phi_{\bar\lambda}(X_i'(\bar p)) = \sum_{j=2}^{n} a_{ij}(\varphi_{\bar\lambda}(\bar p)) \frac{\partial}{\partial x_j}, \qquad i = 1, 2, \dots, m, \tag{1}$$

for $\bar p \in U_{\bar\lambda}$. We define an $(n-1)$-dimensional submanifold W of $U_{\bar\lambda}$ by

$$W = \varphi_{\bar\lambda}^{-1}(\{(0, x_2, x_3, \dots, x_n) \in \varphi_{\bar\lambda}(U_{\bar\lambda})\}).$$

Then $p \in W$ and by construction, $X_2', X_3', \dots, X_m'$ are C^r tangent vector fields of $T(W)$ on W, and the vectors

$$X_2'(p'), X_3'(p'), \dots, X_m'(p'), \qquad p' \in W$$

are linearly independent. Thus, the $X_2', X_3', \dots, X_m'$ define the $(m-1)$-dimensional C^r plane field $\mathscr{D}^{m-1}$ on W. Now $[X_i', X_j'](p') \in \mathscr{D}^m(p')$ for each $p' \in W$, but from the equation (1) for $\Phi_{\bar\lambda}(X_i'(p')$ we see that

$$[X_i', X_j'](p') \in \mathscr{D}^{m-1}(p').$$

Hence $\mathscr{D}^{m-1}$ is involutive. By the induction assumption $\mathscr{D}^{m-1}$ is completely integrable in W, and so there exists a $C^{r'}$ codimension $n-m$ foliation $\mathscr{F}'$ of W such that $\mathscr{D}(\mathscr{F}') = \mathscr{D}^{m-1}$.

Let $(U'_{\lambda'}, \varphi'_{\lambda'})$ be a foliated chart of $\mathscr{F}'$ about p in W. Then for a leaf $L'_{\alpha'}$ of $\mathscr{F}'$, the image of a connected component of $L'_{\alpha'}$ under $\varphi'_{\lambda'}$ is of the form

$$\{ (y_1, y_2, \dots, y_{n-1}) \in \varphi'_{\lambda'}(U'_{\lambda'});\, y_m = c_m, y_{m+1} = c_{m+1}, \dots, y_{n-1} = c_{n-1} \}.$$

Choosing $U_{\bar{\lambda}}$ and $(U'_{\lambda'}, \varphi'_{\lambda'})$ suitably and identifying $\varphi_{\bar{\lambda}}(W)$ and $\varphi'_{\lambda'}(U'_{\lambda'})$, we may regard y_i as a C^r function of $x_1, x_2, \dots, x_n$, $i = 1, 2, \dots, n-1$.

We now define $z_1, z_2, \dots, z_n$ by

$$z_1 = x_1, \qquad z_{i+1} = y_i, \qquad i = 1, 2, \dots, n-1. \tag{2}$$

Then,

$$\left| \frac{\partial(z_1, z_2, \dots, z_n)}{\partial(x_1, x_2, \dots, x_n)}(\varphi_{\bar{\lambda}}(p)) \right| \neq 0.$$

Hence, in some neighborhood of p we may replace $(x_1, x_2, \dots, x_n)$ by $(z_1, z_2, \dots, z_n)$, that is, we may choose $(U_\mu, \varphi_\mu) \in \mathscr{S}$ with $p \in U_\mu \subset U_\lambda$ such that $\varphi_\mu(p')$ can be written as

$$\varphi_\mu(p') = (z_1, z_2, \dots, z_n), \qquad p' \in U_\mu,$$

where $z_1, z_2, \dots, z_n$ are given in (2). Denote by $Y_1, Y_2, \dots, Y_m$ the restriction of the vector fields $X_1, X'_2, X'_3, \dots, X'_m$ to U_μ, and for $p' \in U_\mu$ set

$$\Phi_\mu(Y_i(p')) = \sum_{j=1}^{n} b_{ij}(\varphi_\mu(p')) \frac{\partial}{\partial z_j}, \qquad i = 1, 2, \dots, m.$$

By the choice of $Y_1, Y_2, \dots, Y_m$ and $z_1, z_2, \dots, z_n$, we have

$$b_{11} = 1, \qquad b_{1j} = 0, \qquad j = 2, 3, \dots, n.$$

By our assumption, there exist C^{r-2} functions c_i^j, $i, j = 1, 2, \dots, m$, on U_μ such that

$$[Y_1, Y_i](p') = \sum_{j=1}^{m} c_i^j(p') Y_j(p').$$

By the definition of the bracket product, we get

$$\begin{aligned} \Phi_\mu([Y_1, Y_i](p')) &= \sum_{l=1}^{n} \left(\sum_j \left(b_{1j} \frac{\partial b_{il}}{\partial z_j} - b_{ij} \frac{\partial b_{1l}}{\partial z_j} \right) (\varphi_\mu(p')) \right) \frac{\partial}{\partial z_l} \\ &= \sum_{l=1}^{n} \frac{\partial b_{il}}{\partial z_1}(\varphi_\mu(p')) \frac{\partial}{\partial z_l}. \end{aligned} \tag{3}$$

On the other hand, we have

$$\begin{aligned} \Phi_\mu([Y_1, Y_i](p')) &= \sum_{j=1}^{m} c_i^j(p') \Phi_\mu(Y_j(p')) \\ &= \sum_{l=1}^{n} \left(\sum_{j=1}^{m} c_i^j(p') b_{jl}(\varphi_\mu(p')) \right) \frac{\partial}{\partial z_l}. \end{aligned} \tag{4}$$

From (3) and (4) we obtain a system of linear equations in $b_{1l}, b_{2l}, \dots, b_{ml}$:

$$\frac{\partial b_{il}}{\partial z_1}(\varphi_\mu(p')) = \sum_{j=1}^{m} c_i^{\,j}(p') b_{jl}(\varphi_\mu(p')), \qquad i = 1, 2, \dots, m.$$

From the definition of $(y_1, y_2, \dots, y_{n-1})$, we get

$$b_{il}(0, z_2, z_3, \dots, z_n) = 0,$$

where $l = m+1, m+2, \dots, n$. Hence keeping $z_2, z_3, \dots, z_n$ fixed, we must have

$$b_{il} = 0, \qquad i = 1, 2, \dots, m; \; l = m+1, m+2, \dots, n$$

by the uniqueness of the solutions of ordinary differential equations. Hence the $\Phi_\mu(Y_1), \Phi_\mu(Y_2), \dots, \Phi_\mu(Y_m)$ restricted to $\varphi_\mu(U_\mu)$ are expressed in terms of the unit vectors $\partial/\partial z_1, \partial/\partial z_2, \dots, \partial/\partial z_m$. Hence if we define a C^r codimension $n-m$ foliation $\mathscr{F}_\mu$ of U_μ by

$$\mathscr{F}_\mu = \{ \varphi_\mu^{-1}(z_1, z_2, \dots, z_m, c_{m+1}, c_{m+2}, \dots, c_n); \; c_{m+i} \text{ a constant, } i = 1, 2, \dots, n-m \},$$

then $\mathscr{D}(\mathscr{F}_\mu) = \mathscr{D}^m | U_\mu$. Hence we have shown that $\mathscr{D}^m$ is completely integrable.

In Theorem 7.10, we restate Theorem 7.8 in terms of differential forms. To do this we need the following

LEMMA 7.9. *Let ω be a C^r one-form on M^n. Let X and Y be C^r vector fields on M^n. Then the following identity holds:*

$$d\omega(X, Y) = X(\omega(Y)) - Y(\omega(X)) - \omega([X, Y]). \tag{5}$$

PROOF. It is enough to show that equation (5) holds at a point $p \in M^n$. Let $(U_\lambda, \varphi_\lambda)$ be a chart about p, and set

$$(\Phi_\lambda^*)^{-1}\omega = \sum_{i=1}^{n} f_i\, dx_i, \qquad \Phi_\lambda(X) = \sum_{i=1}^{n} a_i \frac{\partial}{\partial x_i}, \qquad \Phi_\lambda(Y) = \sum_{i=1}^{n} b_i \frac{\partial}{\partial x_i},$$

where f_i, a_i and b_i are C^r functions defined on $\varphi_\lambda(U_\lambda)$, $i = 1, 2, \dots, n$. By Theorem 7.3 (iii)

$$\begin{aligned} &d\omega(p)(X(p), Y(p)) \\ &= \left(\sum_{i,j} \frac{\partial f_i}{\partial x_j}(\varphi_\lambda(p))\, dx_j \wedge dx_i \right) \left(\sum_{i=1}^{n} a_i(\varphi_\lambda(p)) \frac{\partial}{\partial x_i}, \sum b_i(\varphi_\lambda(p)) \frac{\partial}{\partial x_i} \right) \\ &= \sum_{j,i} a_j b_i \left(\frac{\partial f_i}{\partial x_j} - \frac{\partial f_j}{\partial x_i} \right) (\varphi_\lambda(p)), \end{aligned} \tag{6}$$

(cf. Lemma 7.1). On the other hand, we have

$$\begin{aligned} X(\omega(Y))(p) &= \left(\sum_{j=1}^{n} a_j \frac{\partial}{\partial x_j}\left(\sum_{i=1}^{n} b_i f_i\right)\right)(\varphi_\lambda(p)) \\ &= \sum_{i,j}\left(a_j f_i \frac{\partial b_i}{\partial x_j} + a_j b_i \frac{\partial f_i}{\partial x_j}\right)(\varphi_\lambda(p)). \end{aligned} \tag{7}$$

Similarly,

$$Y(\omega(X))(p) = \sum_{i,j}\left(b_j f_i \frac{\partial a_i}{\partial x_j} + b_j a_i \frac{\partial f_i}{\partial x_j}\right)(\varphi_\lambda(p)). \tag{8}$$

We also have

$$\begin{aligned} \omega(p)([X, Y](p)) &= \sum_{i=1}^{n} f_i \left(\sum_{j=1}^{n}\left(a_j \frac{\partial b_i}{\partial x_j} - b_j \frac{\partial a_i}{\partial x_j}\right)\right)(\varphi_\lambda(p)) \\ &= \sum_{i,j}\left(a_j f_i \frac{\partial b_i}{\partial x_j} - b_j f_i \frac{\partial a_i}{\partial x_j}\right)(\varphi_\lambda(p)). \end{aligned} \tag{9}$$

We combine equations (6), (7), (8) and (9) to conclude the proof.

THEOREM 7.10 (the Frobenius theorem, second form). *Let $\mathscr{D}^{n-q}$ be a C^r $(n-q)$-tangent plane field on an n-dimensional C^s manifold M^n, where $r \geq 2$. Then $\mathscr{D}^{n-q}$ is completely integrable if and only if for each point $p \in M^n$, there exist a neighborhood U and C^r one-forms $\omega_1, \omega_2, \dots, \omega_q$ on U which satisfy the following.*

(i) *The tangent plane defined by $\omega_1, \omega_2, \dots, \omega_q$ is $\mathscr{D}^{n-q}|U$.*

(ii) *$d\omega_i$ can be written as:*

$$d\omega_i = \sum_{j=1}^{q} \omega_j \wedge \theta_{ij}, \qquad i = 1, 2, \dots, q,$$

where θ_{ij} is a C^{r-1} one-form defined on U, $i, j = 1, 2, \dots, q$.

PROOF. We show first that the condition is sufficient. Let $(U_\lambda, \varphi_\lambda)$ be a chart of M^n about p such that $U_\lambda \subset U$. Choose C^r one-forms ω_{q+1}, $\omega_{q+2}, \dots, \omega_n$ on U_λ such that $\omega_1(p'), \omega_2(p'), \dots, \omega_n(p')$ form a basis in $(T_{p'}(M))^*$ for $p' \in U_\lambda$. Choose C^r vector fields $X_1, X_2, \dots, X_n$ on U_λ such that

$$(\Phi_\lambda^*)^{-1}(\omega_1(p')), (\Phi_\lambda^*)^{-1}(\omega_2(p')), \dots, (\Phi_\lambda^*)^{-1}(\omega_n(p'))$$

form the dual basis of $\Phi_\lambda(X_1(p')), \Phi_\lambda(X_2(p')), \dots, \Phi_\lambda(X_n(p'))$ for each $p' \in U_\lambda$. The vector fields $X_{q+1}, X_{q+2}, \dots, X_n$ generate $\mathscr{D}^{n-q}$ on U_λ. Since $\omega_i(X_j) = \delta_{ij}$ in U_λ, by Lemma 7.9 the following holds in U_λ:

$$d\omega_i(X_j, X_l) = -\omega_i([X_j, X_l]). \qquad i = 1, 2, \dots, q; \; j, l = q+1, q+2, \dots n. \tag{10}$$

By our assumption, the left-hand side of (10) is zero, so we get

$$-\omega_i([X_j, X_l]) = 0, \qquad i = 1, 2, \ldots, q; \quad j, l = q+1, q+2, \ldots, n.$$

Thus $\mathscr{D}^{n-q}$ is involutive on U_λ and hence on M^n. Therefore by Theorem 7.8 $\mathscr{D}^{n-q}$ is completely integrable.

We next show that the condition is necessary. Suppose that $\mathscr{D}^{n-q}$ is completely integrable. Let $\mathscr{F}$ be a $C^{r'}$ codimension q foliation of M^n such that $\mathscr{D}(\mathscr{F}) = \mathscr{D}^{n-q}$, $r' = r+1$ or r. Pick a foliated chart $(U_\lambda, \varphi_\lambda)$ of $\mathscr{F}$ about p. Then for a leaf L_α of $\mathscr{F}$, a connected component of $\varphi_\lambda(L_\alpha \cap U_\lambda)$ is of the form

$$\{(x_1, x_2, \ldots, x_n) \in \varphi_\lambda(U_\lambda); x_{n-q+i} = c_i,\ i = 1, 2, \ldots, q\}.$$

Hence, ω is written in terms of C^{r-1} functions c_{ij} defined on $\varphi_\lambda(U_\lambda)$ as

$$(\Phi_\lambda^*)^{-1}\omega_i = \sum_{j=n-q+1}^{n} c_{ij}\,dx_j, \qquad i = 1, 2, \ldots, q. \tag{11}$$

Since $\omega_1, \omega_2, \ldots, \omega_q$ are linearly independent, they satisfy

$$dx_j = \sum_{i=1}^{q} e_{ji}(\Phi_\lambda^*)^{-1}\omega_i, \qquad j = n-q+1, n-q+2, \ldots, n, \tag{12}$$

for some C^{r-1} functions e_{ij} defined on $\varphi_\lambda(U_\lambda)$. From (11) and (12), we get

$$\begin{aligned}(\Phi_\lambda^*)^{-1}\,d\omega_i &= \sum_{j=n-q+1}^{n}\left(\sum_{l=1}^{n}\frac{\partial c_{ij}}{\partial x_l}\,dx_l\right)\wedge dx_j \\ &= \sum_{j=n-q+1}^{n}\left(\sum_{i=1}^{q} e_{ji}(\Phi_\lambda^*)^{-1}\omega_i\right)\wedge\left(-\sum_{l=1}^{n}\frac{\partial c_{ij}}{\partial x_l}\,dx_l\right).\end{aligned}$$

Hence we have the desired result.

We proved in the Frobenius Theorem that for an involutive C^r plane field there exists a codimension $n-m$ foliation such that $\mathscr{D}(\mathscr{F}) = \mathscr{D}^m$. In general this foliation is of class C^r; however, the leaves of $\mathscr{F}$ are C^{r+1} manifolds (see Example A in §16).

For $q = 1$, Theorem 7.10 has a corollary given in the following

THEOREM 7.11. *Let M^n be a compact C^s manifold and let $\mathscr{D}^{n-1}$ be a C^r $(n-1)$-plane field on M^n, $r \geq 2$. Let the family $\{\omega_\sigma; \sigma \in \Sigma\}$ be a C^{r-1} one-form over some open cover $\{V_\sigma; \sigma \in \Sigma\}$ of M^n which defines the plane-field $\mathscr{D}^{n-1}$ (Lemma 7.7). Then $\mathscr{D}^{n-1}$ is completely integrable if and only if there exists a C^{r-1} differential form θ such that*

$$d\omega_\sigma = \omega_\sigma \wedge \theta$$

on each V_σ, $\sigma \in \Sigma$. *Further the necessary condition is equivalent to*

$$d\omega_\sigma \wedge \omega_\sigma = 0, \qquad \sigma \in \Sigma.$$

PROOF. That the condition is sufficient is a part of Theorem 7.10. Next assuming that $\mathscr{D}^{n-1}$ is completely integrable we wish to construct θ. By Theorem 7.10, there exist an open cover $\{U_i; i = 1, 2, \ldots, u\}$ of M^n and C^{r-1} one-forms θ_i on U_i, $i = 1, 2, \ldots, u$, such that

$$d\omega_\sigma = \omega_\sigma \wedge \theta_i, \qquad \sigma \in \Sigma,$$

on $V_\sigma \cap U_i$. Now let μ_i, $i = 1, 2, \ldots, u$, be a partition of unity subordinate to $\{U_i\}$, and set

$$\theta = \sum_{i=1}^{u} \mu_i \theta_i.$$

Then we have

$$\begin{aligned} \omega_\sigma \wedge \theta &= \sum_{i=1}^{u} \omega_\sigma \wedge \mu_i \theta_i = \sum_{i=1}^{u} \mu_i (\omega_\sigma \wedge \theta_i) \\ &= \sum_{i=1}^{u} \mu_i \, d\omega_\sigma = d\omega_\sigma. \end{aligned}$$

Moreover, if there exists θ such that $d\omega_\sigma \wedge \theta$, then clearly $d\omega_\sigma \wedge \omega_\sigma = 0$. Conversely if $d\omega_\sigma \wedge \omega_\sigma = 0$, $\sigma \in \Sigma$, then there exists θ_σ such that

$$d\omega_\sigma = \omega_\sigma \wedge \theta_\sigma$$

on V_σ. We construct a C^{r-1} one-form θ on M^n such that $d\omega_\sigma = \omega_\sigma \wedge \theta$ using the method above.

CHAPTER VIII

Cobordisms of Foliations

§29. Foliated cobordisms

The concept of cobordisms was introduced by Pontrjagin, Rochlin, Thom, and others in the 1950s, as a tool which provides a crude but a geometrically easy-to-handle classification of differentiable manifolds. It has come to play an important role in differential topology. Even when we add more structures to manifolds, the concept retains its utility. In this section we introduce cobordisms of foliated C^r manifolds.

Let M_1^n and M_2^n be C^s manifolds, and let $\mathcal{F}_1$ and $\mathcal{F}_2$ be their respective C^r codimension q foliations, $0 \le q \le n$, $0 \le r \le s$. We say that the foliated manifolds or foliations $(M_1^n, \mathcal{F}_1)$ and $(M_2^n, \mathcal{F}_2)$ are C^r *diffeomorphic* if there exists a C^r diffeomorphism

$$f: M_1^n \to M_2^n,$$

which maps leaves of $\mathcal{F}_1$ onto leaves of $\mathcal{F}_2$. For each leaf L_α of $\mathcal{F}_1$ $f(L_\alpha)$ is a leaf of $\mathcal{F}_2$. In this case, we write

$$(M_1^n, \mathcal{F}_1) = (M_2^n, \mathcal{F}_2).$$

We say that f is a *foliation-preserving* $\mathbf{C}^r$ *diffeomorphism* and write

$$f: (M_1^n, \mathcal{F}_1) \to (M_2^n, \mathcal{F}_2).$$

Notice that if $L'_{\alpha'}$ is a leaf of $\mathcal{F}_2$, then $f^{-1}(L'_{\alpha'})$ is a leaf of $\mathcal{F}_1$.

If M_1^n and M_2^n are oriented and $f: M_1^n \to M_2^n$ is an orientation-preserving C^r diffeomorphism taking a leaf of $\mathcal{F}_1$ onto a leaf of $\mathcal{F}_2$, then we write $(M_1^n, \mathcal{F}_1) = (M_2^n, \mathcal{F}_2)$ and say that $(M_1^n, \mathcal{F}_1)$ and $(M_2^n, \mathcal{F}_2)$ are C^r *diffeomorphic.*

In order to define a cobordism of $(M^n, \mathcal{F})$, it is necessary to modify the definition given in §16 of foliations of manifolds with boundaries. Let $(M^n, \mathcal{S}^{(r)})$ be an n-dimensional C^s manifold with boundary (see §16 for the definition of $\mathcal{S}^{(r)}$) and let $\mathcal{F} = \{L_\alpha; \alpha \in A\}$ be the family of arcwise-connected subsets L_α of M^n. We say that $\mathcal{F}$ is a C^r *codimension* q *foliation transverse to the boundary* if $\mathcal{F}$ satisfies the following.

($\hat{\mathcal{F}}_{\mathrm{I}}$) If $\alpha, \beta \in A$, $\alpha \ne \beta$, then $L_\alpha \cap L_\beta = \varnothing$.

($\hat{\mathcal{F}}_{\mathrm{II}}$) $\bigcup_{\alpha \in A} L_\alpha = M$.

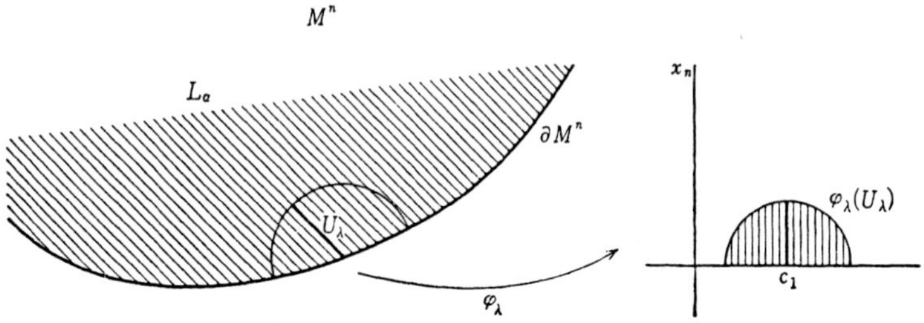

FIGURE 8.1

($\hat{\mathscr{F}}_{\mathrm{III}}$) Let $p \in M^n$. There exists a chart $(U_\lambda, \varphi_\lambda) \in \mathscr{S}^{(r)}$ about p with $\varphi_\lambda: U_\lambda \to \mathbf{R}^n_+$, such that for L_α, $\alpha \in A$, with $U_\lambda \cap L_\alpha \neq \varnothing$, each arcwise connected component of $\varphi_\lambda(U_\lambda \cap L_\alpha)$ is of the form

$$\{(x_1, x_2, \dots, x_n) \in \varphi_\lambda(U_\lambda); x_1 = c_1, x_2 = c_2, \dots, x_q = c_q\},$$

(see Figure 8.1). Here $c_1, c_2 \dots, c_q$ are constants depending on the connected components.

All the notation and terminology introduced in §16 carry over in this section: We denote by $(M^n, \mathscr{F})$ a manifold M^n with a foliation $\mathscr{F}$ transverse to the boundary or a foliation $\mathscr{F}$ of a manifold M^n transverse to the boundary, we call an arcwise-connected component L_α a leaf, a chart $(U_\lambda, \varphi_\lambda)$ in ($\hat{\mathscr{F}}_{\mathrm{III}}$) again a foliated chart, and so forth.

Let $(M^n, \mathscr{F})$ be a C^r codimension q foliation transverse to the boundary. Then a leaf L_α of $\mathscr{F}$ with $L_\alpha \cap \partial M^n \neq \varnothing$ intersects ∂M^n transversely, and if for $p \in L_\alpha \cap \partial M^n$, $(U_\lambda, \varphi_\lambda)$ is a foliated chart about p, then the connected components of $\varphi_\lambda(U_\lambda \cap L_\alpha \cap \partial M^n)$ are $(n-1-q)$-dimensional C^r submanifolds of $\varphi_\lambda(U_\lambda \cap \partial M^n)$ (Figure 8.1). Let a family $\partial\mathscr{F}$ be defined by

$$\partial\mathscr{F} = \{\text{the connected components of } L_\alpha \cap \partial M^n; \alpha \in A\};$$

the elements of $\partial\mathscr{F}$ are the arcwise-connected subsets of ∂M^n, and $\partial\mathscr{F}$ clearly satisfies ($\mathscr{F}_{\mathrm{I}}$) and ($\mathscr{F}_{\mathrm{II}}$) of §16 with respect to ∂M^n. Moreover, set

$$\partial\mathscr{S}^{(r)} = \{(U_\lambda \cap \partial M^n, \varphi_\lambda | U_\lambda \cap \partial M^n); (U_\lambda, \varphi_\lambda) \in \mathscr{S}^{(r)}\},$$

and consider the the submanifold $(\partial M^n, \partial\mathscr{S}^{(r)})$. If $(U_\lambda, \varphi_\lambda) \in \mathscr{S}^{(r)}$ is a foliated chart of $\mathscr{F}$ then for $L_\alpha \cap \partial M^n$ with $(U_\lambda \cap \partial M^n) \cap (L_\alpha \cap \partial M^n) \neq \varnothing$, a connected component of $\varphi_\lambda(U_\lambda \cap L_\alpha \cap \partial M^n)$ turns out to be a connected component of the set

$$\{(x_1, x_2, \dots, x_{n-1}) \in \varphi_\lambda(U_\lambda \cap \partial M^n); x_1 = c_1, x_2 = c_2, \dots, x_q = c_q\}.$$

Therefore $\partial\mathscr{F}$ is a C^r codimension q foliation of a C^s manifold ∂M^n without boundary of dimension $n-1$.

Thus we have obtained $(\partial M^n, \partial\mathscr{F})$ from $(M^n, \mathscr{F})$. When M^n is oriented, we give its boundary ∂M^n the induced orientation as defined in §11.

Let $\mathscr{F}_1$ and $\mathscr{F}_2$ be C^r codimension q foliations of n-dimensional closed oriented C^s manifolds M_1^n and M_2^n respectively. We say that the foliated manifolds $(M_1^n, \mathscr{F}_1)$ and $(M_2^n, \mathscr{F}_2)$ are *foliated cobordant,* or simply *cobordant,* if there exists an $(n+1)$-dimensional compact oriented C^s manifold W^{n+1} with a C^r codimension q foliation $(W^{n+1}, \hat{\mathscr{F}})$ transverse to the boundary such that

$$\partial W^{n+1} = M_1^n \cup (-M_2^n),$$

where $-M_2^n$ is M_2^n with the reversed orientation, and

$$(\partial W^{n+1}, \partial \hat{\mathscr{F}}) = (M_1^n, \mathscr{F}_1) \cup (-M_2^n, \mathscr{F}_2).$$

We call $(W^{n+1}, \hat{\mathscr{F}})$ a *foliated cobordism* between $(M_1^n, \mathscr{F}_1)$ and $(M_2^n, \mathscr{F}_2)$.

The cobordant relation on the C^r codimension q foliations of the closed oriented C^s n-manifolds is clearly an equivalence relation. An equivalence class under cobordisms is called a *foliated cobordism class.* The family of all such foliated cobordism classes is denoted by $\mathscr{F}\Omega^r_{n,q}$ and the cobordism class of a foliated manifold $(M^n, \mathscr{F})$ by $[(M^n, \mathscr{F})]$.

Let $\mathscr{F}_0 = \{\{x\} \times S^{n-q} ; x \in S^q\}$. Then $(S^q \times S^{n-q}, \mathscr{F}_0)$ is a C^r codimension q foliation of $S^q \times S^{n-q}$, which is also a bundle foliation. If we define a family $\hat{\mathscr{F}}_0$ of subsets in $S^q \times D^{n-q+1}$ by $\hat{\mathscr{F}}_0 = \{\{x\} \times D^{n-q+1} ; x \in S^q\}$, then $(S^q \times D^{n-q+1}, \hat{\mathscr{F}}_0)$ is a C^r codimension q foliation transverse to the boundary. If we orient the manifolds $S^q \times S^{n-q}$ and $S^q \times D^{n-q+1}$ in a suitable way, we will have

$$(\partial(S^q \times D^{n-q+1}), \partial\hat{\mathscr{F}}_0) = (S^q \times S^{n-q}, \mathscr{F}_0).$$

We now make $\mathscr{F}\Omega^r_{n,q}$ into an abelian group , defining the sum of two elements

$$[(M_1^n, \mathscr{F}_1)], [(M_2^n, \mathscr{F}_2)] \in \mathscr{F}\Omega^r_{n,q}$$

by

$$[(M_1^n, \mathscr{F}_1)] + [(M_2^n, \mathscr{F}_2)] = [(M_1^n \cup M_2^n, \mathscr{F}_1 \cup \mathscr{F}_2)].$$

The unit element of $\mathscr{F}\Omega^r_{n,q}$ is $[(S^q \times S^{n-q}, \mathscr{F}_0)]$. The inverse element of $[(M^n, \mathscr{F})]$ is $[(-M^n, \mathscr{F})]$, because if we set $\mathscr{F} \times I = \{L_\alpha \times I ; L_\alpha \in \mathscr{F}\}$, then $(M^n \times I, \mathscr{F} \times I)$ is a C^r codimension q foliation of $(M^n \times I)$ transverse to the boundary such that

$$(\partial(M^n \times I), \partial(\mathscr{F} \times I)) = (M^n, \mathscr{F}) \cup (-M^n, \mathscr{F}),$$

with a suitable orientation on $M^n \times I$. The group $\mathscr{F}\Omega^r_{n,q}$ is called the *n-dimensional foliated C^r codimension q cobordism group.*

Two closed oriented manifolds M_1^n and M_2^n are said to be cobordant if there exists a compact oriented $(n+1)$-dimensional C^r manifold W^{n+1} such that $\partial W^{n+1} = M_1^n \cup (-M_2^n)$. This is an equivalence relation whose equivalence classes are called *cobordism classes*. The set of cobordism classes over n-dimensional manifolds is denoted by Ω_n ; Ω_n is an abelian group—in the same way that $\mathcal{F}\Omega^r_{n,q}$ is—and is called the *n dimensional C^r cobordism group*. The equivalence class of M^n is denoted by $[M^n]$.

We now define a map

$$f\colon \mathcal{F}\Omega^r_{n,q} \to \Omega_n$$

by $f([(M^n, \mathcal{F})]) = [M^n]$; then f is evidently a homomorphism. We shall show in §31 that this map is not one-to-one. Already in the case in which $n = 3, q = 1, r = \infty$, the cardinality of $f^{-1}(0)$ is at least the continuum.

It goes without saying that foliated manifolds are more complicated than their counterparts without foliations. Thus as the reader will see in the following example a complete classification of any sort is almost impossible; however, the cobordism group $\mathcal{F}\Omega^r_{n,q}$ provides one convenient means of classifying foliated manifolds.

Let M^n be a closed oriented C^s manifold and $\mathcal{F}$ a C^r codimension one foliation of M^n. Let $\mathcal{F}'$ be the C^{r-1} codimension $n-1$ foliation of M^n transverse to $\mathcal{F}$ constructed for Theorem 4.2 in §16. Applying the method used in the proof of Lemma 6.6 to a leaf L_α of $\mathcal{F}'$, we obtain a simple closed C^r curve in M^n:

$$l\colon S^1 \to M^n,$$

which intersects the leaves of $\mathcal{F}$ transversely. For this l we can construct a C^r embedding of $D^{n-1} \times S^1$:

$$g\colon D^{n-1} \times S^1 \to M^n,$$

such that $g|\{0\} \times S^1 = l$, and for each $y \in S^1$, $g(D^{n-1} \times \{y\})$ is contained in a leaf of $\mathcal{F}$ (see Figure 8.2 (i)). Denote by $\mathcal{F}_1$ the restriction of $\mathcal{F}$ to the submanifold $g(D^{n-1} \times S^1)$ and by $\mathcal{F}_2$ the restriction of $\mathcal{F}$ to the submanifold $M^n - g(\operatorname{Int} D^{n-1} \times S^1)$. Then $\mathcal{F}_1$ is a foliation of $g(D^{n-1} \times S^1)$ and $\mathcal{F}_2$ a foliation of $M^n - g(\operatorname{Int} D^{n-1} \times S^1)$; each foliation is of class C^r and is of codimension one transverse to the boundary. We construct a smooth codimension one foliation $\mathcal{F}'$ of $D^{n-1} \times S^1$ using the map f of (1) in §16 and following the method of construction for $D^2 \times S^1$ in Example B verbatim: the leaves of $\mathcal{F}'$ consist of $S^{n-2} \times S^1$ and the sets

$$L_\alpha = \{ (x_1, x_2, \dots, x_{n-1}, \exp 2\pi(\alpha + f((x_1^2 + x_2^2 + \cdots + x_{n-1}^2)^{1/2}))i);$$
$$(x_1, x_2, \dots, x_{n-1}) \in D^{n-1} \}.$$

The image of the leaves of $\mathcal{F}'$ under g defines a C^r codimension one foliation of $g(D^{n-1} \times S^1)$, which will be denoted by $g(\mathcal{F}')$. We obtain $g(\mathcal{F}')$ from $\mathcal{F}_1$ by letting each leaf $g(D^{n-1} \times \{x\})$ spiral towards the boundary $g(S^{n-2} \times S^1)$ of $g(D^{n-1} \times S^1)$ (Figure 8.2 (ii)). In a like manner we

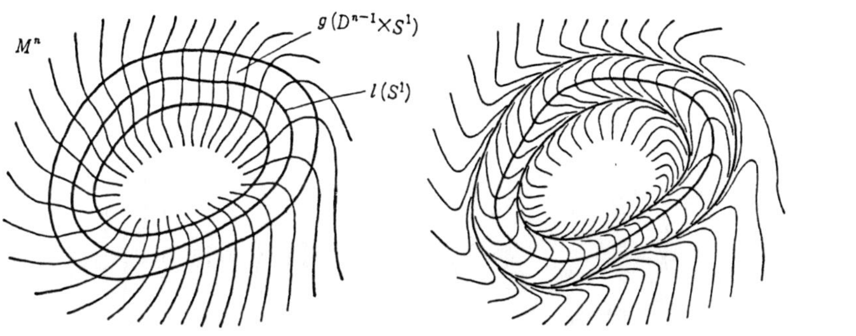

FIGURE 8.2

modify $\mathcal{F}_2$ by letting the leaves spiral towards $g(S^{n-2}\times S^1)$ in some neighborhood of the boundary $g(S^{n-2}\times S^1)$ and adding $g(S^{n-2}\times S^1)$ as a leaf to obtain a C^r codimension one foliation $\bar{\mathcal{F}}_2$ of $M^n - g(\operatorname{Int} D^{n-1}\times S^1)$. If we let $\bar{\mathcal{F}} = g(\mathcal{F}')\cup\bar{\mathcal{F}}_2$, then $\bar{\mathcal{F}}$ is a C^r codimension one foliation of M^n. We note that $\bar{\mathcal{F}} \neq \mathcal{F}$ since $\bar{\mathcal{F}}$ has a compact leaf, namely $g(S^{n-2}\times S^1)$, which is not a leaf of $\mathcal{F}$. In this way one can construct a new foliation $\bar{\mathcal{F}}$ from any foliation $\mathcal{F}$ of M^n.

We shall show next, however, that the above $\mathcal{F}$ and $\bar{\mathcal{F}}$ are cobordant. We consider first a C^r codimension one foliation $\hat{\mathcal{F}} = \{L_\alpha \times I;\, L_\alpha \in \mathcal{F}\}$ of the manifold $M^n\times I$, which is transverse to the boundary. If we define a curve

$$\hat{l}\colon S^1 \to M^n\times I$$

by

$$\hat{l}(y) = (l(y),\, 1), \qquad y\in S^1,$$

then $\hat{l}$ is an embedding. Set $D^n_- = \{(x_1, x_2 \dots, x_n)\in D^n;\, x_n \le 0\}$. For the curve $\hat{l}$, there exists a C^r map:

$$\hat{g}\colon D^n_-\times S^1 \to M^n\times I,$$

which satisfies the following.

(i) $\hat{g}|(D^n_- - D^{n-1})\times S^1$ is a C^r embedding.

(ii) $\hat{g}|(D^n_-\cap \operatorname{Int} D^n)\times S^1$ is a C^r embedding.

(iii) $\hat{g}(D^n_-\times\{y\})$ is contained in a leaf of $\hat{\mathcal{F}}$, $y\in S^1$.

(iv) For $x\in D^{n-1}$, $\hat{g}((x, y)) = (g(x, y), 1)$, (see Figure 8.3 (i)).

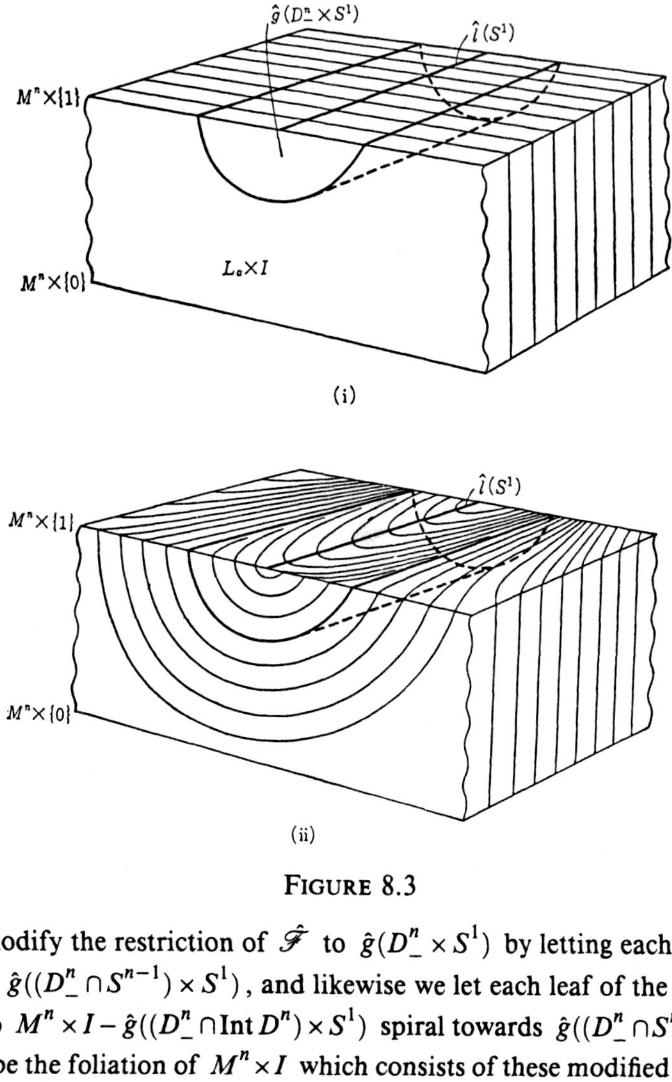

FIGURE 8.3

We modify the restriction of $\hat{\mathscr{F}}$ to $\hat{g}(D^n_- \times S^1)$ by letting each leaf spiral towards $\hat{g}((D^n_- \cap S^{n-1}) \times S^1)$, and likewise we let each leaf of the restriction of $\hat{\mathscr{F}}$ to $M^n \times I - \hat{g}((D^n_- \cap \operatorname{Int} D^n) \times S^1)$ spiral towards $\hat{g}((D^n_- \cap S^{n-1}) \times S^1)$. Let $\hat{\mathscr{F}}_1$ be the foliation of $M^n \times I$ which consists of these modified leaves and the leaf $\hat{g}((D^n_- \cap S^{n-1}) \times S^1)$. Then $\hat{\mathscr{F}}_1$ is a C^r codimension one foliation of $M^n \times I$ transverse to the boundary. Moreover

$$\hat{\mathscr{F}}_1 | M^n \times \{0\} = \mathscr{F}, \qquad \hat{\mathscr{F}}_1 | M^n \times \{1\} = \bar{\mathscr{F}},$$

so we have

$$(\partial(M^n \times I), \partial\hat{\mathscr{F}}_1) = (M^n, \mathscr{F}) \cup (-M^n, \bar{\mathscr{F}}).$$

Hence, $\mathscr{F}$ and $\bar{\mathscr{F}}$ are cobordant.

Let M^n be an n-dimensional C^s manifold with boundary. A map

$$\mathscr{D}^m : M^n \to \bigcup_{p \in M^n} G^m_p,$$

which assigns to each $p \in M^n$ an m-dimensional subspace of $T_p(M^n)$ is called a C^r *m-plane field* on M^n *transverse to the boundary* if it satisfies the following.

(i) If $p \in \partial M^n$, $\mathscr{D}^m(p) \cap T_p(\partial M^n)$ is an $(m-1)$-dimensional subspace of the tangent space $T_p(\partial M^n)$.

(ii) For each $p \in M^n$, there exist a neighborhood U of p and C^r vector fields $X_1, X_2, \ldots, X_m$ on U such that $X_1(p'), X_2(p'), \ldots, X_m(p')$ form a basis in $\mathscr{D}^m(p')$, $p' \in U$.

Let $\mathscr{F} = \{L_\alpha; \alpha \in A\}$ be a C^r codimension q foliation of M^n transverse to the boundary with $r \geq 1$. By assigning to each point p in M^n the $(n-q)$-dimensional tangent plane $T_p(L_\alpha)$ which consists of all vectors of $T_p(M^n)$ tangent to L_α, $p \in L_\alpha$, we obtain a C^{r-1} $(n-q)$-plane field on M^n transverse to the boundary; we denote this plane field by $\mathscr{D}(\mathscr{F})$.

Let $\mathscr{D}^{n-q}$ be a C^r $(n-q)$-plane field on M^n transverse to the boundary, and suppose that there exists a $C^{r'}$ codimension q foliation of M^n transverse to the boundary such that $\mathscr{D}(\mathscr{F}) = \mathscr{D}^{n-q}$, where $r' = r+1$ or r. We then say that $\mathscr{D}^{n-q}$ is *completely integrable.*

The theorems of Frobenius—Theorems 7.8 and 7.10—are directly applicable to a necessary and sufficient condition that an $(n-q)$-dimensional tangent plane field $\mathscr{D}^{n-q}$ transverse to the boundary be completely integrable. Further, if M^n is compact and $q = 1$, then Theorem 7.11 holds verbatim.

§30. Foliated cobordism invariants (Godbillon-Vey numbers)

Let M^n be a compact arcwise connected C^s manifold with or without boundary, $s \geq 4$. Let $\mathscr{F}$ be a C^r codimension one foliation of M^n, which is transverse to the boundary if $\partial M^n \neq \varnothing$, $r \geq 4$. By Lemma 7.7 there exists a C^{r-1} one-form $\{\omega_\sigma; \sigma \in \Sigma\}$ over an open cover $\{V_\sigma; \sigma \in \Sigma\}$ which defines the C^{r-1} $(n-1)$-plane field $\mathscr{D}(\mathscr{F})$. By Theorem 7.11, we may choose a C^{r-2} one-form θ on M^n such that for each ω_σ,

$$d\omega_\sigma = \omega_\sigma \wedge \theta,$$

on V_σ. We then define a C^{r-3} three-form Γ on M^n by

$$\Gamma = \theta \wedge d\theta,$$

which is called a *Godbillon-Vey form of* $\mathscr{F}$. Here we assumed that $r \geq 4$; however, we can define Γ when $r = 3$ if the plane field $\mathscr{D}(\mathscr{F})$ is actually differentiable of class C^3.

LEMMA 8.1. *A Godbillon-Vey form* Γ *is closed and differentiable of class* C^{r-3}.

PROOF. From $d\omega_\sigma = \omega_\sigma \wedge \theta$, we get

$$0 = d(d\omega_\sigma) = d\omega_\sigma \wedge \theta - \omega_\sigma \wedge d\theta = \omega_\sigma \wedge \theta \wedge \theta - \omega_\sigma \wedge d\theta = -\omega_\sigma \wedge d\theta.$$

Hence, there exists a C^{r-3} one-form η_σ on V_σ such that

$$d\theta = \omega_\sigma \wedge \eta_\sigma.$$

Therefore, on V_σ we must have:

$$d\Gamma = d(\theta \wedge d\theta) = d\theta \wedge d\theta = \omega_\sigma \wedge \eta_\sigma \wedge \omega_\sigma \wedge \eta_\sigma = 0.$$

LEMMA 8.2. *Let θ' be a C^{r-2} one-form different from θ such that for each ω_σ, $d\omega_\sigma = \omega_\sigma \wedge \theta'$ on V_σ. Define a three-form Γ' by $\Gamma' = \theta' \wedge d\theta'$. Then $\Gamma - \Gamma'$ is exact and differentiable of class C^{r-3}.*

PROOF. Since $d\omega_\sigma = \omega_\sigma \wedge \theta = \omega_\sigma \wedge \theta'$, we have

$$\omega_\sigma \wedge (\theta' - \theta) = 0.$$

Thus there exists a C^{r-2} function f_σ defined on V_σ such that

$$\theta' - \theta = f_\sigma \omega_\sigma.$$

When $V_\sigma \cap V_\nu \neq \varnothing$, for some $\sigma, \nu \in \Sigma$, we have that $\omega_\sigma = \pm\omega_\nu$ on $V_\sigma \cap V_\nu$; so $f_\sigma = \pm f_\nu$. Hence on V_σ,

$$\begin{aligned}
\theta' \wedge d\theta' &= (\theta + f_\sigma \omega_\sigma) \wedge d(\theta + f_\sigma \omega_\sigma) \\
&= (\theta + f_\sigma \omega_\sigma) \wedge (d\theta + df_\sigma \wedge \omega_\sigma + f_\sigma \, d\omega_\sigma) \\
&= \theta \wedge d\theta - d(f_\sigma \theta \wedge \omega_\sigma) + f_\sigma^2 \omega_\sigma \wedge d\omega_\sigma \\
&= \theta \wedge d\theta - d(f_\sigma \theta \wedge \omega_\sigma).
\end{aligned}$$

But we have

$$f_\sigma \theta \wedge \omega_\sigma = f_\nu \theta \wedge \omega_\nu$$

on $V_\sigma \cap V_\nu$. Hence the two-forms $f_\sigma \theta \wedge \omega_\sigma$, $\sigma \in \Sigma$, define a C^{r-2} one-form, say η, on M^n. Therefore we have

$$\theta \wedge d\theta - \theta' \wedge d\theta' = d\eta.$$

LEMMA 8.3. *Suppose that a C^{r-1} one-form $\{\omega'_{\sigma'}; \sigma' \in \Sigma'\}$ over an open cover $\{V'_{\sigma'}; \sigma' \in \Sigma'\}$ also defines the C^{r-1} plane field $\mathscr{D}(\mathscr{F})$. Define a one-form θ'' by*

$$d\omega'_{\sigma'} = \omega'_{\sigma'} \wedge \theta'', \qquad \sigma' \in \Sigma',$$

and let Γ'' be a three-form defined by

$$\Gamma'' = \theta'' \wedge d\theta''.$$

Then $\Gamma - \Gamma''$ is exact and differentiable of class C^{r-3}.

PROOF. Let $\Sigma'' = \{V_\sigma \cap V_{\sigma'}; \sigma \in \Sigma, \sigma' \in \Sigma'\}$. Then Σ'' is an open cover of M^n. Let $\omega_{\sigma''}$ and $\omega'_{\sigma''}$ be the restrictions of ω_σ and $\omega'_{\sigma'}$ respectively to $V''_{\sigma''} = V_\sigma \cap V'_{\sigma'}$. Replacing $\{\omega_\sigma; \sigma \in \Sigma\}$ by $\{\omega_{\sigma''}; \sigma'' \in \Sigma''\}$, and the open cover $\{\omega'_{\sigma'}; \sigma' \in \Sigma'\}$ by $\{\omega'_{\sigma''}; \sigma'' \in \Sigma''\}$ if necessary, we may assume that $\Sigma = \Sigma'$.

By assumption, there exists a C^{r-1} function g_σ on V_σ such that

$$\omega'_\sigma = g_\sigma \omega_\sigma.$$

The value of g_σ is never zero on V_σ. If we now replace $\omega'_{\sigma'}$ by $-\omega'_{\sigma'}$, we retain the equality

$$d(-\omega'_{\sigma'}) = (-\omega'_{\sigma'}) \wedge \theta''.$$

Hence by choosing the appropriate sign of $\pm\omega'_{\sigma'}$ in each arcwise connected component of $V_{\sigma'}$ and replacing $\{\omega'_{\sigma'}; \sigma' \in \Sigma\}$ by $\{\pm\omega'_{\sigma'}; \sigma' \in \Sigma\}$, we may assume that g_σ is a positive-valued function. Since $g_\sigma = \pm g_\nu$ on $V_\sigma \cap V_\nu$ when $V_\sigma \cap V_\nu \neq \varnothing$, the functions g_σ, $\sigma \in \Sigma$, define a C^{r-1} function g on M^n provided that we choose each g_σ, $\sigma \in \Sigma$, to be positive-valued.

By definition,

$$\omega'_\sigma = g\omega_\sigma,$$

on V_σ, so we have

$$\begin{aligned} d\omega'_\sigma &= dg \wedge \omega_\sigma + g\, d\omega_\sigma \\ &= \omega_\sigma \wedge (-dg + g\theta) \\ &= \omega'_\sigma \wedge \left(\frac{-dg}{g} + \theta\right). \end{aligned}$$

Hence we may take $-d(\log g) + \theta$ for θ''. But then we have

$$d\theta'' = d(-d(\log g) + \theta) = d\theta.$$

Thus,

$$\begin{aligned} \theta'' \wedge d\theta'' &= (-d(\log g) + \theta) \wedge d\theta \\ &= d((-\log g)\, d\theta) + \theta \wedge d\theta. \end{aligned}$$

Lemmas 8.1, 8.2, and 8.3 imply that the closed C^{r-3} three-form Γ is well-defined by a C^r foliation $\mathscr{F}$ up to the exact C^{r-3} three-forms.

Given a compact C^r n-manifold M^n we denote by $H^q_D(M^n, \mathbf{R})$ the quotient space of the vector space consisting of all closed C^{r-1} q-forms by the subspace of the exact C^{r-1} q-forms, where $r \geq 2$. We call $H^q_D(M^n, \mathbf{R})$ the q-dimensional *de Rham cohomology group* and its elements q-dimensional *de Rham cohomology classes.* As we saw above, a codimension one foliation $\mathscr{F}$ determines the three-dimensional de Rham cohomology class $[\Gamma]$, called the *Godbillon-Vey characteristic class* of $\mathscr{F}$.

EXAMPLE 1. Let E be a closed C^r manifold, $\pi: E \to S^1$ a C^r bundle and $\mathscr{F} = \{\pi^{-1}(y); y \in S^1\}$ the C^r codimension one bundle foliation defined by π, $r \geq 4$. If we let y be a coordinate representing the points of the interval $[0, 1]$ and regard S^1 as $[0, 1]$ with $\{0\}$ and $\{1\}$ identified, dy is a C^∞ one-form on the unit circle S^1. Then $\mathscr{D}(\mathscr{F})$ is the C^{r-1} $(n-1)$-plane field defined by $\pi^* dy$, and since $d(dy) = 0$ we have

$$d(\pi^* dy) = 0.$$

Hence we can take zero for θ, so the Godbillon-Vey form Γ of a bundle foliation is zero.

EXAMPLE 2. We shall calculate the Godbillon-Vey form of the Reeb foliation $\mathscr{F}_R$ of S^3 (Example B, §16). Represent the points of D^2 by the polar coordinates (r, ψ), $0 \leq r \leq 1$, $0 \leq \psi \leq 2\pi$, and let $(r, \psi, y) \in D^2 \times S^1$ with $y \in S^1$. Let $f : (-1, 1) \to \mathbf{R}$ be the function (1) in Example B in §16. Further let

$$h : [0, 1] \to [0, 1]$$

be a monotone decreasing C^∞ function which takes on the values one near zero and zero near one (cf. Lemma 1.5). Define a C^∞ one-form ω on S^3 by

$$\omega = h(r)\left(dy - \frac{df(r)}{dr}\,dr\right) + (1 - h(r))\left(\left(1 \middle/ \frac{df(r)}{dr}\right) dy - dr\right),$$

on $\operatorname{Int} D_1^2 \times S_1^1$, and by

$$\omega = -\,dr, \quad \text{on } \partial D_1^2 \times S_1^1.$$

On $D_2^2 \times S_2^1$, ω is defined with the opposite sign.

The C^∞ two-plane field $\mathscr{D}(\mathscr{F}_{\mathbf{R}})$ on S^3 is defined by ω. Via a simple calculation, we get

$$d\omega = -\omega \wedge \frac{d}{dr}\left(\log\left(h(r) + (1 - h(r)) \middle/ \frac{df(r)}{dr}\right)\right) dr.$$

Hence the Godbillon-Vey form Γ of the Reeb foliation is zero.

Let M^3 be a closed oriented three-dimensional C^s-manifold and let $\mathscr{F}$ be a C^r codimension one foliation of M^3 with $s, r \geq 4$. We saw above that the foliation $\mathscr{F}$ defines a C^{r-3} three-form Γ. The integral of Γ over M^3:

$$\int_{M^3} \Gamma,$$

is called the *Godbillon-Vey number* and is denoted by $\text{G.V.}(M^3, \mathscr{F})$. Recall that the Γ is well defined by $\mathscr{F}$ up to the exact C^{r-s} three-forms. Now let

$$\Gamma' = \Gamma + d\Theta, \qquad \Theta \text{ a } C^{r-2} \text{ 2-form on } M^3.$$

Then, Stokes theorem—Theorem 7.6—implies that

$$\int_M \Gamma' = \int_M \Gamma + \int_M d\Theta = \int_M \Gamma.$$

Thus the Godbillon-Vey number of $\mathscr{F}$ is well defined.

The next theorem shows that a Godbillon-Vey number is a foliated cobordism invariant.

THEOREM 8.4. *Let M_1^3 and M_2^3 be closed oriented C^s manifolds and let $(M_1^3, \mathscr{F}_1)$ and $(M_2^3, \mathscr{F}_2)$ be C^r codimension one foliations, $r \geq 4$. Suppose that $(M_1^3, \mathscr{F}_1)$ and $(M_2^3, \mathscr{F}_2)$ are foliated cobordant. Then their Godbillon-Vey numbers are equal:*

$$\text{G.V.}(M_1^3, \mathscr{F}_1) = \text{G.V.}(M_2^3, \mathscr{F}_2).$$

PROOF. Let $(W^4, \mathscr{F})$ be a cobordism between $(M_1^3, \mathscr{F}_1)$ and $(M_2^3, \mathscr{F}_2)$. Let

$$(\partial W^4, \partial\mathscr{F}) = (M_1^3, \mathscr{F}_1) \cup (-M_2^3, \mathscr{F}_2).$$

Let $\{U_\sigma\}_{\sigma=1}^u$ be an open cover of W^4 and $\{\omega_\sigma\}_{\sigma=1}^u$ a C^{r-1} one-form over $\{U_\sigma\}$ which defines $\mathscr{D}(\mathscr{F})$. Among the open sets U_σ, $\sigma = 1, 2, \ldots, u$, let $U_{i_1}, U_{i_2}, \ldots, U_{i_{u'}}$ be those with $U_\sigma \cap M_1^3 \neq \varnothing$ and $U_{j_1}, U_{j_2}, \ldots, U_{j_{u''}}$ those with $U_\sigma \cap M_2^3 \neq \varnothing$. Then $\{M_1^3 \cap U_{i_k}\}_{k=1}^{u'}$ and $\{M_2^3 \cap U_{j_k}\}_{k=1}^{u''}$ are open covers of M_1^3 and M_2^3 respectively. If ω'_{i_k} denotes the restriction of the one-form ω_{i_k} to $M_1^3 \cap U_{i_k}$ and ω''_{j_k} the restriction of ω_{j_k} to $M_2^3 \cap U_{j_k}$, then $\{\omega'_{i_k}\}_{k=1}^{u'}$ defines $\mathscr{D}(\mathscr{F}_1)$ and $\{\omega''_{j_k}\}_{k=1}^{u''}$ defines $\mathscr{D}(\mathscr{F}_2)$

By Theorem 7.11, $\{\omega_\sigma\}_{\sigma=1}^u$ determines a C^{r-2} one-form θ on W^4. Let $\Gamma = \theta \wedge d\theta$ be the Godbillon-Vey form of $\mathscr{F}$. Let

$$\imath_1^*\Gamma\colon M_1^3 \to W^4, \qquad \imath_2^*\colon M_2^3 \to W^4$$

be the inclusion maps and denote

$$\theta_1 = \imath_1^*\theta, \qquad \theta_2 = \imath_2^*\theta.$$

Then as in Theorem 7.11, θ_1 and θ_2 are C^{r-2} one-forms on M_1^3 and M_2^3 which are defined by $\{\omega'_{i_k}\}_{k=1}^{u'}$ and $\{\omega''_{j_k}\}_{k=1}^{u''}$ respectively. The Godbillon-Vey forms of $\mathscr{F}_1$ and $\mathscr{F}_2$ are

$$\Gamma_1 = \theta_1 \wedge d\theta_1, \qquad \Gamma_2 = \theta_2 \wedge d\theta_2.$$

Since $\imath_1^*\Gamma = \Gamma_1$ and $\imath_2^*\Gamma = \Gamma_2$, we have

$$\begin{aligned} \text{G.V.}(M_1^3, \mathscr{F}_1) - \text{G.V.}(M_2^3, \mathscr{F}_2) &= \int_{M_1^3} \Gamma_1 - \int_{M_2^3} \Gamma_2 \\ &= \int_{\partial W^4} \imath^*\Gamma = \int_{W^4} d\Gamma \\ &= 0, \end{aligned}$$

where $\imath\colon \partial W^4 \to W^4$ is the inclusion map.

In the next section we shall show that the Godbillon-Vey number is indeed a meaningful invariant.

§31. Foliated cobordisms of S^3 (theorem of Thurston)

We consider the upper half plane $H = \{z = x + yi; y > 0\}$ of $\mathbf{C}$. We identify H with $\mathbf{R}_+^2 - \mathbf{R}^1 = \{(x, y); y > 0\}$ by associating to each

$z = x + yi$ the ordered pair (x, y). We denote by $\mathrm{SL}(2, \mathbf{R})$ the set of all real matrices $\left(\begin{smallmatrix} a & b \\ c & d \end{smallmatrix}\right)$ with $ad - bc = 1$, and turn $\mathrm{SL}(2, \mathbf{R})$ into a group with the matrix multiplication. Given an element $\left(\begin{smallmatrix} a & b \\ c & d \end{smallmatrix}\right)$ of $\mathrm{SL}(2, \mathbf{R})$, we define a linear transformation $\tau \colon \mathbf{C} \to \mathbf{C}$ by $\tau(z) = \frac{az+b}{cz+d}$; it is evident that we have $\tau(H) = H$, and so the restriction of τ to H is a linear transformation

$$\tau : H \to H.$$

The equation of a circle or a line in $\mathbf{C}$ is given by

$$Az\bar{z} + Bz + \bar{B}\bar{z} + C = 0, \tag{1}$$

where A, C are real numbers and $B\bar{B} - AC > 0$. Now put $w = \frac{az+b}{cz+d}$; then $z = \frac{-dw+b}{cw-a}$, and substituting this in (1), we obtain

$$A'w\bar{w} + B'w + \bar{B}'\bar{w} + C' = 0, \tag{2}$$

where

$$A' = Ad^2 + Cc^2 - (B + \bar{B})cd, \qquad B' = Bad + \bar{B}bc - Abd - Cac,$$
$$C' = Ab^2 + Ca^2 - (B + \bar{B})ab.$$

Since A', C' are real numbers and $B'\bar{B}' - A'C' > 0$, the equation (2) also represents a circle or a line: the linear transformation τ maps a circle or a line onto a circle or a line. In particular if B is real in (1), the center of the circle (1) is on the real axis as well as the center of the circle (2).

The intersection of a circle centered on the real line and the upper half plane H is called a *circular arc in* H. The intersection of H and a line parallel to the imaginary axis, a circle centered at ∞, is also called a circular arc. Suppose we orient a circular arc in H as the arc directed from P to P' in Figure 8.4; it will be called a *circular arc in* H *starting at* P. If L and L' are oriented circular arcs in H, it should be obvious that some orientation-preserving linear transformation defined by an element of $\mathrm{SL}(2, \mathbf{R})$ maps L onto L'.

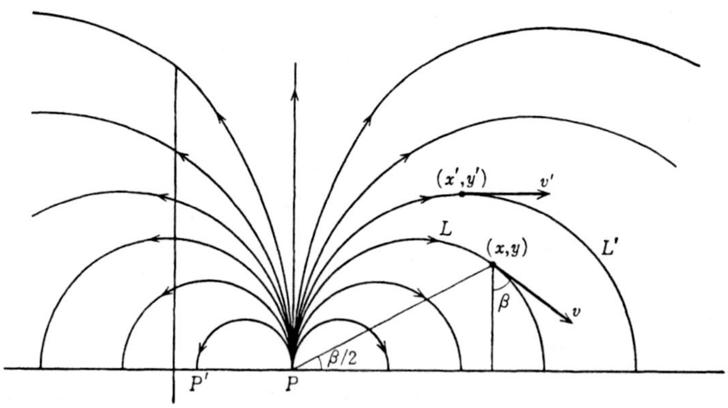

FIGURE 8.4

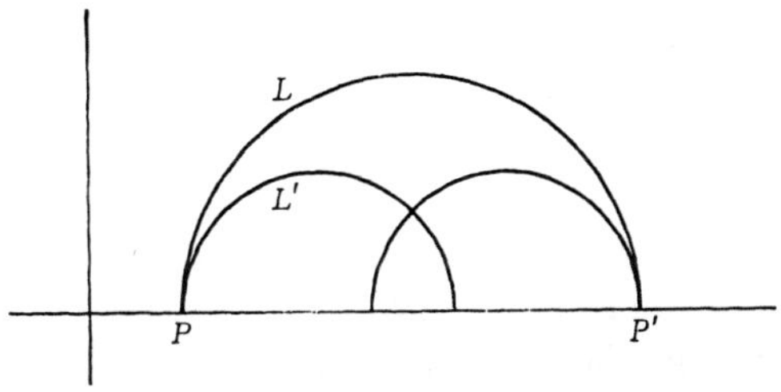

FIGURE 8.5

Let a map

$$r' : H \to H$$

be defined by $r'(z) = 1/\bar{z}$. Set $S^1 = \{z; |z| = 1\}$. Then each point of $H \cap S^1$ is left fixed by r'. The map r' is the *reflection* with respect to $H \cap S^1$. Let L be a circular arc in H and τ a linear transformation mapping $H \cap S^1$ onto L, then $\tau \circ r' \circ \tau^{-1}$ leaves each point of L fixed. We say that $\tau \circ r' \circ \tau^{-1} : H \to H$ is a reflection with respect to L.

Now introduce the Riemannian metric

$$ds^2 = \frac{1}{y^2}(dx^2 + dy^2) \tag{3}$$

to the two-dimensional C^∞ manifold H (cf. Appendix VIII, Note 1). Circular arcs in H turn out to be geodesics with respect to this metric. Taking circular arcs in H to be 'lines' we obtain a model for the non-Euclidean geometry of Lobachevsky. We define 'lines' L and L' to be parallel if they approach arbitrarily close to the same point P on the real axis. With this definition all but the parallel postulate of the axioms for Euclidean geometry are satisfied: given a 'line' L and a point Q not on L there is more than one 'line' passing through Q and parallel to L (Figure 8.5).

Let $\tau : H \to H$ be a linear transformation defined by an element of the group $\mathrm{SL}(2, \mathbf{R})$. Since τ maps a circular arc in H onto a circular arc in the half plane H, we have

$$\langle \tau_* u, \tau_* v \rangle_{\tau(z)} = \langle u, v \rangle_z, \qquad u, v \in T_z(H),$$

where $\langle\ ,\ \rangle_z$ is the inner product defined by the Riemannian metric (3). Hence the Riemannian metric (3) is invariant under linear transformations defined by elements of $\mathrm{SL}(2, \mathbf{R})$.

An element in the tangent space $T(H)$ of H is given by (z, v), $v \in T_z(H)$ or by (x, y, v), where $z = x + yi$. The correspondence

$$(x, y, v) \mapsto ((x, y), v) \tag{4}$$

is a C^∞ diffeomorphism of C^∞ manifolds $T(H)$ and $H \times \mathbf{R}^2$.

Denote by E the tangent sphere bundle of H:

$$\{(x, y, v) \in T(H); \| v \| = 1\},$$

and by $\hat{\pi}$ the C^∞ projection

$$\hat{\pi}: E \to H, \qquad \hat{\pi}(x, y, v) = (x, y).$$

The total space E is a C^∞ three-manifold diffeomorphic to $H \times S^1$ under the correspondence (4). We give $H \times S^1$ the Riemannian metric consisting of the metric (3) on H and the standard Riemannian metric on S^1.

Let L and L' be circular arcs in H starting at P, and let $(x, y) \in L$ and $(x', y') \in L'$ (Figure 8.4). Now consider vectors

$$v \in T_{(x,y)}(H), \qquad v' \in T_{(x',y')}(H),$$

tangent to L and L', pointed with the orientations of L and L' respectively. We then say that v and v' are *parallel* (Figure 8.4). We define a relation $\sim$ on E:

$$(x, y, v) \sim (x', y', v')$$

if and only if v and v' are parallel; $\sim$ is an equivalence relation. If L_α, $\alpha \in A$, denotes an equivalence class under $\sim$, then $\mathscr{F} = \{L_\alpha; \alpha \in A\}$ is a C^∞ codimension one foliation of E. This will be verified below where we define a C^∞ one-form ω (6) on E which satisfies $d\omega \wedge \omega = 0$ in (8).

Pick a point P on the real axis—including ∞—and consider the family of all circular arcs in H starting at P (Figure 8.4). For a circular arc L in this family the set of all tangent vectors of L of length one which is pointed in the same direction as L is a leaf L_α. Hence, we might use $A = \mathbf{R} \cup \{\infty\}$ as the index set for $\mathscr{F}$ letting the coordinate of P be $\alpha \in \mathbf{R} \cup \{\infty\}$. A leaf L_α gives rise to the following equation.

Let $(x, y, v) \in L_\alpha$. If β is the angle between the imaginary axis in the negative direction and v pointed with the orientation of L, as shown in Figure 8.4, then

$$\tan\frac{\beta}{2} = \frac{y}{x - \alpha}. \tag{5}$$

Expressing v as the angle β we get the functional equation (5) for L_α.

Applying the exterior derivative d to (5) we get

$$d\beta + \frac{2\sin^2(\beta/2)}{y}\,dx - \frac{\sin\beta}{y}\,dy = 0.$$

Thus if we define a C^∞ one-form ω on E by

$$\omega = d\beta + \frac{2\sin^2(\beta/2)}{y}\,dx - \frac{\sin\beta}{y}\,dy, \tag{6}$$

the tangent plane field defined by ω is $\mathscr{D}(\mathscr{F})$. Let τ be a linear transformation defined by an element of $\mathrm{SL}(2, \mathbf{R})$. We recall that τ maps a circular arc in H onto a circular arc in H and that the Riemannian metric (3) is invariant under τ. If we denote the restriction of $\tau_*: T(H) \to T(H)$ to E again by τ_*, we have a C^∞ diffeomorphism $\tau_*: E \to E$. We can verify that if $L_\alpha \in \mathcal{F}$ then $\tau_*(L_\alpha) \in \mathcal{F}$, that is, $\mathcal{F}$ is SL(2, **R**)-invariant, and that ω is also SL(2, **R**)-invariant:

$$(\tau_*)^*\omega = \omega$$

(cf. Appendix VIII, Note 2).

We shall next calculate the Godbillon-Vey form Γ with respect to ω. The exterior derivative $d\omega$ of ω is:

$$d\omega = -\frac{2\sin^2(\beta/2)}{y^2}\,dy \wedge dx + \frac{\sin\beta}{y}\,d\beta \wedge dx - \frac{\cos\beta}{y}\,d\beta \wedge dy.$$

Hence, setting

$$\theta = \frac{\sin\beta}{y}\,dx - \frac{\cos\beta}{y}\,dy, \tag{7}$$

we get $d\omega = \omega \wedge \theta$. In particular we have

$$d\omega \wedge \omega = 0. \tag{8}$$

Therefore,

$$\Gamma = \theta \wedge d\theta = \frac{-1}{y^2}\,dx \wedge dy \wedge d\beta.$$

The negative of Γ, $-\Gamma$, is the volume element of H associated with the Riemannian metric (3). It is again routine to verify that θ is SL(2, **R**)-invariant.

Let K be a 'convex polygon' in H whose sides are segments of circular arcs in H (Figure 8.6). Let $s_1, s_2, \dots, s_q$ denote the 'sides' of K and $p_1, p_2, \dots, p_q$ the 'vertices' of K. Let K', s_i', and p_i' be the reflections of K, s_i, and p_i respectively through the circular arc containing s_1, $i = 1, 2, \dots, q$. Here $s_1 = s_1'$, $p_1 = p_1'$, and $p_q = p_q'$ (Figure 8.6). It is easy to see that for each i, there is a linear transformation

$$\tau_i : s_i \to s_i',$$

defined by an element of SL(2, **R**), realizing the correspondence $s_i \to s_i'$ under this reflection.

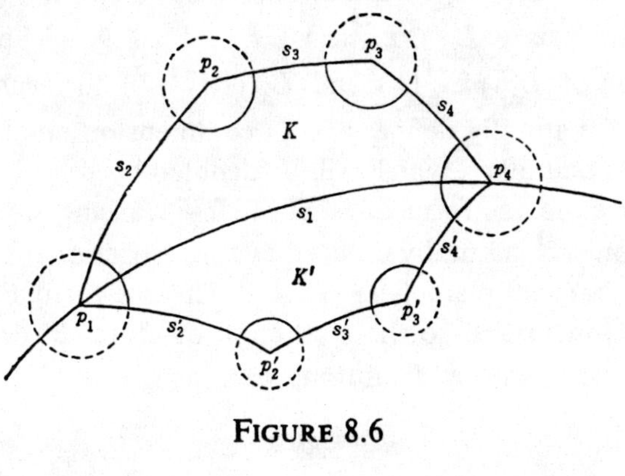

FIGURE 8.6

In the following we consider, the distance in H induced by the Riemannian metric (3). Let $\epsilon > 0$ be sufficiently small, and take ϵ-neighborhoods $U_\epsilon(p_i)$ and $U_\epsilon(p_i')$ of p_i and p_i' respectively, $i = 1, 2, \dots, q$. Set

$$\hat{K} = K \cup K' - \bigcup_{i=1}^{q} U_\epsilon(p_i) - \bigcup_{i=1}^{q} U_\epsilon(p_i'),$$

and $\hat{\pi}^{-1}(\hat{K}) = \hat{E}$. Further label

$$\hat{K} \cap s_i = \hat{s}_i, \qquad \hat{K} \cap s_i' = \hat{s}_i',$$

and denote by $\hat{\tau}_i: \hat{s}_i \to \hat{s}_i'$, the restriction of $\tau_i: s_i \to s_i'$ to $\hat{s}_i$. Let V^2 be a C^∞ two-manifold obtained from $\hat{K}$ by identifying $\hat{s}_i$ and $\hat{s}_i'$ under $\hat{\tau}_i$; the manifold V^2 is S^2 with q holes:

$$V^2 = S^2 - \bigcup_{i=1}^{q} \operatorname{Int} D_i^2.$$

Let M^3 be the C^∞ three-manifold obtained from $\hat{E}$ by identifying $\hat{\pi}^{-1}(\hat{s}_i)$ and $\hat{\pi}^{-1}(\hat{s}_i')$ under

$$(\hat{\tau}_i)_*: \hat{\pi}^{-1}(\hat{s}_i) \to \hat{\pi}^{-1}(\hat{s}_i').$$

Let

$$\kappa: \hat{E} \to M^3$$

be the quotient map. Finally let

$$\pi': M^3 \to V^2$$

be the induced projection map of $\hat{\pi}: E \to H$. Then π' is a C^∞ bundle whose fiber is S^1. Since V^2 is a compact two-manifold with boundary, it follows that π' is a product bundle, and $M^3 = V^2 \times S^1$ (cf. Appendix VIII, Note 3).

Because $\mathscr{F}$ is $SL(2, \mathbf{R})$-invariant, the map $(\hat{\tau}_i)_*: \hat{\pi}^{-1}(\hat{s}_i) \to \hat{\pi}^{-1}(\hat{s}_i')$ sends $L_\alpha \cap \hat{\pi}^{-1}(\hat{s}_i)$ onto $L_{\alpha'} \cap \hat{\pi}^{-1}(\hat{s}_i')$, $L_\alpha, L_{\alpha'} \in \mathscr{F}$. Therefore, through the identification of $L_\alpha \cap \hat{\pi}^{-1}(\hat{s}_i)$ and $L_{\alpha'} \cap \hat{\pi}^{-1}(\hat{s}_i')$ for each i, the restriction of $\mathscr{F}$ to $\hat{E}$ naturally defines a C^∞ codimension one foliation of M^3 transverse to the boundary, which will be denoted by $\bar{\mathscr{F}}$.

Likewise, since the one-form ω is $SL(2, \mathbf{R})$-invariant, we have the differential form $\bar{\omega}$ on M^3 naturally induced by the restriction of ω to $\hat{E}$, which defines the C^∞ tangent plane field $\mathscr{D}(\bar{\mathscr{F}})$. The one-form θ defined by (7) is also $SL(2, \mathbf{R})$-invariant, so the restriction of θ to $\hat{E}$ defines a smooth one-form $\bar{\theta}$ on M^3, and by definition,

$$d\bar{\omega} = \bar{\omega} \wedge \bar{\theta}.$$

Let $U_{2\epsilon}(p_i)$ and $U_{2\epsilon}(p_i')$ be the 2ϵ-neighborhoods of p_i and p_i' respectively. We consider the 'polar coordinates' with origins at p_i and p_i' respectively, and represent points of $U_{2\epsilon}(p_i)$ or $U_{2\epsilon}(p_i')$ by (r, ψ). Here $0 \le r < 2\epsilon$. We now set

$$P_i = \{ (r, \psi, v); (r, \psi) \in U_{2\epsilon}(p_i) - U_\epsilon(p_i), v \in T_{(r,\psi)}(H), \| v \| = 1 \},$$
$$P_i' = \{ (r, \psi, v); (r, \psi) \in U_{2\epsilon}(p_i') - U_\epsilon(p_i'), v \in T_{(r,\psi)}(H), \| v \| = 1 \}.$$

Then $P_i, P_i' \subset \hat{E}$. For $(r, \psi, v) \in P_i$—or $(r, \psi, v) \in P_i'$—we choose a vector $\bar{v} \in T_{p_i}(H)$—or $\bar{v} \in T_{p_i'}(H)$—such that $\bar{v}$ is parallel to v and points in the same direction as v. Denote by $\gamma_i(r, \psi, v)$—or $\gamma_i'(r, \psi, v)$—the angle between $\bar{v}$ and the imaginary axis in the negative direction. We then have obtained C^∞ functions

$$\gamma_i : P_i \to \mathbf{R}, \qquad \gamma_i' : P_i' \to \mathbf{R}.$$

The leaves of $\mathcal{F}$ restricted to P_i—or P_i'—are defined by $\gamma_i = (\text{constant})$—or $\gamma_i' = (\text{constant})$.

If we denote by δ_i the difference of the angle between $\bar{v} \in T_{p_i}(H)$ and the imaginary axis in the negative direction, and the angle between the vector $(\hat{\tau}_i)_*(v) \in T_{p_i'}(H)$ and the imaginary axis in the negative direction, by the definition of γ_i and γ_i' we get

$$\gamma_i'((\hat{\tau}_i)_*(r, \psi, v)) = \gamma_i(r, \psi, v) + \delta_i, \qquad i = 1, 2, \dots, q,$$

where $(r, \psi, v) \in P_i$. Therefore, the restriction of $d\gamma_i$ to $P_i \cap \hat{\pi}^{-1}(\hat{s}_i)$ and the restriction of $d\gamma_i'$ to $P_i' \cap \hat{\pi}^{-1}(\hat{s}_i')$ are both $(\hat{\tau}_i)_*$-invariant, and there exists a C^∞ one-form defined on the set $\kappa((P_i \cup P_i') \cap \hat{E})$, induced naturally by $d\gamma_i$ and $d\gamma_i'$; we shall denote this one-form by $d\hat{\gamma}_i$, $i = 1, 2, \dots, q$. Since on $\kappa((P_i \cup P_i') \cap \hat{E})$ the one-form $d\hat{\gamma}_i$ defines $\mathcal{D}(\mathcal{F})$ which is also defined by $\bar{\omega}$, there exists a C^∞ function

$$g : \kappa((P_i \cup P_i') \cap \hat{E}) \to \mathbf{R} - \{0\},$$

such that $d\hat{\gamma}_i = g\bar{\omega}$.

Now let $h : [\epsilon, 2\epsilon] \to [0, 1]$ be a monotone decreasing C^∞ map such that

$$h(t) = \begin{cases} 1, & \epsilon \le t \le 4\epsilon/3, \\ 0, & 5\epsilon/3 \le t \le 2\epsilon \end{cases}$$

(cf. Lemma 1.5). Define a C^∞ one-form ω' on M^3 by

(i) $\omega' = \bar{\omega}$, on $M^3 - \bigcup_{i=1}^q \kappa((P_i \cup P_i') \cap \hat{E})$,

(ii) $\omega'(\kappa(r, \psi, v)) = (1 - h(r))\bar{\omega}(\kappa(r, \psi, v)) + h(r)\, d\hat{\gamma}_i(\kappa(r, \psi, v))$, on $\kappa((P_i \cup P_i') \cap \hat{E})$.

It is evident from the definition of ω', the one-forms ω' and $\bar{\omega}$ define the same C^∞ tangent plane field on M^3, namely $\mathcal{D}(\mathcal{F})$.

Calculating $d\omega'$ on $\kappa((P_i \cup P_i') \cap \hat{E})$, we get

$$\begin{aligned}
d\omega' &= -\frac{dh}{dr} dr \wedge \bar{\omega} + (1-h)\, d\bar{\omega} + \frac{dh}{dr} dr \wedge d\hat{\gamma}_i \\
&= \bar{\omega} \wedge \left(\frac{dh}{dr} dr\right) + (1-h)\bar{\omega} \wedge \bar{\theta} + \bar{\omega} \wedge \left(-g\frac{dh}{dr} dr\right) \\
&= \bar{\omega} \wedge \left((1-g)\frac{dh}{dr} dr + (1-h)\bar{\theta}\right) \\
&= \frac{1}{(1-h)+gh}\omega' \wedge \left((1-g)\frac{dh}{dr} dr + (1-h)\bar{\theta}\right).
\end{aligned}$$

Thus setting

$$\theta' = \frac{1}{(1-h)+gh}\left((1-g)\frac{dh}{dr} dr + (1-h)\bar{\theta}\right) \tag{9}$$

we conclude that

$$d\omega' = \omega' \wedge \theta'.$$

Recall that $d\bar{\omega} = \bar{\omega} \wedge \bar{\theta}$ on $M^3 - \bigcup_{i=1}^q \kappa((P_i \cup P_i') \cap \hat{E})$. Thus putting $\theta' = \bar{\theta}$, we establish

$$d\omega' = \omega' \wedge \theta'$$

on $M^3 - \bigcup_{i=1}^q \kappa((P_i \cup P_i') \cap \hat{E})$.

We now look at the difference between $\bar{\Gamma} = \bar{\theta} \wedge d\bar{\theta}$ and $\Gamma' = \theta' \wedge d\theta'$, the Godbillon-Vey forms with respect to $\bar{\omega}$ and ω'. We have that $\bar{\Gamma} = \Gamma'$ on the set $M^3 - \bigcup_{i=1}^q \kappa((P_i \cup P_i') \cap \hat{E})$. On $\kappa((P_i \cup P_i') \cap \hat{E})$, put

$$\bar{\Gamma} - \Gamma' = g_i\, dr \wedge d\psi \wedge d\beta, \qquad i = 1, 2, \dots, q.$$

A simple calculation with formula (9) shows that as $\epsilon \to 0$, g_i is bounded. Hence

$$\lim_{\epsilon \to 0} \int_{M^3} \bar{\Gamma} - \Gamma' = 0. \tag{10}$$

We shall next modify $\mathscr{F}$, which is a C^∞ foliation of M^3 transverse to the boundary, to construct a C^∞ foliation of M^3 such that each connected component of ∂M^3 is a leaf according to the definition given in §16. First we define a monotone decreasing function $\bar{h}\colon [\epsilon, 2\epsilon] \to \mathbf{R}$, such that

$$\begin{aligned}
&\bar{h}(t) = 1, && \epsilon \le t \le 10\epsilon/9, \\
&0 < \bar{h}(t) < 1, && 10\epsilon/9 < t < 11\epsilon/9, \\
&\bar{h}(t) = 0, && 11\epsilon/9 \le t \le 2\epsilon.
\end{aligned}$$

Define a C^∞ one-form ω'' on M^3 as follows:

(i) $\omega'' = \omega'$, on $M^3 - \bigcup_{i=1}^q \kappa((P_i \cup P_i') \cap \hat{E})$,

(ii) $\omega''(\kappa(r, \psi, v)) = (1 - \bar{h}(r))\omega'(\kappa(r, \psi, v)) + \bar{h}(r)\, dr$, on $\bigcup_{i=1}^q \kappa((P_i \cup P_i') \cap \hat{E})$. (11)

Denote by $\mathscr{D}$ the C^∞ tangent plane field defined by ω''. Then on

$$M^3 - \bigcup_{i=1}^{q} \kappa(((U_{4\epsilon/3}(p_i) - U_\epsilon(p_i)) \cup (U_{4\epsilon/3}(p_i') - U_\epsilon(p_i'))) \cap \hat{E}),$$

the plane-field $\mathscr{D}$ is same as the C^∞ tangent plane field defined by ω'; so it is completely integrable. On the other hand, on

$$\kappa(((U_{4\epsilon/3}(p_i) - U_\epsilon(p_i)) \cup (U_{4\epsilon/3}(p_i') - U_\epsilon(p_i'))) \cap \hat{E}),$$

we have

$$\omega'' = (1-\bar{h})\, d\hat{y}_i + \bar{h}\, dr, \quad \text{and so} \quad d\omega'' = \frac{-d\bar{h}}{dr}\, dr \wedge d\hat{y}_i;$$

consequently we get

$$\omega'' \wedge d\omega'' = 0.$$

Hence by the theorem of Frobenius, the one-form ω'' is again completely integrable. Thus ω'' determines a C^∞ codimension one foliation of M^3, which will be denoted by $\mathscr{F}'$. By the definition of ω'' in (11), each connected component of ∂M^3 is a leaf of $\mathscr{F}'$.

We recall that $M^3 = V^2 \times S^1$ and $V^2 = S^2 - \bigcup_{i=1}^{q} D_i^2$; hence we have that $D^2 = V^2 \cup D_1^2 \cup D_2^2 \cup \cdots \cup D_{q-1}^2$ and

$$D^2 \times S^1 = (V^2 \cup D_1^2 \cup \cdots \cup D_{q-1}^2) \times S^1 = M^3 \cup (D_1^2 \times S^1) \cup \cdots \cup (D_{q-1}^2 \times S^1).$$

Foliate each $D_i^2 \times S^1$ by the Reeb foliation, $i = 1, 2, \ldots, q-1$; then together with $\mathscr{F}'$ of M^3, we have obtained a C^∞ codimension one foliation of $D^2 \times S^1$, which will be denoted by $\mathscr{F}''$. Finally, as in Example B in §16, let $S^3 = (D^2 \times S^1) \cup (S^1 \times D^2)$ and foliate $D^2 \times S^1$ by $\mathscr{F}''$ and $S^1 \times D^2$ by the Reeb foliation. The result is a C^∞ codimension one foliation which we shall denote by $\mathscr{F}(K, \epsilon)$.

Let us calculate the Godbillon-Vey number of $(S^3, \mathscr{F}(K, \epsilon))$. In the smooth foliation defined by ω'', each connected component of ∂M^3 is a leaf. On each of the $D_i^2 \times S^1$, $i = 1, 2, \ldots, q$, as well as on $S^1 \times D^2$, we have defined the C^∞ one-form ω of Example 2 in §30 which in turn defines the Reeb foliation. Putting all these forms together we obtain a C^∞ one-form ω_ϵ on S^3 which defines $\mathscr{F}(K, \epsilon)$. Using the one-form ω_ϵ we calculate the Godbillon-Vey form Γ_ϵ on S^3. But first notice that by the calculation in Example 2 in §30, $\Gamma_\epsilon = 0$ on each $D_i^2 \times S^1$, $i = 1, 2, \ldots, q$ as well as on $S^1 \times D^2$. Since $\omega_\epsilon = \omega''$ on $M^3 = V^2 \times S^1$, we calculate Γ_ϵ as follows. Consider an open cover of M^3:

$$\{M^3 - \bigcup_{i=1}^{q} \kappa(((\overline{U_{10\epsilon/9}(p_i) - U_\epsilon(p_i)}) \cup (\overline{U_{10\epsilon/9}(p_i') - U_\epsilon(p_i')})) \cap \hat{E}),$$

$$\kappa(((U_{11\epsilon/9}(p_i) - U_\epsilon(p_i)) \cup (U_{11\epsilon/9}(p_i') - U_\epsilon(p_i'))) \cap \hat{E}),$$

$$i = 1, 2, \ldots, q\},$$

and let μ and the μ_i, $i = 1, 2, \dots, q$, be a partition of unity subordinate to this cover.

Define θ_1'' on

$$M^3 - \bigcup_{i=1}^{q}(((\overline{U_{10\epsilon/9}(p_i) - U_\epsilon(p_i)}) \cup (\overline{U_{10\epsilon/9}(p_i') - U_\epsilon(p_i')})) \cap \hat{E})$$

by

$$\theta_1'' = \frac{1}{1-\bar{h}}\frac{d\bar{h}}{dr}dr + \theta',$$

where θ' was defined in (9). Then we get $d\omega'' = \omega'' \wedge \theta_1''$. On

$$\kappa(((U_{11\epsilon/9}(p_i) - U_\epsilon(p_i)) \cup (U_{11\epsilon/9}(p_i') - U_\epsilon(p_i'))) \cap \hat{E}),$$

we have $\omega'' = (1-\bar{h})\,d\hat{\gamma}_i + \bar{h}\,dr$, so

$$d\omega'' = \frac{-d\bar{h}}{dr}dr \wedge d\hat{\gamma}_i.$$

Putting

$$\theta_{2,i}'' = \frac{-1}{\bar{h}}\frac{d\bar{h}}{dr}d\hat{\gamma}_i;$$

we get $d\omega'' = \omega'' \wedge \theta_{2,i}''$. Hence, setting

$$\theta'' = \mu\theta_1'' + \sum_{i=1}^{q}\mu_i\theta_{2,i}'',$$

we have established that

$$d\omega'' = \omega'' \wedge \theta''.$$

It follows then on M^3 we have $\Gamma_\epsilon = \theta'' \wedge d\theta''$. By direct calculation, verbatim from the calculation of $\bar{\Gamma} - \Gamma'$, we obtain

$$\lim_{\epsilon\to 0}\int_{M^3}\Gamma' - \Gamma_\epsilon = 0.$$

Combining this with the earlier result (10) we get

$$\lim_{\epsilon\to 0}\int_{M^3}\Gamma_\epsilon = \lim_{\epsilon\to 0}\int_{M^3}\bar{\Gamma}.$$

But notice that

$$\lim_{\epsilon\to 0}\int_{M}\bar{\Gamma} = \int_{\hat{\pi}^{-1}(K\cup K')}\frac{-1}{y^2}\,dx\,dy\,d\beta.$$

Setting $a(K\cup K') = \int_{\hat{\pi}^{-1}(K\cup K')} -1/y^2\,dx\,dy\,d\beta$, we see that

$$\text{G.V.}(S^3, \mathcal{F}(K, \epsilon)) = \int_{S^3}\Gamma_\epsilon = \int_{M^3}\Gamma_\epsilon$$

converges to $a(K\cup K')$ as $\epsilon \to 0$. But $a(K\cup K')$ is the negative of the area of $K\cup K'$, so it is different from zero.

Let us compare the foliations $\mathcal{F}(K, \epsilon)$ and $\mathcal{F}(K, \epsilon')$. Let $\tilde{\mathcal{F}}$ be a codimension one foliation of $S^3 \times I$ transverse to the boundary such that for each $t \in I$, $\{L_\alpha \cap (S^3 \times \{t\}); L_\alpha \in \tilde{\mathcal{F}}\}$ is the foliation $\mathcal{F}(K, \epsilon + t(\epsilon' - \epsilon))$ of $S^3 \times \{t\}$. We notice that by choosing a suitable $\bar{h}$ to define $\mathcal{F}'$, we can get a C^∞ foliation $\tilde{\mathcal{F}}$. Evidently we have

$$(\partial(S^3 \times I), \partial\tilde{\mathcal{F}}) = (S^3, \mathcal{F}(K, \epsilon)) \cup (-S^3, \mathcal{F}(K, \epsilon')).$$

Hence the foliations $(S^3, \mathcal{F}(K, \epsilon))$ and $(S^3, \mathcal{F}(K, \epsilon'))$ are cobordant. Thus by Theorem 8.4,

$$\mathrm{G.V.}(S^3, \mathcal{F}(K, \epsilon)) = \mathrm{G.V.}(S^3, \mathcal{F}(K, \epsilon')).$$

We therefore conclude that the Godbillon-Vey number $\mathrm{G.V.}(S^3, \mathcal{F}(K, \epsilon))$ is equal to $a(K \cup K')$ independent of ϵ.

Since the value of $a(K \cup K')$ varies continuously in K, by Theorem 8.4 we have proved the following

THEOREM 8.5 (Thurston). *There exist at least continuum many smooth codimension-one foliations of S^3 each having a different Godbillon-Vey number. Therefore, the cardinality of $\mathcal{F}\Omega^\infty_{3,1}$, the cobordism group of the three-dimensional smooth codimension-one foliations, is at least the continuum.*

The three-dimensional cobordism group Ω_3 is known to be zero.

Appendices

I

The following are the four papers of Poincaré cited in §1, Chapter I. Each paper is entitled *Mémoire sur les courbes définies par une équation différentielle* [P1, P2, P3, P4]. Poincaré's development continues in a paper published about 10 years later [P5] and establishes the concept of topology in its five *Compléments*.

The papers *Mémoire sur les courbes définies par une équation différentielle* introduced a branch of mathematics, today called the theory of dynamical systems. For the theory of dynamical systems see *Geometrical methods in the theory of ordinary differential equations*, Arnold [Ar]; and *Differential equations, dynamical systems, and linear algebra*, Hirsch and Smale [HS].

The topic of Chapter I is the introduction of the works by Denjoy [De] and Siegel [Si]. We have changed some parts of the proofs to make them adaptable to the Poincaré-Bendixson theorem.

NOTE 1. For theorems on the existence and uniqueness of the solutions of ordinary differential equations, see [Ar, HS].

NOTE 2. This follows from the compactness of the torus. Consider the torus obtained by identifying the opposite sides of the square (Figure 1.5), then the conclusion follows from the fact that a continuous function defined on the square is of bounded variation.

NOTE 3. The family $\{\overline{\{\varphi(t, p); -\infty < t \leq s\}}; s \in (-\infty, 0]\}$ of compact sets in T has the finite intersection property; so the intersection of its members is a nonempty closed set (see §8).

NOTE 4. For the Diophantine approximation see for example *Fundamentals of Diophantine geometry*, Lang [L2].

NOTE 5. As in Note 2, consider the torus obtained from the square by identifying its opposite edges; it is thus sufficient to consider a sequence in the square.

NOTE 6. For closed surfaces and their classification according to their Euler numbers, see any introductory text book on topology. *Surface topology*, Firby and Gardiner [FG] is readable and entertaining for a reader who intends not to become a topologist. See also *Topology from a differentiable viewpoint*, Milnor [M], where the reader will find the necessary information on how to calculate Euler numbers and mapping degrees as well.

II

We presented definitions and basic concepts of differentiable manifolds, limiting our scope to what is needed in this book. For greater depth, read any basic differential topology text, for instance *Differential topology*, Guillemin and Pollack [GP]; *Differential topology*, Hirsch [H].

NOTE 1. For the implicit function and inverse function theorems, see any introductory analysis book, for instance Lang [L1].

NOTE 2. In Theorem 2.6 we assumed the compactness of M^n to simplify our argument; however, M^n need not be compact. Here is why. By the definition of manifolds, M^n is a locally compact Hausdorff space and has a countable basis, so M^n is paracompact: for an open cover $\mathfrak{U}$ of M^n there exists a locally finite open cover which is a subdivision of $\mathfrak{U}$. Hence we can construct $\{\mu\sigma\,;\,\sigma\in\Sigma\}$ in the same way as in the proof of Theorem 2.6.

NOTE 3. We shall first prove the following.

LEMMA. *Let f be a C^∞ function defined on a convex neighborhood V of the origin O of $\mathbf{R}^n$ such that $f(0)=0$. Then there exist C^∞ functions g_i, $i=1,2,\dots,n$, defined on V satisfying the following*:

(i) $f(x_1, x_2, \dots, x_n) = \sum_{i=1}^n x_i g_i(x_1, x_2, \dots, x_n)$,
$(x_1, x_2, \dots, x_n)\in V$,

(ii) $g_i(0) = (\partial f/\partial x_i)(0)$, $i=1,2,\dots,n$.

PROOF. $f(x_1, x_2, \dots, x_n)$ can be written as follows:

$$f(x_1, x_2, \dots, x_n) = \int_0^1 \frac{df(tx_1, tx_2, \dots, tx_n)}{dt}\,dt$$

$$= \int_0^1 \sum_{i=1}^n \frac{\partial f}{\partial x_i}(tx_1, tx_2, \dots, tx_n)x_i\,dt.$$

Hence, setting

$$g_i(x_1, x_2, \dots, x_n) = \int_0^1 \frac{\partial f}{\partial x_i}(tx_1, tx_2, \dots, tx_n)\,dt, \qquad i=1,2,\dots,n,$$

we see that the g_i satisfy the conditions.

Now choose U_λ, $U_\lambda\subset U$, such that $\varphi_\lambda(p)=0$ and $\varphi_\lambda(U_\lambda)$ is convex. Applying the above lemma to $f\circ\varphi_\lambda^{-1}(x_1, x_2, \dots, x_n) - f(p)$, we get

$$f\circ\varphi_\lambda^{-1}(x_1, x_2, \dots, x_n) - f(p) = \sum_{i=1}^n x_i g_i(x_1, x_2, \dots, x_n). \tag{1}$$

Here we have

$$g_i(0) = \frac{\partial(f\circ\varphi_\lambda^{-1})}{\partial x_i}(0),$$

so applying the lemma to $g_i(x_1, x_2, \ldots, x_n) - (\partial(f \circ \varphi_\lambda^{-1})/\partial x_i)(0)$, we obtain the expression

$$g_i(x_1, x_2, \ldots, x_n) = \frac{\partial(f \circ \varphi_\lambda^{-1})}{\partial x_i}(0) + \sum_{j=1}^{n} x_j f_{ij}(x_1, x_2, \ldots, x_n), \qquad (2)$$

where each f_{ij} is a C^∞ function defined on $\varphi_\lambda(U_\lambda)$. From (1) and (2) we get the desired equation.

NOTE 4. Apply the remark in Note 2.

NOTE 5. It is obvious that the metric axioms (ii) and (iii) are satisfied; however, to show that (i) is satisfied it is necessary to consider geodesics.

III

In Chapter III we discussed three results concerning dynamical systems. The Poincaré-Bendixson theorem in §13 was taken from a paper of Bendixson [Be] which followed Poincaré's work cited in I. This theorem plays a vital role in the proof of Novikov's theorem in §25.

We introduce Schweitzer's paper [Sc] in §14. Not only is Schweitzer's result intimately related to the C^1 vector field of Denjoy in §6, but it also reappears in Chapter 6 as a mechanism to remove compact leaves. See [Se] for the Seifert conjecture.

The discussion on Wilson's C^∞ vector field in §15 which should be compared with Schweitzer's work in §14, comes from [W].

NOTE 1. See Milnor [M] concerning the relation between nonsingular vector fields and Euler numbers.

NOTE 2. Let M_1, M_2 be C^r manifolds, $r \geq 1$, and define the projection $\pi_i : M_1 \times M_2 \to M_i$, $i = 1, 2$, by $\pi_i(x_1, x_2) = x_i$, $x_1 \in M_1$ and $x_2 \in M_2$. Then for $v_1 \in T_{x_1}(M_1)$ there exists a unique $v \in T_{(x_1, x_2)}(M_1 \times M_2)$ such that $(\pi_1)_*(v) = v_1$, $(\pi_2)_*(v) = 0$. Denote v by v_1. Similarly for $v_2 \in T_{x_2}(M_2)$ we can define $v_2 \in T_{(x_1, x_2)}(M_1 \times M_2)$.

IV

We defined foliations in Chapter 4. Bundle foliations defined by C^r bundles form a special category of foliations. In comparing bundle foliations with more general types of foliations one obtains a clear-cut insight into the role played by foliations in the field of geometry.

NOTE 1. Let $(M, \mathscr{S})$, $(M', \mathscr{S}')$ be manifolds and $C^r(M, M')$ the set of C^r maps from M into M'. For $f \in C^r(M, M')$, choose $(U_\lambda, \varphi_\lambda) \in \mathscr{S}$ and $(U_{\lambda'}', \varphi_{\lambda'}') \in \mathscr{S}'$ such that $f(U_\lambda) \subset U_{\lambda'}'$. We say that $g \in C^r(M, M')$ is in a neighborhood of f if for a compact set K contained in U_λ, the maps $\varphi_{\lambda'}' \circ f \circ \varphi_\lambda^{-1}$ and $\varphi_{\lambda'}' \circ g \circ \varphi_\lambda^{-1}$ are within some approximation up to the rth partial derivatives in $\varphi_\lambda(K)$. The set of all such neighborhoods generates a topology, called the C^r topology on $C^r(M, M')$.

NOTE 2. Let M^n be a C^r manifold, $r \geq 2$. Suppose that a Riemannian metric is defined on M^n. Let p be a point of M^n. If v is an element of the tangent space $T_p(M^n)$ with $\| v \|$ sufficiently small, there exists a unique geodesic $g\colon [0, 1] \to M^n$ such that $g(0) = p$ and $(dg/dt)(0) = v$. We define the map $\operatorname{Exp}(v)$ by

$$\operatorname{Exp}(v) = g(1).$$

NOTE 3. Consult for instance *Topology*, Dugundji [Du] about Lebesgue numbers.

V

In Chapter V we introduced the main results from a monograph of Reeb [R]. Our definition of coherent systems of charts differs slightly from that given in Reeb's book. A more recent result on the stability theorem is due to Thurston [T2].

NOTE 1. See any elementary text book on algebraic topology for fundamental groups; for instance, Algebraic Topology, Massey [Ma].

NOTE 2. See Reeb [Re], (A,III,c).

VI

In writing §23, Chapter VI, we used Nishimori's notes: in §25 we introduced Novikov [N]; this was done by mainly following Haefliger [Ha].

NOTE 1. A universal covering of a $L'_{\alpha(t)}$ is a covering whose base space is $L'_{\alpha(t)}$ and whose total space is connected with trivial fundamental group. The existence of universal coverings is known. Let $\pi\colon \tilde{L}'_{\alpha(t)} \to L'_{\alpha(t)}$ be the projection and $g\colon A \to L'_{\alpha(t)}$ be a map, where A is a topological space. A map $\tilde{g}\colon A \to \tilde{L}'_{\alpha(t)}$ is called a *lift* of g if $\pi \circ \tilde{g} = g$.

VII

The examples in Chapter I indicate the significance of the value r of C^r in foliations. In Chapter VII all differential forms investigated are of class C^r; for example, the Frobenius Theorem was stated in conjunction with C^r foliations. *A comprehensive Introduction to differential geometry*, vol. I, Spivak [Sp], is a good reading here.

VIII

In Chapter VIII we presented Thurston's work [T1], filling in computational details. This presentation used notes from a Milnor seminar by Nishimori and Mizutani as a guide.

NOTE 1. In general, suppose a Riemannian metric $\{\langle\ ,\ \rangle_p ; p \in M^n\}$ is given on M^n. For a chart $(U_\lambda, \varphi_\lambda)$ of M^n, let $(x_1, x_2, \dots, x_n) \in \varphi_\lambda(U_\lambda)$ and define $g_{ij}\colon U_\lambda \to \mathbf{R}$ by

$$g_{ij}(p') = \left\langle (\Phi_\lambda^*)^{-1}\left(\frac{\partial}{\partial x_i}\right), (\Phi_\lambda^*)^{-1}\left(\frac{\partial}{\partial x_j}\right)\right\rangle_{p'}, \qquad p' \in U_\lambda.$$

We then define a Riemannian metric on U_λ by

$$ds^2 = \sum_{i,j} g_{ij}\, dx_i\, dx_j.$$

NOTE 2. Let τ be the linear transformation sending z to $(az+b)/(cz+d)$; τ maps an oriented circular arc L in H onto an oriented circular L' in H. Let $(x, y, v) \in L_\alpha$ and v a tangent vector to L at $(x, y) \in L$ with $\| v \| = 1$, and put

$$\tau_*(x, y, v) = (x', y', v').$$

Then we see that v' is a tangent vector to L' at $(x', y') \in L'$, and since a Riemannian metric is invariant under τ we have that $\| v' \| = 1$. Therefore $(x', y', v') \in L_{\tau(\alpha)}$, and τ maps the leaf L_α onto the leaf $L_{\alpha(\tau)}$. Thus $\mathscr{F}$ is invariant under τ.

We shall next show that $(\tau_*)^*\omega = \omega$. For simplicity let

$$\tau(x) = x', \qquad \tau(y) = y', \qquad \tau(\alpha) = \alpha', \qquad \tau_*(\beta) = \beta',$$

where α is the coordinate of P in Figure 8.4.

From the above remark, just as we have $\tan(\beta/2) = y/(x-\alpha)$, we get

$$\tan\frac{\beta'}{2} = \frac{y'}{x'-\alpha'}.$$

Further we evidently have

$$x' = \frac{(ax+b)(cx+d)+acy^2}{(cx+d)^2+c^2y^2}, \qquad y' = \frac{y}{(cx+d)^2+c^2y^2},$$

$$\alpha' = \frac{a\alpha+b}{c\alpha+d}.$$

We break up the correspondence $(x, y, \beta) \mapsto (x', y', \beta')$ into three stages

$$(x, y, \beta) \mapsto (x, y, \alpha) \mapsto (x', y', \alpha') \mapsto (x', y', \beta').$$

Since $\alpha = x - y\cot\beta/2$, we have with respect to the first correspondence

$$\frac{\partial x}{\partial \beta} = \frac{\partial y}{\partial \beta} = 0, \qquad \frac{\partial \alpha}{\partial \beta} = \frac{(x-\alpha)^2+y^2}{2y},$$

with respect to the second correspondence

$$\frac{\partial x'}{\partial \alpha} = \frac{\partial y'}{\partial \alpha} = 0, \qquad \frac{\partial \alpha'}{\partial \alpha} = \frac{1}{(c\alpha+d)^2},$$

and with respect to the third correspondence

$$\frac{\partial x'}{\partial \alpha'} = \frac{\partial y'}{\partial \alpha'} = 0, \qquad \frac{\partial \beta'}{\partial \alpha'} = \frac{2y'}{(x'-\alpha')^2+y'^2}.$$

Further we obtain by direct computation

$$(x' - \alpha')^2 + y'^2 = \frac{(x-\alpha)^2 + y^2}{((cx+d)^2 + c^2y^2)(c\alpha + d)^2}.$$

Hence we have established

$$\frac{\partial \beta'}{\partial \alpha'} = \frac{2y(c\alpha+d)^2}{(x-\alpha)^2 + y^2}.$$

Using the above result we show that $(\tau_*)_*(\partial/\partial\beta) = \partial/\partial\beta'$

$$\begin{aligned}
(\tau_*)_*\left(\frac{\partial}{\partial\beta}\right) &= \frac{\partial x'}{\partial\beta}\frac{\partial}{\partial x'} + \frac{\partial y'}{\partial\beta}\frac{\partial}{\partial y'} + \frac{\partial\beta'}{\partial\beta}\frac{\partial}{\partial\beta'} \\
&= \frac{\partial\beta'}{\partial\beta}\frac{\partial}{\partial\beta'} = \left(\frac{\partial\beta'}{\partial\alpha}\frac{\partial\alpha}{\partial\beta} + \frac{\partial\beta'}{\partial x}\frac{\partial x}{\partial\beta} + \frac{\partial\beta'}{\partial y}\frac{\partial y}{\partial\beta}\right)\frac{\partial}{\partial\beta'} \\
&= \frac{\partial\beta'}{\partial\alpha}\frac{\partial\alpha}{\partial\beta}\frac{\partial}{\partial\beta'} = \left(\frac{\partial\beta'}{\partial\alpha'}\frac{\partial\alpha'}{\partial\alpha} + \frac{\partial\beta'}{\partial x'}\frac{\partial x'}{\partial\alpha} + \frac{\partial\beta'}{\partial y'}\frac{\partial y'}{\partial\alpha}\right)\frac{\partial\alpha}{\partial\beta}\frac{\partial}{\partial\beta'} \\
&= \frac{\partial\beta'}{\partial\alpha'}\frac{\partial\alpha'}{\partial\alpha}\frac{\partial\alpha}{\partial\beta}\frac{\partial}{\partial\beta'} \\
&= \frac{\partial}{\partial\beta'}.
\end{aligned}$$

As shown earlier, $\mathcal{F}$ is invariant under τ_*, so there exists a C^∞ map $f\colon E \to \mathbf{R}$ such that $(\tau_*)^*\omega = f\omega$. Hence it follows that

$$((\tau_*)^*\omega)\left(\frac{\partial}{\partial\beta}\right) = (f\omega)\left(\frac{\partial}{\partial\beta}\right) = f.$$

On the other hand, from the above calculation we get

$$((\tau_*)^*\omega)\left(\frac{\partial}{\partial\beta}\right) = \omega\left((\tau_*)_*\left(\frac{\partial}{\partial\beta}\right)\right) = \omega\left(\frac{\partial}{\partial\beta'}\right) = 1,$$

so we must have that $f = 1$. Hence $(\tau_*)^*\omega = \omega$.

NOTE 3. The classification space of C^r S^1 bundles is the complex projective space $\mathbf{C}P^N$, where N is sufficiently large. But a map $V^2 \to \mathbf{C}P^N$ is always homotopic to zero, so a C^r bundle whose base is V^2 with the fiber S^1 is the product bundle.

We recommend Bott [Bo] or Tamura and Mizutani [MT] for further reading by interested readers. We also believe that Lawson [La] will shed some light on an overall picture of the foliation theory (see the translator's afterword).

References

[Ar] V.I. Arnold, *Geometrical methods in the theory of ordinary differential equations*, Springer-Verlag, New York, 1983; English transl.

[Be] I. Bendixson, *Sur les courbes définies par les équations différentielles*, Acta Math. **24** (1901), 1–88.

[Bo] R. Bott, *Lectures on characteristic classes and foliations*, Lectures on Algebraic and Differential Topology, Lecture Notes in Math. vol. 279, Springer-Verlag, New York, 1972.

[CN] C. Camacho and A. Lins Neto, *Geometric theory of foliations*, Birkhäuser, Boston, 1985; English transl.

[De] A. Denjoy, *Sur les courbes définies par les équations différentielles à la surface du tore*, J. Math. Pures Appl. **11** (1932), 333–375.

[Du] J. Dugundji, *Topology*, Allyn and Bacon, Boston, 1966.

[FG] P. A. Firby and C. F. Gardiner, *Surface topology*, Ellis Horwood, Chichester, 1982.

[F] D. Fuks, *Foliations*, J. Soviet Math. **18** (1982), 255–291; English transl.

[GP] V. Guillemin and A. Pollack, *Differential topology*, Prentice-Hall, Englewood Cliffs, 1974.

[G] C. Godbillon, *Feuillettages, études géométriques*, Prog. Math., vol. 98, Birkhäuser Verlag, Basel, 1991.

[Ha] A. Haefliger, *Travaux de Novikov sur les feuilletages*, Séminaire Bourbaki 1971/72, no 412.

[HH1] G. Hector and U. Hirsch, *Introduction to the geometry of foliations, Part* A, Aspects of Mathematics, vol. E1, Vieweg, Braunschweig, 1981.

[HH2] ——, *Introduction to the geometry of foliations, Part* B, Aspects of Mathematics, vol. E3, Vieweg, Braunschweig, 1983.

[H] M. Hirsch, *Differential topology*, Graduate Texts in Math., Springer-Verlag, New York, 1976.

[HS] M. Hirsch and S. Smale, *Differential equations, dynamical systems and linear algebra*, Academic Press, New York, 1974.

[L1] S. Lang, *Analysis* I, Addison-Wesley, Reading, 1968.

[L2] ——, *Fundamentals of Diophantine geometry*, Springer Verlag, New York, 1983.

[La] H. B. Lawson, *Foliations*, Bull. Amer. Math. Soc. **80** (1974), 369–418.

[Ma] W. Massey, *Albegraic topology: an introduction*, Harcourt, Brace & World, New York, 1967.

[M] J. Milnor, *Topology from a differentiable viewpoint*, University Press of Virginia, Charlottesville, 1965.

[MT] T. Mizutani and I. Tamura, *On the existence of foliations*, Sūgaku **25** (1973), 197–296. (Japanese)

[Ni] T. Nishimori, *A vector field on* S^3 *without closed orbits* (*after P. Schweitzer*) in Comprehensive Studies in Dynamical Systems, Lecture Series, vol. 173, R.I.M.S., University of Kyoto, 1973. (Japanese)

[N] S. P. Novikov, *Topology of foliations*, Trans. Moscow. Math. Soc. **14** (1967), 268–304.

[P1] H. Poincaré, *Mémoire sur les courbes définies par une équation différentielle*, J. Math. Pures Appl. 3e série **7** (1881), 375–424.

[P2] ——, *Mémoire sur les courbes définies par une équation différentielle*, J. Math. Pures Appl. 3e série **8** (1882), 251–296.

[P3] ——, *Mémoire sur les courbes définies par une équation différentielle*, J. Math. Pures Appl. 4e série **1** (1885), 167–244.

[P4] ——, *Mémoire sur les courbes définies par une équation différentielle*, J. Math. Pures Appl. 4e série **2** (1886), 151–217.

[P5] ——, *Analysis situs*, Jour. Ecole Polytechn. 2e série **1** (1895), 1–121.

[R] G. Reeb, *Sur certains propriétés topologiques des variétés feuilletées*, Actualités Sci. Indust., vol. 1183, Hermann, Paris, 1952.

[Re] B. Reinhart, *Differential geometry of foliations*, A Series of Modern Surveys in Mathematics, vol. 99, Springer Verlag, New York, 1983.

[Sc] P. Schweitzer, *Counterexamples to Seifert conjecture and opening closed leaves of foliations*, Ann. of Math. **100** (1974), 386–400.

[Se] H. Seifert, *Closed integral curves in 3-space and isotopic two-dimensional deformations*, Proc. Amer. Math. Soc., **1** (1950), 287–302.

[Si] C. L. Siegel, *Note on differential equations on the torus*, Ann. of Math. **46** (1945), 423–428.

[Sp] M. Spivak, *A Comprehensive Introduction to Differential Geometry, Part* 1, Publish or Perish, Boston, 1970.

[Ta] I. Tamura, *Topologiya sloenii* (D. V. Anosov, ed.), Mir, Moscow, 1979; Russian transl.

[T1] W. Thurston, *Non-cobordant foliations of* S^3, Bull. Amer. Math. Soc. **78** (1972), 511–514.

[T2] ——, *A generalization of the Reeb stability theorem*, Topology **13** (1974), 347–352.

[W] F. Wilson, *On the minimal sets of non-singular vector fields*, Ann. of Math. **84** (1966), 529–536.

Translator's Afterword

The *Topology of Foliations* was first published in Japanese in 1976 and a Russian translation with substantial revisions appeared in 1979. When Tamura agreed to have the book translated into English in 1990, he intended to make the following revisions.

1. Rewrite Chapters I and II, making them more concise;
2. change the references, in particular to cite only those papers mentioned in the text itself;
3. write a new preface;
4. replace the list of recommended books, the standard texts in Japanese, with an equivalent list in English.

Unfortunately, Tamura became seriously ill before the work of translation began, and passed away early in 1991. His premature death greatly disturbed his many mathematical friends in Japan and elsewhere in the world. The illness presented technical problems for the translation, because I had expected Tamura's comments and assistance at every stage in the work.

Takashi Tsuboi, one of Tamura's distinguished students, kindly agreed to write the new preface, for which I am extremely grateful.

The new reference list is quite short. A reader interested in the subject matter of this book may wish to continue in further depth with some of the following literature:

1. Godbillon [G] has complete references, but unfortunately they are given in chronological order making it somewhat difficult to follow a specific topic; this book does not touch on cobordisms;
2. Fuks [F];
3. Reinhart [Re];
4. Hector-Hirsch [HH1, HH2].

For those capable of reading Russian, the Russian translation of the present book [Ta], edited by Anosov, may be quite interesting. Anosov added numerous footnotes and supplemented many of the calculations, making the book truly readable for "the reader with a background of mathematics at the lower division level in college." The Camacho book [CN], written in 1979 and translated into English in 1985, is equivalent in spirit to the present book.

There is a large degree of overlap, but an agreeable difference of perspective, and each book touches on topics not included in the other.

I would like to thank T. Hashiguchi, T. Inaba, K. Masuda, S. Matsumoto, Y. Mitsumatsu, T. Mizutani, H. Nakayama, T. Nishimori, A. Sato, T. Tsuboi, N. Tsuchiya, and K. Yano for reading the translation or helping with many matters associated with it.

Kiki Hudson
La Jolla, 1991

Subject Index

Recent Titles in This Series

(Continued from the front of this publication)

(See the AMS catalogue for earlier titles)